Space Systems Failures

Disasters and Rescues of Satellites, Rockets and Space Probes

David M. Harland and Ralph D. Lorenz

Space Systems Failures

**Disasters and Rescues of Satellites,
Rockets and Space Probes**

 Springer

Published in association with
Praxis Publishing
Chichester, UK

David M. Harland
Space Historian
Kelvinbridge
Glasgow
UK

Dr Ralph D. Lorenz
Lunar and Planetary Laboratory
University of Arizona
Tucson
Arizona
USA

SPRINGER–PRAXIS BOOKS IN SPACE EXPLORATION
SUBJECT *ADVISORY EDITOR*: John Mason B.Sc., M.Sc., Ph.D.

ISBN 0-387-21519-0 Springer Berlin Heidelberg New York

Springer is a part of Springer Science + Business Media (*springeronline.com*)

Library of Congress Control Number: 2005922815

Cover design: Jim Wilkie
Project Copy Editor: Alex Whyte
Typesetting: BookEns Ltd, Royston, Herts., UK

Printed in Germany on acid-free paper

Table of contents

List of illustrations

Authors' preface

Space utilisation in applications such as communications, navigation and remote sensing, and to a certain extent space exploration, is a multibillion dollar business. As such, the stakes are high and competition for lucrative contracts is fierce. Industry, and indeed government and the various other participating entities, are reluctant to document in open sources the details, or even the fact of failures. It is not unknown for a company to bolster its public image by listing projects in which it participated, even where its own contribution to a project failed – the eventual success of a project is usually remembered longer than the effort required to overcome early crises. This is, perhaps, as it should be, because space is no different from any other field of human endeavour. In as much as we relied in researching this book on failures documented in the public domain, the reader must recognise that this survey is far from complete. Although our theme is failures, it is *not* our intent to point fingers, or blame participants. Indeed, mention in this book may be considered in some sense to be an acknowledgment of integrity, in that these failures have been documented in order that others may learn. We must emphasise that this book is not a catalogue of doom. Many failures turned out to be recoverable, and it was during such efforts that spacecraft engineers demonstrated that they had the 'right stuff'. There are examples of missions being recovered weeks after what had seemed to be a hopeless failure, and useful discoveries made about systems and operating techniques that would never have been learned if the failure had *not* occurred. Sadly, there are also examples of hard-won lessons being neglected.

A truism is that "engineering is the art of doing with one dollar that which any damn fool can do with two". Many of the failures described herein are the result of the intense schedule pressure, allowing little time for testing. In other cases, financial pressures have forced design compromises that are less robust than those that the designers might have preferred. This is not to say that applying more money to a project necessarily enhances reliability, but it must be recognised that in general there is an increase in risk associated with cost-cutting. This was explicitly recognised by NASA administrator Dan Goldin in the early 1990s, with the introduction of the 'faster-better-cheaper' strategy. He went so far as to say that if failures did not occur, then NASA was not trying hard enough. Some programmes were introduced with the specific intent of testing newer technologies, with an acceptance of an increased degree of risk in order to improve the ratio of performance to cost. To some extent, this vision stalled, in that after several failures had occurred, the tradition of risk-aversion reasserted itself. A more sanguine long-term view might have permitted a few more high-risk attempts, in order to achieve a better determination of the real reliability of lower-cost space systems. Several examples of space systems failure in this book appear directly as results of various cost-saving measures. The failure of Mars Observer in 1993 can be attributed to the ostensibly sensible approach of adapting Earth-orbiting satellite designs for deep space missions. But (in this case at least) this approach failed to take account of the different mission duration and thermal environment. Other failures can be attributed to overly lean design or operations teams, in which insufficient time, or oversight, or inexperienced (and therefore cheaper) staffing allowed problems to pass unnoticed. The task is to find the appropriate balance.

We have split this book into two parts, the first dealing with launch vehicles and the second with satellites and space probes, and have included citations to enable readers to refer to our sources for further information. While all these systems share some failure possibilities, because the development of launch vehicles has a set of specific problems, pressures and motivations, these are treated in a historical fashion. Spacecraft failures are discussed in the context of the various engineering disciplines in order to compare how similar failures can occur on widely different kinds of spacecraft.

David M. Harland
and
Ralph D. Lorenz

January 2005

Acknowledgements

Many colleagues from the space science and engineering community have confided interesting stories, only some of which we have been able to document. In particular we thank Andrew Ball of the Planetary and Space Sciences Research Institute at The Open University, James Garry of the University of Leiden, Marc Rayman of the Jet Propulsion Laboratory, and Bill Ailor of the Aerospace Corporation for reference material. We are also grateful to various agencies for making publicly available information on failures in the hope that others may learn from their experience, in particular the NASA Public Lessons Learned on-line database, the NASA Office of Logic Design, and the Goddard Space Flight Center. We benefited, too, from the reportage by Craig Covault, Tim Furniss, Phillip Clark and Henry Spencer. Sharon Ng, who was supported by the Arizona Space Grant Program, assisted early on in the search for material in the excellent library of the University of Arizona. We are grateful to Zibi Turtle for critical feedback of the manuscript, and of course to Clive Horwood of Praxis for his enthusiastic support throughout.

As regards the illustrations, we are grateful to the corporations involved in the space business, in particular Boeing, International Launch Services, Sea Launch, Lockheed Martin, Space Systems/Loral, Surrey Satellite Technology, the Orbital Sciences Corporation and PanAmSat; to Bart Hendrickx for our cover picture; to Jonathan McDowell for some archive pictures; and to Corby Waste for his paintings of Mars missions for JPL.

Part One
Launch vehicles

1

The missiles

THE FIRST SPACE LAUNCHER

After successfully testing its own hydrogen bomb in 1953, the Soviet Union set out to outflank the American air defences by building a ballistic missile with which to threaten Washington from a securely defended base on Soviet territory. It was recognised that the most effective configuration for a multi-stage missile would be to stack the stages one on top of the other, but the shut down and jettisoning of one stage and the ignition of a large liquid rocket engine in flight were beyond the state of the art. Sergei Korolev, the leading Soviet rocket engineer, therefore opted for 'parallel staging' in which 'boosters' would augment the main engine to lift the missile into the air and subsequently be released. In that way he would need only to perfect the art of discarding the boosters without disturbing the main stage. To lob the 3-tonne warhead a distance of 8,000 kilometres, Korolev strapped four conical boosters around a core stage, each of which contained a turbopump driven by hydrogen peroxide steam that fed kerosene and liquid oxygen into an engine that had four combustion chambers. Because these engines were fixed, the missile was steered by small vernier engines. The construction of a launch pad on the Kazakh steppe began in 1955.[1] In May 1957 the first flight failed when the vehicle exploded in the act of staging. Similar failures in June and July were followed by a successful flight in August that not only staged but achieved the assigned range. On 4 October 1957, to demonstrate the power of this rocket, the Soviets used it to put the world's first artificial satellite, Sputnik, into orbit. As it was his seventh rocket design, Korolev referred to this missile as the *Semyorka*, which in Russian means 'number seven'.[2] With an upper stage, it inserted Vostok 1 into orbit on 12 April 1961, with Yuri Gagarin on board. A more powerful stage enabled it to send small probes to the Moon and to the planets Mars and Venus. It was later used to launch the Soyuz crew ferries and Progress cargo ships that served the Salyut and Mir space stations, and is still in use today serving the International Space Station. It has truly been the 'workhorse' of the Space Age.

The launch of a *Semyorka* rocket on 12 April 1961 carrying Vostok 1 and Yuri Gagarin.

THOR

In 1946, as the US Army fired off the V-2 missiles that it had captured from Germany, the Naval Research Laboratory ordered a study of a high-performance sounding rocket to be its successor. This led to the issuing of a contract to the Glenn L. Martin Company to develop the Viking, utilising the XLR-10 rocket motor that (as in the case of the engine of the V-2) burned alcohol and oxygen.

On 1 August 1955 President Dwight Eisenhower told the Navy to manage the Vanguard project to place a satellite into orbit during the International Geophysical Year, which was to run from mid 1957 to the end of 1958. On 7 October the Navy gave Martin the contract to upgrade the Viking with General Electric's X-405 kerosene-burning motor, and to install two upper stages to enable it to function as a launch vehicle. For the second stage, Aerojet supplied an AJ-10 motor based on the Aerobee-Hi sounding rocket that used hypergolic propellants. The third stage was a 33-KS-2800 solid rocket by the Grand Central Rocket Company. In the event, the goal of being the first to place a satellite into orbit was lost to the Soviets on 4 October 1957. The attempt to even the score using the new Vanguard on 6 December was a fiasco, with the rocket losing thrust after a few seconds, falling back, and exploding. The next attempt failed after 57 seconds when the vehicle lost control, but the third launch on 17 March 1958 succeeded.[3] As the second stage had the flight control apparatus, after its main engine shut off it controlled its attitude using helium jets while it coasted, until the rotator on its nose spun the third stage to ensure that this would remain stable following its release. After a repeat performance on 17 February 1959, the second stage was upgraded with an AJ-10-142, and the third stage was superseded by a fibreglass-reinforced plastic X-248 Altair I solid rocket that had been developed by the Hercules Powder Company in collaboration with the Allegheny Ballistics Laboratory.[4]

Meanwhile, in December 1955 the Air Force had issued a contract to the Douglas Aircraft Company to develop the Thor intermediate-range ballistic missile utilising the Rocketdyne MB-3-I kerosene and oxygen engine. On 20 September 1957, after a very rapid development, the missile successfully flew to its assigned range.[5,6] After installing the Vanguard's upper stages on a Thor in a configuration called Thor–Able, on 17 August 1958 the Air Force tried to send a probe to the Moon. The Thor rose smoothly from Pad 17 at Canaveral but its turbopump seized at T + 17 seconds and the vehicle exploded. On trying again on 11 October, the second stage shut down prematurely due to an

The attempt to launch Vanguard on 6 December 1957 failed when the vehicle lost thrust, toppled back onto the pad and exploded.

Thor–Delta models (left to right): Thor–Delta-B, 21 December 1963 with TIROS 8; Thor–Delta-C1, 25 May 1966 with Explorer 32; Thor–Delta-D, 19 August 1964 with Syncom 3; Thor–Delta-E1, 11 January 1967 with Intelsat 2F2; Thor–Delta-G, 14 December 1966 with Biosat 1; Thor–Delta-M6, 13 March 1971 with Explorer 43.

accelerometer programming error, the third stage was unable to make up the shortfall, and Pioneer 1 peaked at an altitude of 113,000 kilometres before falling back. The accelerometer system was improved to ensure that the next probe was given the correct velocity. On 8 November, however, the third stage failed to ignite. Given the state of the technology at that time, these were very ambitious missions.

It was fortunate that the Thor had matured so rapidly, because the Air Force planned to fit it with an Agena upper stage equipped for orbital reconnaissance. Built by the Lockheed Aircraft Company, the Agena used the XLR-81 hypergolic engine that had been developed for the rocket-powered pod that was to have been released by the B-58 Hustler strike aircraft to deliver a nuclear warhead. The first launch of the Thor–Agena in January 1959 for the Discoverer series failed, but on 28 February the Agena attained low polar orbit for the first of a series of flights to test a capsule for returning film from a spacecraft, and this was subsequently utilised operationally with successive generations of 'KH' cameras.[7,8]

ATLAS

The prototype hydrogen bomb that America detonated in May 1951 was too bulky to be dropped by an aircraft, exceeding the capacity of even the B-36, at that time the largest bomber in the world, but in the expectation of a much more compact device eventually becoming available Convair was hired to design a ballistic missile with a range of 10,000 kilometres. In 1953 (the year that the Soviets tested their own hydrogen bomb) it was realised that a breakthrough in miniaturisation would soon facilitate the manufacture of a warhead that weighed only 1 tonne, and the development of this missile, named Atlas, became a national priority. The same logic that led Korolev to adopt parallel staging led Convair to place three thrust chambers in line, but in this case the auxiliaries were mounted one on each side of the central 'sustainer', and the engines were to be gimballed to steer the missile. An innovative feature was that the body of the missile was a thin-skinned structure that relied on pressurisation to maintain its rigidity. The engines burned kerosene and oxygen fed from common tanks. Two verniers, one on each side, perpendicular to the line of the main engines, could be independently pivoted on that axis, and after the sustainer shut down these were to adjust the velocity to refine the range.

The fact that the auxiliary engine of the Rocketdyne MA-1 power plant became available prior to the sustainer prompted the test of missiles with only the side chambers. On 11 June 1957 first vehicle was destroyed by the range safety officer at an altitude of only 10,000 feet, and on 25 September the second was blown up after three minutes. In both cases the problem involved the propellant feed system. The test on 17 December, however, was successful, and the vehicle fell into the Atlantic Ocean 1,000 kilometres downrange. Although photographs were published in newspapers, the Air Force refused to confirm that the test had taken place! Eight flights of this Atlas-A configuration were sufficient to verify that the airframe maintained its integrity in flight, and that the guidance system functioned. It was an excellent start. The Atlas-B had all three chambers. Only 16 per cent of the lift-off thrust was provided by the sustainer. After 120 seconds the side chambers shut off and were jettisoned along with the skirt around the base of the missile, and the sustainer burned for a further 150 seconds. During staging, the propellant flow to the side chambers had to be terminated without disrupting the sustainer, and the hardware had to be released without disturbing the vehicle. This represented a real engineering challenge, and it took 10 test flights in 1958 and 1959 to rectify the problems. Meanwhile, the Atlas-C tested systems that were not in these initial vehicles. In 1959 the Atlas-D equipped with the MA-2 engine became the first intercontinental ballistic missile to enter service with the Air Force; however, deployment was slow and there were only a dozen or so by the end of 1960. Having finished the development phase that demonstrated the technology, plans were set in motion to re-engineer the missile so that the Atlas-E would be not only more powerful but also more operationally flexible. The MA-3 incorporated a number of refinements, including a simplified igniter, and further increased the thrust. Ironically, having perfected the staging with the Atlas-B, Convair was baffled by the Atlas-E's tendency to explode during this operation. An investigation established that propellant was draining from the severed pipes, igniting in the sustainer's exhaust, and sending a shock wave up into the engine compartment that fractured the still pressurised pipes, resulting in an explosion. The remedy was to fit valves on the engine end as well as the tankage end of the pipes. It was later realised that this flaw was in the original design, but had been benign since the shape of the earlier pipes had not dumped the unused propellant. As a result, switching from the developmental to the operational missile took much longer than expected. The Atlas-F could be stored vertically in a silo, but had to be raised to the surface prior to firing. By 1962, there were 30 Atlas-D, 30 Atlas-E and 70 Atlas-F vehicles in service, but when the Minuteman solid-propellant missile entered service in 1963 it was decided to phase out the Atlas. Although the schedule did not call for the final missile to be decommissioned until the end of 1967, by mid-1965 they had all been returned to the manufacturer for upgrading with the MA-5 power plant (there was no MA-4) to serve as launch vehicles.[9] Meanwhile, in June 1961 Convair had become part of the General Dynamics Corporation.

Testing times

The Atlas-E was first tested on 11 October 1960 from Canaveral, and was a partial success in that it achieved a peak altitude of 1,000 kilometres. On 30 November the second had a hydraulics failure; on 24 January 1961 the third was disabled by a flight control failure; on 24 February the fourth peaked at 1,500 kilometres; on 14 March the fifth had a propellant problem; and on 25 March the sixth suffered an electrical failure. Two successes were followed on 7 June by the first test flight from Vandenberg, which was a total loss. On 23 June there was a flight control failure, then two successes, followed by a propulsion failure on 9 September. After a further two successful flights, on 10 November 1961 the sustainer engine shut down 15 seconds after lift-off and the missile had to be destroyed by the range safety officer. The first test of the Atlas-F was on 9 August 1961 and after two successes that flew the planned trajectory peaking at an altitude of 1,400 kilometres, on 12 December 1961 the third suffered a guidance failure and peaked at 1,000 kilometres and on 21 December the fourth had a hydraulics problem and peaked at only 500 kilometres. The test on 9 April 1962 was a total failure. The first Atlas-F test at Vandenberg on 1 August 1962 was a success, but the next on 10 August was a total loss. After four successes, the test on 14 November 1962 reached only 300 kilometres. After another success, the missile on 1 March 1963 overperformed and reached an altitude of 1,800 kilometres. Another two successes were followed by a total failure on 24 March and another success. An accident during propellant loading for an operational test at Walker Air Force Base on 1 June caused an explosion that destroyed the missile. A launch at Vandenberg the next day was a total loss, as was a test on 4 October, but that on 18 December 1963 succeeded. Three missiles were then lost in pad mishaps during operational readiness tests on 13 February, 9 March and 14 May 1964. And so the programme continued, with a progressively improving success rate.[10,11]

Converted launchers

Once the Atlas-F had been established as a reliable vehicle, the Air Force started to fire it from Vandenberg to conduct a variety of experiments and to launch satellites, in most cases successfully. In an incident on 13 April 1975 a lack of deluge water allowed a gel of kerosene and oxygen to collect in the flame bucket and explode as the vehicle lifted off, damaging an engine sufficiently to prompt it to fail in flight. On 29 May 1980, the sustainer engine suffered an internal fuel leak as it ignited, causing it to underperform by about 20 per cent. This not only slowed the rate of acceleration, it left the vehicle fat on propellant, with the result that the sustainer had to burn for 50 seconds longer than planned. Since the payload had no communication with the vehicle, at T + 375 seconds, oblivious to the fact that it was still under thrust, the Star 37 stage separated and fired its motor, blowing off the top of the vehicle's oxygen tank. The payload, NOAA B, an Advanced-TIROS meteorological satellite, survived, but the resultant highly elliptical orbit was useless

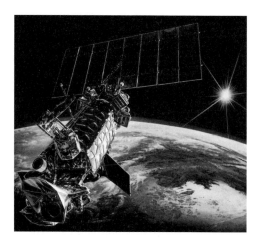

An artist's impression of a DMSP weather satellite.

and the mission was declared a total loss. A replacement was deployed by the next (and penultimate) launch on 23 January 1981. After a long hiatus, the final missile was dispatched on 6 February 1985. In the course of 101 launches the Atlas-F had suffered 15 failures, most of them in the early days. The first attempt to use an Atlas-E to orbit a satellite, in this case a 'White Cloud' Navy Ocean Surveillance Satellite, was on 9 December 1980, on which occasion a hydraulics problem in one of the auxiliary engines caused it to fail a fraction of a second before it was to have shut down and been jettisoned. The momentary asymmetric thrust flipped the vehicle around, and the sustainer drove it back through its own exhaust trail until the range safety officer destroyed it. Although the development of the Atlas-E as a missile was protracted, its later value as a launcher was illustrated by the fact that the final vehicle was retrieved from storage at the Smithsonian Air and Space Museum and used to place the DMSP 13 weather satellite into polar orbit for the Department of Defense on 24 March 1995.[12,13]

Atlas–Agena

The decommissioned Atlas-D missiles were fitted with Agena upper stages for use as space launchers. As the Agena was stretched to carry more propellant, and its engine was made both more powerful and capable of being restarted in space, it became available in A, B and D models (the planned C model was cancelled). The Atlas–Agena-A employed the MA-2 power plant.[14] Two of four launches failed. On the first launch, on 26 February 1960, the Agena with the first satellite for the Missile Defense Alarm System (MIDAS) failed to separate from its booster, but the second was inserted into orbit on 24 May. On 11 October the Agena with the first of the Satellite and Missile Observation System (SAMOS) satellites failed, but the second was launched on 31 January 1961. The Atlas–Agena-B had the MA-3 power plant.[15] Of 28 launches between 12 July 1961 and 21 March 1965, eight suffered problems. On 23 August 1961, on the second mission, the Agena attained low 'parking orbit' and then failed to restart to make the burn that was to insert the Ranger 1 spacecraft into a highly elliptical orbit. The next launch on 9 September with SAMOS 3 was a total loss when the vehicle exploded on the pad. MIDAS 4 was successfully dispatched on 21 October, but on 18 November Ranger 2 was stranded in parking orbit when an inoperative roll gyroscope prevented the Agena restarting. After SAMOS 4 was lost on 22 November, SAMOS 5 was successfully placed into orbit on 22 December but failed to return its film capsule. On 26 January 1962 the Agena with Ranger 3 was able to restart to leave parking orbit, but a guidance fault made the probe miss the Moon by some by 37,000 kilometres. After a string of successes, Mariner 1 was lost on 22 July when the Atlas flew off course and at T + 293 seconds was destroyed by the range safety officer. Although this is often cited as the exemplar of the loss of a mission to a software error, in fact, as is often the case, two independent faults had interacted fatally.[16] The guidance antenna on the Atlas performed below specifications. When the received signal became weak and noisy, the vehicle lost its lock on the ground reference that supplied steering commands. In the event that radio guidance was lost, the autopilot was supposed to reject the spurious signals from the antenna and proceed on its stored program. However, at this point a second fault took effect. Due to an oversight, a hyphen had been omitted from the program, and this had the effect of allowing the flawed signals to command the vehicle to veer left and drop its nose. It was found that the hyphen had always been missing, but had been benign since there had been no radio guidance failure on the previous flights. Mariner 2 was successfully dispatched on 27 August and became the first spacecraft to return data about another planet when it flew past Venus on 14 December 1962 at a range of 35,000 kilometres. There were another two successes, then MIDAS 6 was lost on 17 December 1962 and MIDAS 8 on 12 June 1963. The Atlas–Agena-D had the MA-5 power plant.[17,18] Of 27 launches between 12 July 1963 and 7 April 1978 there were only two failures. On 5 November 1964 Mariner 3 was trapped when the shroud failed to separate. An investigation found that the new lightweight shroud had bonded to the probe. However, it was able to be revised in time to dispatch Mariner 4 on 28 November, a few days before the 'window' for a mission to Mars closed, and the pictures that it snapped during a 10,000-kilometre flyby on 15 July 1965 revealed the surface of the planet to be scarred by large impact craters. The other incident was on 4 December 1971, when an Atlas was lost with a Canyon communications intelligence-gathering payload.

In addition, in June 1962 NASA ordered a variant of the Agena on which the

Atlas models (left to right): Atlas-D, 20 February 1962 with John Glenn in a Mercury capsule; Atlas–Able, 26 November 1959 with Pioneer P3 (lost when the shroud collapsed after 45 seconds); Atlas–Agena-A, 26 February 1960 with MIDAS 1; Atlas–Agena-B, 9 September 1961 with SAMOS 3; Atlas–Agena-D, 30 September 1965 with the 22nd KH-7 satellite.

engine could be fired as many as six times to enable it to undertake orbital manoeuvres while docked to a manned Gemini spacecraft. This Gemini–Agena Target Vehicle also had a pair of thrusters mounted in 'saddle bag' fashion on each side of the main engine. The first of these vehicles lifted off from Pad 14 at Canaveral on 25 October 1965. After the Agena was released, it coasted for 50 seconds and then fired its thrusters for ullage to settle the propellants in their tanks prior to firing the main engine for orbital insertion. The telemetry data ceased a fraction of a second after the engine ignited.[19] Discovering what had gone wrong from the momentary telemetry was a daunting task. The multiple restart capability had been provided by replacing the solid-charge starter that generated the gas to drive the turbine with one that created the gas by vaporising liquid drawn from a tank. This was not a factor. However, in the Agena-D, which could be restarted only once, the flow of oxidiser was initiated prior to feeding fuel since trials had shown that this improved the engine's performance. But because NASA intended to use the vehicle for extensive orbital manoeuvring, it had rejected this 'waste' of oxidiser, and demanded that the fuel and oxidiser be injected simultaneously. Despite successful ground tests, the fuel in this case had been injected into the chamber *ahead* of the oxidiser, creating a combustion instability in which the fuel-rich mix had burned explosively, and this 'hard start' had created a pressure transient that blew the engine apart.[20] This therefore serves as an excellent example of how a seemingly straightforward modification can give rise to complications. The next flight on 16 March 1966 was successful, and when Gemini 8 rendezvoused with the vehicle a few hours later Neil Armstrong and Dave Scott became the first people to see another spacecraft in orbit.[21]

A GATV as seen from Gemini 8.

The target vehicle for the next mission, Gemini 9, was lost on 17 May 1966 when, 10 seconds before the Atlas was to shed its side-mounted auxiliaries, one of the engines gimballed hard over and forced it to nose-dive into the Atlantic. As a contingency, it was decided to launch an *ad hoc* target comprising the docking system stabilised by a bolted-on attitude control system of the type used by the Gemini spacecraft. This Augmented Target Docking Adapter was placed directly into the rendezvous orbit on 1 June, but the telemetry during the ascent was ambiguous in that although the command to jettison the aerodynamic shroud had been issued, the confirmation that this had occurred was not received. When Tom Stafford and Gene Cernan drew along-

The ATDA with its fouled shroud.

side the vehicle, they found that the shroud had partially deployed – the pyrotechnics had separated its base from the vehicle and split it lengthwise, but a thin metal strap was preventing it from slipping off the aperture of the conical docking collar. Stafford likened the fouled shroud to the jaws of "an angry alligator". When the pictures taken by the astronauts were examined by the engineers, the fault was identified. The Agena shrouds were made by Douglas Aircraft, and the Gemini–Agena Target Vehicles were built by Lockheed, whose pad technicians were trained to install the shrouds. However, the improvised Augmented Target Docking Adapter had been made by McDonnell (which had built the Gemini spacecraft and the docking system) and in this case the shroud had been fitted by McDonnell's technicians, the Lockheed supervisor had been called away, and his verbal instructions for finishing the job had been misunderstood.

Atlas–Centaur

The Centaur was a powerful upper stage that used a pair of RL-10 hydrogen-burning engines. On the first test of the Atlas–Centaur on 8 May 1962, the shroud collapsed early in the ascent.[22] On 27 November 1963 the Centaur achieved orbit, but on 30 June 1964 it suffered a hydraulic failure. The flight on 11 December 1964 succeeded, but on 2 March 1965 the vehicle exploded on the pad. After two successes it made its operational début on 30 May 1966 by dispatching Surveyor 1 to land on the Moon. On 10 August 1968, on its first attempt to insert a satellite into geosynchronous transfer orbit, the Centaur failed to reignite, stranding ATS 4 in parking orbit. On 30 November 1970 the shroud failed to separate, trapping OAO B. On 8 May 1971 a faulty gyroscope circuit in the autopilot prevented the Centaur gimballing its engines, with the result that Mariner 8 was dumped in the Atlantic.[23] Intelsat 406 was lost on 20 February 1975 when the Atlas suffered an electrical failure during separation, and Intelsat 4A5 was lost on 30 September 1977 when one of the turbopumps in the Atlas failed at $T + 55$ seconds.[24]

When the Atlas-G with its stretched core was introduced on 9 June 1984, the Centaur exploded as a result of an oxygen leak caused by an anomalously violent separation, and Intelsat 509 was lost.[25] The first FLTSATCOM satellite for the Navy was put into geosynchronous transfer orbit by an Atlas–Centaur in 1978, but when the fifth satellite in this series was launched on 6 August 1981 the fibreglass

Atlas–Centaur, 26 May 1977 with Intelsat 4A5.

shroud collapsed.[26] Although the Centaur put the payload into geosynchronous transfer orbit and the Star 37F apogee motor circularised the orbit, the satellite had been badly damaged by the shroud and had to be written off.[27] With the initial FLTSATCOMs surviving beyond their five-year design life, the first of the Block-II series, FLTSATCOM 6, lifted off on 26 March 1987 on the penultimate Atlas-G into driving rain from a 2,500-foot overcast. When it strayed off course at T + 51 seconds the range safety officer destroyed it.[28] John W. Gibb, the manager of the Atlas–Centaur project office, reported that this launch had not violated the long-established rules that allowed a launch in rain, prohibiting doing so only if there was a thunderstorm within 5 miles of the pad or anvil clouds within 3 miles of the pad.[29] In this case, the countdown had been 'held' to await acceptable conditions. As the débris had fallen close offshore it was readily recovered and the investigation, led by John R. Busse, soon found a piece of the aerodynamic shroud with a number of pin-hole punctures burned through it, indicating that the vehicle had been hit several times by lightning.[30] This was confirmed when the digital computer unit was located and showed that the last command gimballed the engines hard over.[31] This was not a software error, the false command was the result of "a single random upset" in the computer's memory induced by the intense electric field of the lightning strike. The erroneous command had been issued at T + 38.3 seconds, and the aerodynamic loads had caused the vehicle to break up at T + 50.7 seconds. The destruction command had finished the job. It was concluded that the launch had been a "vehicle no-trial", because the Atlas had been dispatched into an environment in which it was not designed to survive.[32,33,34] Busse dismissed NASA's claim that the launch had not violated the commit criteria. He said the basic failure was a "missed call", and criticised the lack of coordination between the weather forecasters and the launch team.[35] NASA upgraded its local weather monitoring (particularly of the electric fields) to enable it to make a better assessment of when an acceptable 'window' might occur in poor weather. The loss of the Atlas was in marked contrast to Apollo 12, which was launched on 14 November 1969 into an overcast of stratocumulus in defiance of a rule that no vehicle be launched into a thunderstorm. No sooner had the 360-foot-long Saturn V cleared the 400-foot tower than it vanished into the murk. At T + 36.5 seconds, at an altitude of 6,100 feet, the vehicle was hit by a bolt of lightning that discharged to ground down the exhaust plume. A reverse electrical surge damaged some of the instrumentation in the spacecraft and caused its fuel cell system to disconnect from the power buses, and 16 seconds later a cloud-to-cloud bolt disrupted its inertial guidance. Fortunately, the Saturn V had its own

The launch of Apollo 12 into a thunderstorm, and the resultant lightning strike half a minute later.

power supply and guidance, and flew on unscathed.[36] As this was the first time that Walter Kapryan (the launch director of the Kennedy Space Center) and Gerry Griffin (the flight director in Houston) had supervised an Apollo launch, they promptly became known as "the rookies who launched into a thunderstorm".[37]

THE TITANS

When the Atlas became a priority, the Air Force ordered the parallel development of a completely different design, in case the Atlas proved impractical. The contract went to the Martin Company. By this time, the engineering database had advanced sufficiently to make air-starting a liquid rocket an acceptable challenge, and it was decided to employ an efficient tandem configuration. The first stage had a twin-chamber Aerojet LR-87 engine and the second had a single-chamber LR-91, both burning kerosene and oxygen. The engines were to gimbal to steer the vehicle in powered flight, and verniers were to refine the trajectory of the second stage after its main engine shut down. Development was rapid, and in February 1959 the first stage was tested using a mock-up second stage. Ten full-range flights followed in 1960. Like the Atlas-F, the Titan was to be stored vertically in a silo and raised for firing. Being more advanced, it could have been expected to complement and later supersede the Atlas, but in 1963 the Air Force decided to withdraw both. Unlike the Atlas, the Titan was junked. The Titan II, on which work started in 1960, was a very different story. Although it benefitted from lessons learned with the Titan I, it was a wholly new design and a variety of technological developments simplified its construction. In this case the engines burned hydrazine and nitrogen tetroxide which, being hypergolic, burned on coming into contact, thereby eliminating the igniter. The fact that these propellants were not cryogenic enabled the missile to be stored ready to fire from its silo literally at one-minute's notice. Development was rapid, and after a test programme that began with the début launch in March 1962, the Air Force began the deployment of three 'wings', each of 18 missiles, in 1963.

In addition, the Air Force intended to use the Titan II as a space launcher, and

Titan II and early III models (left to right): Titan II ICBM; Titan II, 8 April 1964 with Gemini 1; Titan IIIA, 11 February 1965 with the Lincoln experimental communications satellite; Titan IIIB, 6 November 1968 with the 17th KH-8 satellite; Titan 34B, 24 April 1981 with Jumpseat 6.

NASA ordered a dozen adapted for use in the Gemini programme in 1964–1966. The Titan IIIA was a 'stretched' Titan II fitted with a third stage. The first test from Pad 20 at Canaveral on 1 September 1964 went well until a pressurisation failure caused the premature shut down of the third stage, known as the Transtage. In 1966, after three successful flights of the Titan IIIA, the Air Force switched over to the Agena-D for the simple reason that the mass-produced stage was more cost-effective for the envisaged missions. The first of the Titan IIIB vehicles was successfully dispatched from SLC-4W at Vandenberg on 29 July 1966. All but two of a total of 54 launches over the next 20 years successfully inserted KH-8 'close look' reconnaissance satellites into low polar orbit.[38] The losses were on 26 April 1967 when the second stage lost thrust, and 26 June 1973 when the Agena failed.

Meanwhile, the Titan IIIC was introduced by augmenting the Titan IIIA with a pair of five-segment solid rocket boosters. The inaugural flight from Pad 40 at Canaveral was a success with a dummy payload. On the second flight on 15 October 1965, however, the Transtage broke up. Nevertheless, all but three of 35 fired from Canaveral between 1966 and 1982 successfully deployed a variety of payloads. The failures were 26 August 1966 when the shroud failed at T + 78 seconds, 20 May 1975 when the Transtage lost attitude control following a power failure to its gyroscopic platform, and 25 March 1978 when a hydraulics pump in the second stage failed. The Titan IIID was essentially the Titan IIIB with five-segment boosters. The first lifted off from SLC-4E at Vandenberg on 15 June 1971, and 22 successfully deployed KH-9 and KH-11 reconnaissance satellites by 1983. The Titan 34B was the Titan IIIB with an enlarged shroud that enclosed both the Agena and either an SDS data-relay or a Jumpseat electronic intelligence-gathering satellite. Of 14 launched from SLC-4W between 1971 and 1987, all but one successfully inserted its payload into an orbit with a high apogee above the northern hemisphere; the failure was on 16 February 1972 (the second flight). Meanwhile, NASA had introduced the Titan IIIE, which

Titan IIIC and Titan IIID models (left to right): Titan IIIC, 18 January 1967 with a dispenser of eight satellites for the Initial Defense Communications Satellite Program; Titan IIIC, 14 December 1975 with the 5th IMEWS (the last of the first model of DSP); Titan IIID, 15 June 1971 with the 1st KH-9 'Big Bird'; Titan IIID, 14 June 1978 with the 2nd KH-11.

was a Titan IIIC with a Centaur upper stage. Although the Centaur failed on the vehicle qualification flight on 11 February 1974, over the next three years six were successfully launched from Pad 41 with a pair of Helios, a pair of Viking, and a pair of Voyager spacecraft.

The Titan 34D was the Titan 34B augmented with longer strap-ons with five-and-a-half segments per booster and either a Transtage or a two-stage IUS instead of the Agena. Its introduction on 30 October 1982 from Pad 40 at Canaveral marked the début of the IUS, which successfully delivered a pair of DSCS communications satellites to geostationary orbit for the Department of Defense. The first Titan 34D to be launched from SLC-4E at Vandenberg was on 20 June 1983. Of 15 launched with a variety of payloads by 1989, all but three were successful. On 28 August 1985 a Titan 34D tumbled and the range safety officer destroyed it.[39,40] Not only was this the first failure for this variant, it was the first such mishap in 18 years of Titan operations at Vandenberg. The investigation concluded that the evidence indicated an oxidiser leak and the consequent failure of a turbopump subassembly when its pinion gear broke as a result of the loss of gear cooling or lubrication. This caused the premature shut down of one of the two engines on the core stage, and because the vehicle required both engines to maintain controlled flight, it tumbled.[41] This incident highlights an important consideration in launch vehicle design – a vehicle with more engines will lose a smaller fraction of thrust if one engine fails, thus (as for aircraft) launchers with more engines are perceived as being more reliable. However, if the vehicle's thrust margin is so low that it requires all of its engines to operate, then the probability of a launch failure will be greater for a many-engined vehicle than for one that has only one engine. By the end of 1985, however, the Air Force was phasing out the Titan, and had only a few left.

A Titan IIIE–Centaur lifts off on 20 August 1976 with Voyager 2.

NASA'S DELTA

On 29 April 1959 (six months after its establishment) NASA ordered 12 Thor–Able launch vehicles, which it renamed the Thor–Delta, and it also asked Douglas to improve the inertial system to provide more accurate guidance during the ballistic coast leading up to the release of the third stage.[42] The maiden launch on 13 May 1960 from Pad 17 at Canaveral failed because the second stage suffered a loss of attitude control, but the next mission on 12 August 1960 successfully placed the Echo passive relay 'balloon' satellite into orbit. The last of the batch was fired on 18 September 1962 and, apart from the first, they all succeeded with a variety of payloads which included the first TIROS weather satellite and Telstar, the first active-repeater communications satellite.[43] By this time, of course, a delighted NASA had ordered another batch of rockets. Fortunately, the potential of the Thor enabled Douglas to exploit the 'building block' approach to undertake a programme of progressive upgrades which were designated alphabetically. Two launches of the Delta-A in October 1962 tested the more powerful MB-3-II first-stage and the AJ-10-118 second-stage motor, which had a restart capability, and on 13 December 1962 the Delta-B introduced

Titan 34D models (left to right): Titan 34D, 30 October 1982, with an IUS carrying the 15th and final DSCS II and the 1st DSCS III; Titan 34D, 10 May 1989 with a Transtage carrying the 6th and final Chalet satellite; Titan 34D, 18 April 1986 with the 20th and final KH-9.

the AJ-10-118A with a longer burn time and better guidance. Nine vehicles of this type were launched with a payloads that included Syncom, the first communications satellite to be placed into geostationary orbit.[44] The only failure was on the final launch, on 19 March 1964, when the third stage generated insufficient thrust and stranded Explorer 20 in the wrong orbit.

In 1963, the Delta-C had the MB-3-III engine and the AJ-10-118D, and introduced the X-258 Altair II third stage. Of 12 fired between 27 November 1963 and 22 January 1969 only one failed. On 25 August 1965 the early ignition of the third stage in the post-boost coast doomed OSO C. Meanwhile, the Air Force had augmented its Thor–Agena by strapping three Castor I (TX-33) solid rockets to the first stage, and this strategy was copied for the Delta-D, which was introduced as the Thrust Augmented Thor (TAT) Delta on 19 August 1964. On 6 April 1965 one put the 'Early Bird' communications satellite in geostationary orbit above the Atlantic.

A Thor–Delta launches Telstar 1 on 10 July 1962. The inset shows a scene from the first transatlantic TV relay.

A TAT Delta launches the Early Bird on 6 April 1965.

Next, the Thor was stretched for greater endurance and augmented by three of the more powerful Castor II (TX-354) strap-on solid rockets, and the second stage was increased in diameter to carry propellant to triple its duration and fitted with the more powerful AJ-10-118E. Termed the Delta-E, six of these Improved TAT Deltas were launched between 6 November 1965 and 20 April 1967, with no failures. A second form utilised the United Technologies FW-4D solid motor as its third stage. Only one of 18 flights between 1 July 1966 and 1 April 1971 had a problem. This was on 26 August 1966, and it was not really a launch vehicle issue as the Intelsat 2F1 satellite stranded itself in geosynchronous transfer orbit when its built-in apogee motor shut down after just 4 seconds.[45] The Delta-F was the FW-4D form of the Delta-E without strap-ons, but there proved to be no requirement for it. The Delta-G was a two-stage version of the Delta-E using the older Castor I strap-ons. It was used in 1966 and 1967 for two recoverable Biosatellites. The Delta-H was the Delta-G without strap-ons, but again there was no need for it. The Delta-J, which was the Delta-E with a Star 37 third stage, was used only once, in 1968. The Delta-K never got off the drawing board.

The next major advance was the Long Tank TAT Delta, in which the first stage was made cylindrical rather than conical, and stretched by 15 feet. The Delta-L, -M and -N differed only in their third stages: the Delta-L had the same third stage as the Delta-E; the Delta-M had the same third stage as the Delta-J, and the Delta-M-6 was the Delta-M with three extra Castor II strap-ons (for a total of six, thereby giving rise to the name Super Six); the Delta-N was a two-stage version of either the Delta-L or the Delta-M, and the Delta-N-6 was a two-stage Delta-M-6. On the maiden launch of the Delta-L on 27 August 1969 a hydraulic oil relief valve leaked due to vibration, resulting in loss of hydraulic pres-

sure which permitted the engine nozzle to develop uncontrolled gimballing and attitude excursions, and at T + 383 seconds the range safety officer destroyed the vehicle and its payload, which was the Pioneer E satellite that was to have been placed into solar orbit to report on the solar wind.[46] When the first Delta-M was launched on 19 September 1968 a pitch-rate system malfunction was noticed about 20 seconds after lift-off, the first stage began to break up at T + 102 seconds, and 6 seconds later the range safety officer destroyed it and its Intelsat 3F1 payload.[47] In the case of the Delta-M on 26 July 1969, the third stage motor's casing either ruptured or there was nozzle failure, with the result that Intelsat 3F5 was stranded in the wrong orbit.[48] When the first Delta-N lifted off from SLC-2 on 21 October 1971 with the ITOS B weather satellite, the second stage suffered an oxygen leak.[49] The pitch and yaw jets pulsed to counteract the force of this venting, maintaining the proper attitude until their gas was expended, whereupon the vehicle tumbled.[50] On 16 July 1973 a two-stage vehicle with ITOS E suffered an attitude control malfunction 270

Delta 1913, 10 June 1973 with Explorer 49.

seconds into the second-stage burn following an abrupt fall in output from the hydraulic pump caused a loss of hydraulic pressure that disabled the thrust vector control system.[51]

Meanwhile, in April 1967 Douglas had merged with McDonnell, to form McDonnell Douglas. Having advanced far through the alphabet to accommodate the recent clutch of very similar variants, it was decided to switch to a numerical system that would provide a more structured scheme. The four digits specified in turn, the first stage, the number of strap-ons, the second stage and the third stage. In the 1000 series (which ran to the

Delta 2313, 19 January 1974 with Skynet 2A.

1410, 1604, 1900, 1910, 1913, 1914 variants) there could be 4, 6 or 9 Castor II strap-ons, the second stage engine was either the AJ-10-118F or the TR-201 (the latter being basically the descent engine of the Apollo Lunar Module with its throttle fixed at 100 per cent), and the third stage (if present at all) was one of several Star 37 solid motors. It introduced the Extended Long Tank configuration that was dubbed the Straight Eight because the upper part of the vehicle continued the 8-foot diameter of the first stage. This time the two-stage form was of use because the increased capability of the second stage enabled it to deliver a heavier satellite than the previous three stage vehicle.[52,53] Eight were launched between 23 September 1972 and 21 June 1975 with a variety of payloads.

The 2000 series marked a major upgrade that introduced a new power plant for the first stage. The Rocketdyne RS-27 combined the control system of the MB-3 with the turbopump, turbine, gas generator, valves and thrust chamber of the H-1 used in the first stage of the Saturn I series.[54] Of 44 launches between 19 January 1974 and 6 October 1981 there were two failures. On the maiden flight (which was a 2313) a fragment of conductive contaminant was shaken loose by vibrations, causing a short circuit in the electronics of the second stage and, as a result, Skynet 2A was released into a low orbit that decayed after several days. The investigation concluded that the insulation coating on the printed circuit boards had been substandard and the contaminant had welded across the exposed ends of component leads.[55] On 20 April 1977 the third stage of a 2914 was released before the rotator on the second stage had spun it up for stability, causing it to tumble.[56] Nevertheless, by any standard, over the years the Delta had been a reliable launcher for small-to-medium payloads.

NOTES

1. *Spaceflight*, November–December 1980, p. 340.
2. The configuration of the *Semyorka* was not publically revealed until it was put on display at the Paris Air Show in 1967.
3. http://ntrs.nasa.gov/archive/nasa/casi.ntrs.nasa.gov/19740072500_1974072500.pdf
4. http://www.tbs-satellite.com/tse/online/lanc_vanguard.html
5. http://www.designation-systems.net/dusrm/app3/b-3.html
6. http://www.designation-systems.net/dusrm/app3/index.html
7. http://www.fas.org/spp/military/program/launch/delta.htm
8. http://www.astronautix.com/lvs/atlgenaa.htm
9. http://www.tbs-satellite.com/tse/online/lanc_atlas.html
10. http://www.astronautix.com/lvs/atlase.htm
11. http://www.astronautix.com/lvs/atlasf.htm
12. *Aviation Week & Space Technology*, 27 March 1995, p. 66.
13. *Aviation Week & Space Technology*, 3 April 1995, p. 29.
14. http://www.astronautix.com/lvs/atlgenaa.htm
15. http://www.astronautix.com/lvs/atlgenab.htm
16. *Far Travelers: The Exploring Machines*, O.W. Nicks, SP-480, NASA, 1985, p. 1.
17. http://www.astronautix.com/lvs/atlgenad.htm
18. http://www.astronautix.com/lvs/atlslv3a.htm

19. *On the Shoulders of Titans – AHistory of Project Gemini*, B.C. Hacker and J.M. Grimwood, NASA SP-4209, 1977, p. 268.
20. *On the Shoulders of Titans – A History of Project Gemini*, B.C. Hacker and J.M. Grimwood, NASA SP-4209, 1977, p. 298.
21. *How NASA Learned to Fly in Space – An Exciting Account of the Gemini Missions*, D.M. Harland, Apogee Books, 2004, p. 152.
22. http://www.tbs-satellite.com/tsc/online/lanc_atlas_centaur.html
23. *Solar System Log*, A. Wilson, Jane's, 1987, p. 64.
24. *Spaceflight*, November–December 1980, p. 351.
25. http://www.astronautix.com/lvs/atlasg.htm
26. http://www.astronautix.com/craft/fltatcom.htm
27. http://www.friends-partners.org/oldfriends/jgreen/fltsatco.html
28. *Aviation Week & Space Technology*, 30 March 1987, p. 20.
29. *Aviation Week & Space Technology*, 6 April 1987, p. 23.
30. *Aviation Week & Space Technology*, 13 April 1987, p. 31.
31. *Aviation Week & Space Technology*, 4 May 1987, p. 21.
32. *Jane's Space Directory*, vol. 10, 1994–1995, p. 260.
33. Note that it was not 26 February 1987, as some say; there was no launch of an Atlas–Centaur on that date; in fact, this was the only such launch in the whole year!
34. *Aviation Week & Space Technology*, 8 August 1988, p. 67.
35. *Aviation Week & Space Technology*, 18 May 1987, p. 25.
36. *Apollo 12 Mission Report*, NASA, MSC-01855, March 1970.
37. *Apollo EECOM – Journey of a Lifetime*, S. Liebergot with D.M. Harland, Apogee Books, 2003, p. 136.
38. http://www.astronautix.com/lvs/titan3b.htm
39. *Aviation Week & Space Technology*, 18 November 1985, p. 26.
40. web.mit.edu/org/s/spacearchitects/Archive/space wars.doc
41. https://www.patrick.af.mil/ heritage/Cape/Cape1/cape1fn.htm
42. *http://www.tbs-satellite.com/tse/online/lanc_thor_delta.html*
43. *Spaceflight*, October 1979, p. 413.
44. http://www.boeing.com/defense-space/space/bss/factsheets/376/syncom/syncom.html
45. http://www76.pair.com/tjohnson/thrfail.txt
46. http://www76.pair.com/tjohnson/thrfail.txt
47. *Aviation Week & Space Technology*, 23 September 1968, p. 27.
48. http://kevinforsyth.net/delta/delta2.htm
49. http://www76.pair.com/tjohnson/thrfail.txt
50. http://kevinforsyth.net/delta/delta2.htm
51. http://kevinforsyth.net/delta/delta2.htm
52. http://www.daviddarling.info/encyclopedia/D/Delta.html
53. http://www.spaceline.org/rocketsum/delta-1000.html
54. *Jane's Space Directory*, vol. 10, 1994–1995, p. 283.
55. http://kevinforsyth.net/delta/delta2.htm
56. http://kevinforsyth.net/delta/delta2.htm

2

The Shuttle

COMMERCIAL SATELLITES

Telstar, the world's first commercial communications satellite, was developed by the American Telephone & Telegraph Company (AT&T) as a private venture, and launched by NASA on a Delta on 10 July 1962. Although the satellite relayed the first television across the Atlantic, large swivelling antennas were required to communicate with it and the 6,000-kilometre apogee restricted its use as a long-distance relay to brief periods.

On 31 August 1962 the US government passed the Communications Satellite Act and ordered the creation of a national consortium to develop and run a system to serve the domestic telecommunications industry.[1] On its establishment in February 1963, this Communications Satellite Corporation (Comsat) considered a constellation of Telstars, but decided to use geostationary relays, as these would provide continuity of service. In 1959 the Hughes Aircraft Company had set out to design such a satellite, and in August 1961 was given a contract by NASA to supply three of them. On 14 February 1963 the first Syncom fell silent towards the end of the 20-second firing of its liquid-propellant rocket to circularise its orbit at the requisite 36,000 kilometres. "The culprit first seemed to be the apogee motor," recalled Harold Rosen, the project leader. But there were other possibilities. "The nitrogen tanks might have had too much pressure and exploded, or the critical wire that powered both the telemetry transmitter and the communications transmitter could have broken suddenly." On 26 July 1963, the second satellite achieved geosynchronous orbit, but was not geostationary because no attempt was made to cancel the inclination of its orbit, with the result that over a 24-hour period it nodded 33 degrees north and south of the equator. The Delta that launched the third satellite on 19 August 1964 flew a series of manoeuvres to insert its payload into geostationary orbit.[2] It arrived above the International Date Line just in time to relay coverage of the Olympic Games in Japan to a fascinated American audience, and the tag 'Live Via Satellite' became the defining symbol of the time. That month, Comsat led the

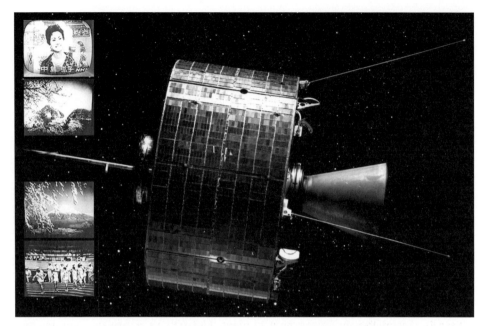

The Syncom satellite that relayed the 1964 Olympic Games.

creation of the International Telecommunications Satellite Consortium (Intelsat) to establish and operate a network of geostationary satellites to relay telecommunications on a global basis, and on 6 April 1965 the first satellite, formally designated Intelsat 1, but more popularly known as the 'Early Bird', was stationed above the Atlantic; it could relay up to 240 voice channels or a single monochrome television channel.

By the mid-1970s Comsat had a fleet of Comstar satellites, and leased transponders to the US domestic market – its main customers being AT&T and General Telephone & Electronics (GTE), both of which ran lucrative businesses relaying voice for a diverse community of users. Intelsat controlled the majority of the international telephone traffic and most transoceanic television relays. Owing to the exponential growth in demand, it made financial sense to pursue the continuous development of successive generations of ever more capable satellites. Having built Syncom, Early Bird and the Comstars, Hughes had a clear lead, and exploited every development in the miniaturisation of electronics to stay ahead. It introduced the 300-kilogram HS-333, a spin-stabilised drum satellite with a conformal array of solar cells and a de-spun antenna that could relay either 1,000 voice channels or one colour television channel. The great advance was the electronics used to shape this C-Band beam to deliver all of the radiated power to a geographical 'footprint' defined by the client, enabling neighbours to share frequencies without interference. In 1972, the US Federal Communications Commission opened the market to satellite broadcasting, and cable television operators eagerly expanded into this new area because it enabled them to reach a vastly enlarged audience without the expense of laying cables. Hughes duly introduced the HS-376 with a variety of options that included the use of the higher-capacity Ku-Band. In addition to members of Intelsat buying satellites for local services, it now became possible for individual companies to purchase them for private use. When Satellite Business Systems (SBS) was spun off by Comsat in 1975, it ordered HS-376 satellites to sell an all-digital service integrating voice, data, video and e-mail to businesses having facilities widely distributed over the continental United States.

A FINAL DELTA

In the early 1970s, McDonnell Douglas set out to upgrade its Delta to carry satellites that were too heavy for the 2000 series, but not so large as to need an Atlas–Centaur. As introduced in 1975, the 3914 was its 2000 series counterpart with Castor IV (TX-526) strap-ons and a Star 37 third stage with a capacity to geosynchronous transfer orbit of 930 kilograms. In 1980 the third stage was replaced by the PAM-D. This Payload Assist Module was developed by McDonnell Douglas for use by Shuttle payloads in need of a perigee kick motor, but the Shuttle was late, and it had been decided to introduce the motor on the Delta – hence the 'D' qualifier. It incorporated a Star 48 motor. Perversely, as the PAM was considered to be part of the payload, the designation had a '0' to indicate the *absence* of a third stage. The 3910/PAM raised the capacity to geosynchronous transfer orbit to 1,090 kilograms. In 1982 a new second stage was introduced using the AJ-10-118K engine that had been developed for the Titan III's Transtage, which had two, as the AJ-10-138. The fact that this new stage had 30 per cent more propellant gave the 3920/PAM a capacity of 1,250 kilograms. The more powerful PAM-D2 (which used a Star 63 motor) was introduced later for heavier satellites.

Only three of the 38 Delta 3000 series dispatched between 13 December 1975 and 24 March 1989 had difficulties. On 13 September 1977 a 3914 carrying OTS 1 exploded at T + 54 seconds, probably due to a burn-through of the casing of one of the strap-ons.[3] A 3914 launched from Canaveral on 7 December 1979 suffered an apogee motor failure that stranded Satcom 3 in geosynchronous transfer orbit. On 3 May 1986 the first stage of a 3914 with GOES G shut down 71 seconds into the planned 223-second firing. This was the first premature shutdown for the RS-27 engine. The telemetry indicated that there had been two power surges immediately beforehand.[4] The wreckage was salvaged, and the electronics that controlled the engine was examined for evidence of a short circuit. McDonnell Douglas built a high-fidelity simulator of the first stage to enable failure scenarios to be assessed, and it was concluded that the fault was a wiring harness.[5,6,7] The investigation found that mechanical damage to wiring by vibration had – as Larry Ross, the chairman of the NASA Delta Review Board put it – prompted "a rather significant electrical fault". The battery current, ordinarily

running about 9.5 to 9.8 amperes, at about 70 and some-odd fraction of seconds into flight, took a very significant excursion on the order of 188 amps.

It did that for about 6 milliseconds. It recovered to normal and stayed that way for about 900 milliseconds. It took another excursion of about the same magnitude, this time for about 13 milliseconds and once again recovered. The impact of that, and the direct cause of the failure, was a reduction in voltage available to relays that are in the aft end of the [stage]. Those relays hold open propellant valves, which feed propellants to the engines. Once sufficient voltage to keep those relay coils energised is lost, even momentarily, it is not possible to reactuate those valves.

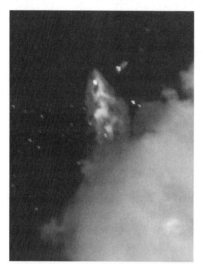

The Delta carrying GOES G is destroyed.

At 71 seconds, the main engine shut down. Without attitude control, while still under the power of the three air-started strap-ons, the vehicle entered a flat spin, and by 77 seconds it had developed sufficient yaw to over-stress the shroud and shear it off, in the process demolishing the payload. The vehicle had completed a full 360-degree rotation before the range safety officer commanded its destruction.[8] Several years earlier, the poly-vinyl chloride insulation on the wiring had been superseded by teflon insulation, and the shape of the wiring harness had allowed the vibration of launch to cause mechanical damage to the insulation. The harness was redesigned.[9] On 26 February 1987 GOES 8 was dispatched, and despite a severe electrical problem with its apogee motor, was success-fully inserted into geostationary orbit to give the National Oceanic and Atmospheric Administration a much needed second operational satellite.[10]

THE NATIONAL SPACE TRANSPORTATION SYSTEM

In 1978, President Jimmy Carter ordered that expendable launch vehicles be phased out in favour of the Space Shuttle which, starting in 1980, was to operate as the National Space Transportation System.[11] The protracted development of the Shuttle prompted the offloading of some commercial satellites, and on 15 November 1980 a Delta 3910/PAM launched SBS 1, the first of the HS-376 satellites. The Shuttle Columbia made its début on 12 April 1981, and after its fourth test flight in June 1982 was declared operational. When launched on 11 November 1982 for mission STS-5, it carried a pair of HS-376s mounted vertically in cradles towards the rear of its payload bay. A satellite launched on an expendable rocket would be spun up and released immediately, but the Shuttle was to deploy its satellites over a period of several days. As soon as the bay doors were opened, a clamshell shade was closed

over each satellite to protect it from
the Sun, and a heater was activated.
The Shuttle faced its bay forward and
tilted one wing downward, so that the
satellites would be aligned with the
velocity vector. Eight hours into the
mission, SBS 3 was spun up by the
turntable in its cradle and then
ejected by a spring – and the National
Space Transportation System was in
the commercial satellite business. The
satellite was on a 2.2-tonne PAM
stage that could insert a 1.25-tonne
satellite into geosynchronous transfer
orbit, and after 45 minutes, while
crossing the equator, the motor fired
for 83 seconds and was released. At
the apogee of the transfer orbit 6
hours later, SBS 3 fired its own
thrusters to circularise. The next
day, an Anik C satellite was similarly
dispatched for Canada's Telesat.

IUS malfunction

When Challenger made its début on 4
April 1983 as STS-6, it carried TDRS
1 for NASA's Tracking and Data

The launch of STS-1.

The deployment of the HS-376 satellite Anik C2.

An IUS carrying a TDRS satellite elevated on its cradle.

Relay System. As this 2.2-tonne satellite was too heavy for a PAM, it used the more powerful IUS, and this 18-tonne stack was carried lengthwise, filling the bay. The first step in bringing the stack to life was to elevate the cradle to 30 degrees. The two-stage IUS was heavily instrumented and the control system of each of its stages was verified. By using a camera installed at the rear of the bay, the astronauts confirmed that the first stage's nozzle could gimbal. Once TDRS 1 was verified by the Air Force's satellite operations facility in Sunnyvale in California, the power umbilical was disconnected, and the elevation increased to 60 degrees. At the appointed time, the ring-clamp holding the IUS in place was unlocked to enable the spring to eject it. Ten minutes later (by which time Challenger had withdrawn) the IUS was activated. After a further 30 minutes, it performed a series of star sightings to update its inertial platform. On crossing the equator it ignited its first-stage motor, which delivered 18,500 kilograms of thrust for 150 seconds for insertion into geosynchronous transfer orbit. The first stage then used its small thrusters to orient the stack and initiate a slow roll to even out thermal stresses during the 6-hour climb. A few minutes before reaching apogee, the first stage re-oriented the stack for the circularisation burn, and separated. The second stage was to have delivered 2,750 kilograms of thrust for 103 seconds but, towards the end of its burn, the oil-filled seal of the gimbal deflated, the nozzle slewed, and the offset thrust induced an end-over-end tumble at a rate of 30 revolutions per minute. As soon as the flight controllers saw this in the telemetry they commanded the release of the satellite, which readily stabilised itself. The 21,700 by 35,550-kilometre orbit had little to recommend it as a communications relay. It was not only non-synchronous, but the fact that the IUS had not yet fully cancelled the initial 28.5-degree inclination left it nodding 3 degrees each side of the equator. The satellite had its own propulsion system, but this was only to enable it to adjust its station, not to make major manoeuvres. Nevertheless, it was able to assume its operating station. The IUS had redundant systems to overcome many faults, but it had no means of recovering from a mechanical failure in the manifold. All Shuttle missions that required the IUS were postponed pending the resolution of the problem, but it soon became evident that this would be a lengthy process.[12] This was bad news for the Air Force, as its most modern satellites had been designed to exploit the Shuttle's cavernous payload bay and were too big to be off-loaded onto the rapidly depleting stock of Titan 34Ds. When the IUS flew again as STS-51C in January 1985 the first-stage underperformed, but the second stage made up the 16-metre-per-second velocity shortfall and delivered its Magnum electronic intelligence-gathering satellite as planned.

Losing two satellites

While the IUS was grounded, the Shuttle continued commercial operations. STS-7 deployed another Anik C for Telesat and Palapa B1 for Indonesia, and STS-8 deployed Insat 1B for India. When STS-41B released Westar 6 for Western Union and Palapa B2 in February 1984, both of their stages

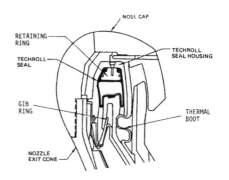

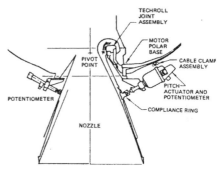

A schematic of the gimbal of the second stage of an IUS.

The deployment of Westar 6.

failed after 10 seconds. The investigation focused on the nozzles of the Star 48B motors. The fact that these were the first of a batch of five recently produced by Thiokol implied a manufacturing fault, and an examination of the remaining motors indicated that the curing of the graphite–epoxy composite wall of the nozzle had trapped bubbles of gas that had caused a burn-through when the motor fired.[13]

Repairing SolarMax

On the next mission in April 1984, Challenger set off to repair SolarMax, which, in November 1980, a few months after its launch, had suffered a power spike that blew the fuses of its attitude control system. The satellite was to supplement the series of Orbiting Solar Observatory satellites by using seven instruments to monitor the Sun through the most active part of its 11-year cycle of sunspots.[14] After the failure, the satellite had been put into a spin to enable the few instruments that did not need a steady view to undertake a restricted programme. If SolarMax could be repaired, it would hopefully survive long enough to follow the Sun through its minimum and the run up to the *next* maximum, and so carry out its original mission. In addition to restoring the attitude control system, the astronauts were to try to repair an instrument that had never met its specification. The rendezvous required Challenger to climb to 560 kilometres, which was much higher than the Shuttle had previously ventured. By the time the orbiter assumed station 100 metres away, the Goddard Space Flight Center had powered down most of SolarMax's systems, and almost cancelled its spin. Its fate rested with George Nelson and James van Hoften. As all the aspects of the repair that could be tested in space had been rehearsed on earlier missions, the recovery was expected to be straightforward. Nelson donned the MMU 'flying backpack' and collected a TPAD, which was a device built to mate with a trunnion pin on the satellite. At the base of SolarMax was the spacecraft bus, which was generic to several satellites. Projecting from each side of the interface ring to the science package was a pair of broad solar panels. As the trunnion pin was on the bus, Nelson had to station himself below the level of the solar panels, which passed over his head as the satellite slowly turned on its main axis at the rate of about one revolution per minute. When the trunnion pin came into view Nelson advanced to slide the collar of the TPAD over it, but when he activated the mechanism it failed to engage the pin.[15] He tried again, imparting more force, but again without result, and this time he caused the satellite to nutate. In frustration, Nelson withdrew several metres to consider what to do. Almost as an afterthought, he reached and grabbed hold of the tip of one of the solar panels as it drifted by – if he could stabilise the satellite, Challenger's robotic manipulator might be able to grasp the pin to capture it directly. However, before the Goddard engineers could warn him off he had turned SolarMax's axial spin into a gyration that prevented it from holding its solar panels facing the Sun. Even with most of its systems powered down, it would soon exhaust its battery. Nelson's efforts to overcome the satellite's tumbling had made matters worse. As Challenger manoeuvred alongside, the efflux from its thrusters further upset the satellite. With his MMU running low on nitrogen propellant, Nelson beat a retreat. With shocking suddenness, it seemed that the satellite that he had set out to rescue had been written off! The Shuttle manoeuvred clear while the Goddard engineers endeavoured to restabilise the satellite and recharge its battery – it was a close call, but they succeeded.[16]

Challenger returned two days later. This time it was decided to attempt to snatch the satellite with the robotic arm, which was achieved without incident. To enable SolarMax to draw power from an umbilical, it was put on a structure at the rear of the bay that had been built to enable the Shuttle to deploy satellites using the same type of multipurpose bus, and had a U-shaped cradle with a tilt-table. If it proved impossible to fix SolarMax, it was to be returned to Earth in this cradle. The following day, van Hoften affixed a foot restraint to the arm and mounted it, and

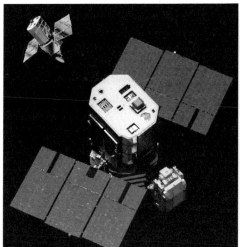

George Nelson manoeuvres to grapple the SolarMax satellite, and its repair once mounted in the Shuttle's payload bay.

Nelson tagged along when Terry Hart swung the arm to the rear of the bay. Nelson inserted another foot restraint into a socket to provide a fixed work platform. The tilt-table was rotated to enable the astronauts to access the satellite's attitude control system. The bus comprised three boxes about 1 metre square, set 120 degrees apart. The box with the electronics of the attitude control system was to be replaced. Fortunately, SolarMax was one of the first satellites designed to be *serviced* in space, and unbolting the box and fitting the replacement proved straightforward. This accomplished the primary objective by restoring the satellite's ability to aim its instruments at the Sun. If they had been running late, the astronauts might have stopped at this point, but as they were ahead of schedule they set out to repair the faulty coronagraph. This would be a more difficult task, because the instrument would have to be taken apart and repaired. First, however, they had to gain access. The satellite was rotated to present its instrument section, a heavy-duty cutter was used to slice through the thermal insulation, and the loose flaps were folded back and held in place by duct tape. Six screws had to be undone to release the hatch beneath. As the screws could damage the instrument, these had to be retrieved, which was an awkward procedure for gloved hands. Since it would not be practical to reuse the screws, another strip of tape was applied to serve as a hinge. The astronauts were making good progress, which was just as well because they had to undo a total of 22 screws to disconnect the instrument's controller! The task of wiring up the new controller fell to Nelson. Fast-action spring clips were used to re-establish the 11 circuits. This done, the hatch was closed, and the insulation was flattened and taped into position. Although this tape would alter the thermal properties of the instrument, the tape would not face the Sun when the spacecraft was operating. Having completed both their primary and secondary tasks, van Hoften and Nelson retreated to the airlock. When SolarMax was released the following day, the coronagraph proved to be fully functional.

A fortnight later, despite the Sun being well past its peak, SolarMax documented the largest flare since 1978. As the solar cycle neared its new peak it inflated the upper layer of the Earth's atmosphere, increasing the rate of decay of SolarMax's orbit. A propellant replenishment was tentatively assigned as a secondary task on a Shuttle mission in 1990, but the satellite re-entered in December 1989, having kept watch on the Sun to the end.

Pad abort and a trio of satellites

Discovery's début mission, as STS-41D, the twelfth flight of the programme, started as an exercise in frustration. After a perfect Flight Readiness Firing on 2 June 1984, NASA set the launch for 22 June, postponed it to 25 June, and then scrubbed it at T–9 minutes due to a fault in one of the general-purpose computers. The next day, the count ran smoothly to the point at which the three hydrogen burning engines were to ignite at 120-millisecond intervals, whereupon the hydraulically activated fuel valve of the first engine failed to open. The second engine was up to 20 per cent thrust before this misfire was diagnosed by the computer, which immediately ordered a shut-down, inhibiting the third engine. The flame fizzled out and the billowing cloud of steam rapidly dispersed. This was the first time that a launch had been scrubbed following engine-start. Although spectacular, it was not an unprecedented situation because the vehicle was in essentially the same state as after an engine test firing. As the engines that had fired required to be refurbished, the launch had to be pushed back by a month. When launched on 30 August, this became the first Shuttle to deploy a trio of satellites. Two rode PAM motors which this time fired properly. The third satellite was the first of a new type. In 1978, the Navy had ordered a series of advanced Syncom satellites. The resulting HS-381 exploited the fact that an early cost–benefit analysis had envisaged charging a fee based on how much of the payload bay's length a satellite occupied. In contrast to the HS-376, which was a 2.8-metre-tall drum of just over 2 metres diameter in order to fit the shroud of a vehicle such as the Delta, the marginally taller drum of the HS-381 was fully 4 metres across. Whereas the HS-376 satellites were mounted on a perigee-kick motor and set upright in the bay, the HS-381 was carried on its side. Although the HS-381 was 7 tonnes, half of this mass was propellant for the in-built motor that was to enable it to attain its operating station. Consequently, this series was not delayed by the grounding of the

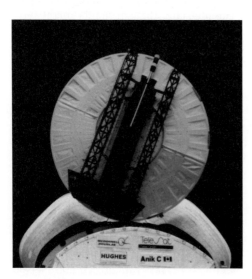

The deployment of Leasat 3.

IUS. As the spring-loaded cradle ejected the satellite 'frisbee style', it set it rotating at 2 revolutions per minute for stability and flipped a lever to activate its sequencer, which immediately elevated the omni antenna to enable Hughes to verify the systems before the satellite set off for geostationary orbit. Since these satellites were to remain the property of Hughes and be leased by the Navy only after they were installed on-station, they were referred to as Leasats.

Retrieving the lost satellites

NASA mounted an ambitious rescue mission on STS-51A in November 1984. After it had deployed another HS-376 and an HS-381, Discovery was to retrieve the satellites stranded in elliptical orbits with 1,000-kilometre apogees in February.[17] After jettisoning their stages, these satellites had used their thrusters to lower themselves, and were now in similar orbits and spaced several hundred kilometres apart. Their retrieval would involve considerably more orbital manoeuvring than on any previous Shuttle mission. Discovery drew to a halt barely 10 metres from Palapa B2. The satellite's 60-rpm spin had been virtually eliminated, leaving only a slow residual roll for stability. Since the HS-376 had not been designed to be manipulated in orbit, and did not have a trunnion pin, recovering it would be more difficult than in the case of SolarMax. To capture it without damaging the conformal solar cells on its drum, a device had been produced to engage the ring at the base that had mated with the PAM stage. Although the designers referred to this as the capture device, its shape had inevitably resulted in the astronauts dubbing it the 'stinger'. It was really several devices in one. Like the TPAD, it had a clamp to attach to the arms of the MMU. A long rod projecting out in front (the stinger itself) was to be inserted into the throat of the satellite's apogee motor. A ring at the base of the stinger was to connect with that on the satellite. There was also a trunnion pin on the side of the device to enable the remote manipulator to take over once the satellite had been stabilised.

Joe Allen manoeuvred into position aft of the satellite, and then eased forward. The stinger's success was evident even before Allen reported it, because he adopt the satellite's spin. Once this motion had been cancelled by the thrusters on the MMU, he manoeuvred into a convenient orientation and Anna Fisher brought the arm in over his shoulder to snare the trunnion pin. With Allen still attached, Fisher eased the satellite into the bay, where Dale Gardner was waiting with shears to remove the rod-like omni antenna that projected from the top of the satellite. The satellite had then

Dale Gardner manoeuvres to capture Westar 6.

to be turned upside down for stowage. A bracket had been built to run over the top of the drum, but this could not be attached. It had been tested on similar satellites at Hughes, but the model was tailored to its customer's requirements, and the feed horn of the folded antenna on top of this one projected a little further than the tool's designers had been told, preventing the clamps from engaging. In the original plan, Gardner was to affix the bracket, Fisher was to raise Allen back out of the bay and release him, Allen would flip the satellite to enable Fisher to grip the trunnion pin on the bridge that Gardner had fitted, Allen would disengage the stinger, and Fisher would line up the satellite and lower it onto the waiting mount. Given that the common bracket clamp (as the bridging bracket was formally known) would not fit, the astronauts set out to improvise. In the revised plan, Allen disengaged, leaving the arm grasping the satellite via the stinger. While Allen stowed the MMU, Gardner placed a portable foot restraint on the sidewall. Allen then stood on this restraint and took hold of the satellite by its still-folded main antenna. When Fisher withdrew the arm, Allen did his best to hold the satellite steady while Gardner disengaged the stinger. Gardner placed a 'shower cap' over the motor nozzle to prevent flakes of solid propellant falling out into the bay when Discovery landed. Gardner then disconnected the A-frame carriage mount from the bay floor and mated it to the ring on the base of the satellite. Throughout this activity (which lasted well over an hour) Allen held the satellite steady. With the mount attached, they pitched Palapa down into the bay and Gardner reconnected the A-frame to the bay floor. In all, the retrieval had taken six hours.

Two days later, Discovery drew up alongside Westar 6, and Allen and Gardner went out to retrieve it. It had been decided to use a revision of the impromptu procedure. This time Gardner donned the MMU and the stinger, and Allen installed a foot restraint on the arm and mounted it. Gardner captured the satellite and flew it to Allen, who grasped it by the antenna. This time, the stinger was promptly disengaged and stowed, together with the MMU. With Gardner on the bay floor, Fisher slowly brought Allen and the satellite down. The fact that Allen had to hold the satellite by its antenna meant that this time the satellite arrived conveniently for fitting the A-frame. With the satellite in place, the omni antenna was cut off to prevent it interfering with the closing of the bay doors. Before they came in, the two men perched on the arm and displayed a placard to a camera advertising two satellites for sale. In recognition that their "extraordinary exertions contributed to the preservation of property", Lloyds of London gave the astronauts its Silver Medal. Both satellites were subsequently sold and relaunched under new names.

Hot-wiring Leasat 3
After deploying Leasat 3 on 13 April 1985, Discovery withdrew to leave the satellite to run through its start-up sequence, but when it failed to issue telemetry it was realised that the lever had failed to activate the sequencer. Rather than leave the satellite dormant, it was decided to extend the mission and re-rendezvous to throw the switch. STS-51D was carrying Senator Jake Garn, who was 'observing' in his capacity as chairman of the subcommittee that oversaw the agency's budget, and this was an excellent opportunity to show that the most effective backup system in space

was a human presence. The initial idea was for David Griggs or Jeff Hoffman to mount the arm and throw the switch. But what if the satellite started in a disturbed state and immediately fired its motor? It was decided to devise a method that did not require an astronaut to be near the satellite when it came to life. The crew was told to get some sleep while the support staff considered the matter, but the astronauts stayed up half the night devising a scheme of their own, and the next day a plastic document cover was cut to construct a flexible 'fly-swatter' with a narrow slit near its tip. Griggs and Hoffman went out and taped this to the end-effector of the arm. Once they were safely back in the airlock, Rhea Seddon brushed the tip of this improvisation against the solar cells coating the surface of the slowly rotating drum, then waited for the lever to come around. After several near misses, the fly-swatter's slit threw the switch. Discovery withdrew 100 metres to observe the satellite's activation, but when the rod of the omni antenna did not swing up on time it was accepted that the hulk had to be abandoned. It was evident that the problem was not a simple switch fault.[18] If a way could be devised to jump-start the derelict satellite it might still be rescued by a later mission.

A detailed analysis of the likely failure modes revealed just such a possibility, and a few months later, after deploying its trio of brand new satellites on STS-51I, Discovery once again drew up alongside Leasat 3.[19] James van Hoften was waiting on the arm, and Mike Lounge eased him out to the satellite, which was still axially stable due to its slow roll. As the lever that had been flipped by the fly-swatter came by, van Hoften reset it. He had taken a bar to span the gap between two of the sockets used to hoist the satellite on the ground. He waited until the attachment points rotated into view, then affixed this capture bar. When it next came around, he grabbed the bar and manually cancelled the satellite's spin, using his arms to absorb the energy. Meanwhile, Bill Fisher (husband of Anna Fisher) had set up station on a foot restraint on the starboard sill. The arm slowly moved van Hoften down towards the bay, and he dragged the 7-tonne satellite with him. Once it was in position, just over the bay, Fisher attached a second handling bar on the opposite side and grasped the satellite. Van Hoften replaced his bar with a trunnion pin assembly, dismounted the arm, and removed the foot restraint from the end-effector. The arm then relieved Fisher of the satellite. It was teamwork all the way. The next task was the installation of the Spun-Bypass Unit (SBU). This contained the timers to activate the satellite. It was to be attached to a point on the side of the drum, and its cables strung out to the circuitry. Although the system had been tested on a satellite in the Hughes factory, doing so in space was not expected to be easy. The job was made more difficult because the HS-381 had not been designed to be worked on by astronauts. The first task was to safe the satellite's pyrotechnics, to ensure that there was no risk of the astronauts triggering some circuitry while disabling the seemingly dead Post-Ejection Sequencer and installing the SBU. All the complex rewiring was completed without incident, and the SBU diagnostic display confirmed that the satellite's battery was in good condition. The next task was to install a Relay Power Unit (RPU) to directly engage the relay to deploy the 2-metre-long rod of the omni antenna and, to everyone's delight, this swung up. This action enabled the satellite to accept radio commands, and started the telemetry. By this point, the astronauts had effectively

James van Hoften spins up Leasat 3.

restored Leasat to a state equivalent to that of a successful deployment. As the two men had been out for almost seven hours, there was no time for redeployment and so the satellite was left on the arm overnight.

There were some uncertainties in activating the satellite after such a long time in a dormant state. If the solid propellant in its integrated perigee-kick motor had grown cold enough to enable cracks to form, it could explode when fired. On the second excursion, a new motor cap was affixed, with a probe to enable the temperature of the interior to be measured. The probe had a telemetry link so that it could monitor later attempts to warm the propellant. Once this was confirmed to be functioning, the arming pins were pulled from the SBU to start its timers. Fisher resumed his station on the sidewall to hold the satellite, the arm disengaged, and van Hoften deleted the trunnion pin. After slipping the manipulator back on the arm, van Hoften mounted it. He then retrieved a third handling bar, which was rather smaller than the capture bar. This was the spin-up bar. Once van Hoften had hold of the satellite Fisher removed his bar, Lounge drew the arm up out of the bay, and van Hoften dragged the satellite with him. However, it could not just be released. When it had been ejected from the bay six months previously, it had been given a 2 revolutions per minute roll. This had to be restored. The 4-metre-diameter, 3-metre-tall drum was the largest satellite thus far encountered by a spacewalker. Even a mighty heave barely made it spin. Two further boosts were required to build it up to the required rate. This was complicated by the fact that the initial impulse had also set the satellite drifting away, and the arm had to follow it to enable van Hoften to grasp the bar each time it came around.[20] Ironically, although Leasat 3 successfully took up its operating station, Leasat 4 (which was one of the satellites released earlier in the flight) failed upon reaching geostationary orbit. The investigation by Hughes concluded that because the signal strength was unacceptably weak, there must be a fault in the cable that linked the transmitter to the antennas.[21]

Abort-To-Orbit
On 29 July 1985 STS-51F became the first mission to undertake an abort during the ascent. As Challenger continued to climb under its own power after jettisoning its SRBs, the centre engine in the SSME cluster overheated. The flight controllers monitoring the situation throughout the 104 per cent phase were relieved to see that it did not stray into the red line prior to being throttled back to 65 per cent for the final phase of the ascent. It continued to overheat, however, and the computer shut it off at T+350 seconds. By this point, with two engines, the Shuttle would have

sufficient energy to attain a low orbit, so an Abort-To-Orbit was ordered, which involved mission commander Gordon Fullerton turning the selector to 'ATO' and pushing a button to enact his selection. The computer throttled up the two remaining engines to 91 per cent, and added 70 seconds to the burn to compensate. However, no sooner had this abort been initiated than a temperature rise was seen in one of the remaining engines. At this point the controllers began to suspect a faulty sensor rather than a genuine problem with the engine, and they recommended that the crew intervene to inhibit the computer from shutting off this engine. Unable to reach orbit on one engine, Challenger would have had to invoke a Transatlantic Abort Landing which, in this case, would be an emergency landing at Zaragoza in Spain. The final 220-kilometre circular orbit was not ideal for the observational programme planned for the telescopes in the payload bay, but any orbit was better than none, and the observing plan was hastily revised.

The loss of Challenger
On 27–28 January 1986 the overnight chill threatened the water pipes on Pad 39, so the valves were opened to allow the water to flow to prevent the pipes freezing. It was so cold that soon the gantry walkways were laced with sheets of ice, and icicles adorned the structure. The next morning, the launch of STS-51L was held for two hours to allow the ice to melt. Nevertheless, when Challenger lifted off at 11:38 EST the temperature was 15°C below that of any previous launch.

At Thiokol, a group of engineers led by Roger Boisjoly had expressed doubts about the resilience of the rubber O-rings that would have to seal the joints of the segments of the SRBs, but Joseph Kilminster, the company's vice president for boosters, was very aware of the imperative to build up the flight rate. Lawrence Mulloy, who managed the Thiokol contract at the Marshall Space Flight Center, asked incredulously: "When do you want me to launch? Next April?" Kilminster overrode his engineers and recommended a launch. In the NASA way, a waiver was issued to recertify the SRBs for this colder temperature. This was regarded as yet another step in the continuing process of stretching the Shuttle's operating envelope. However, although no one was aware of it, the O-ring in the lowest field joint of the righthand SRB was so cold that it failed to seat properly in its groove when the motor was ignited, and let a blast of hot gas pass through the joint. The puff of dense black smoke that issued from the side of the casing was spotted only by the high-speed cameras that filmed each launch from close range for subsequent analysis. When the joint flexed as the casing accommodated the longitudinal stress of acceleration, a rapid series of smaller puffs followed, and a succession of blasts of gas seared further into the O-rings before the joint finally sealed. As the vehicle went supersonic, the thrust was throttled back in order not to over-stress the stack. After the aerodynamic loads had peaked, the SSMEs were increased to 104 per cent, and the geometry of the propellant in the SRBs increased their thrust. A few seconds later, there was a catastrophic explosion. CNN, the only major network to show the launch live, relayed NASA's video feed. As the flight controllers in Houston stared in disbelief at their telemetry displays, which had ceased updating, Stephen Nesbitt, the Public Affairs Officer, told the television audience that it was "obviously a major

malfunction". The loss was all the more shocking because there seemed to have been
no indication of a problem.

In fact, as the SRBs increased thrust at T + 59 seconds, the hot gas had reopened
the breach in the right motor, which emitted a continuous plume similar to a blow
torch. The first indication in the telemetry was at T + 60 seconds, when the internal
pressure of this motor began to depart from nominal. The issue for the flight
controllers was whether this was a temporary decrease in pressure (in which case the
focus of attention would shift to how well the Shuttle was able to correct its
trajectory by swivelling the nominal engines) or whether it marked the onset of a
divergence. The pressure could vary within a narrow band and remain acceptable,
but at T + 63 seconds it left this band. The reason for this under-performance was
not evident. At T + 64 seconds, the plume breached the ET and ignited the resulting
hydrogen leak. The SSMEs gimballed to counter the unwanted vehicle motions. At
T + 72.2 seconds the plume severed the strut connecting the base of the right SRB to
the ET, and as the booster pivoted on its upper strut its rear skirt struck Challenger's
wing. The SSME gimballing increased to 5 degrees per second in an effort to counter
the disturbance resulting from the divergent pitch and yaw rates between the two
SRBs. At T + 73.124 seconds the bottom cap of the ET failed, and dumped the
pressurised hydrogen into its wake. At T + 73.137 seconds the nose of the loose SRB
struck the ET's intertank and fractured the base of the oxygen tank. Although
Challenger survived the resulting detonation of the propellants, it was ripped apart
by the extreme aerodynamic stress.

The Shuttle fleet was grounded during an investigation that interviewed 160

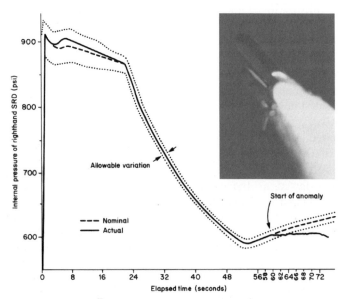

A plot of the pressure in the righthand SRB of STS-51L, and (inset) the plume that
spewed from the field joint that was responsible for the anomaly.

people, examined 6,000 documents, and spun off 35 panels to analyse specific issues in depth. Attention focused on the O-rings. Over the years, all but two of the SRBs had been recovered (those from STS-4 had been lost due to identical parachute failures), and evidence of hot gas flowing past the main O-ring had been detected in nine casings. A broad section of the 75-millimetre-thick insulation that lined the aft nozzle of one of the boosters on STS-8 had been almost completely eroded by the hot efflux. This was attributed to a faulty batch of material. In fact, this had been the first use of an uprated booster. The erosion of the SRBs was to become a matter of some concern, not least because there was so much variation in the damage. When gas burned completely through a narrow arc of the inner O-ring ring and severely eroded the outer ring on one of the SRBs on STS-51B, Boisjoly had 'red flagged' the issue. Thiokol had established a formal study in October 1985, but it was judged that more data was required to characterise the problem, which entailed the inspection of more spent boosters. After the *Report of the Presidential Commission on the Space Shuttle Challenger Accident* was issued on 6 June, NASA redesigned the tang-and-clevis joint to reduce flexure, but in doing so it minimised the modification in order to enable it to upgrade the segments that it had in stock. In addition, a third O-ring was added to each joint, putty was applied to the interior of the joint, an electrical heater was installed to protect the O-rings from chill, and a 'weather strip' was wrapped around the joint to prevent the seepage of rainwater.

The Shuttle had carried 20 commercial communications satellites, which, in view of its operational complexity, was a fair achievement, but after the loss of

Table – Commercial satellites launched by the Shuttle

Satellite	Flight	Supplier	Bus	Operator
SBS 3	STS-5	Hughes	HS-376	SBS, USA
Anik C3	STS-5	Hughes	HS-376	Telesat, Canada
Anik C2	STS-7	Hughes	HS-376	Telesat, Canada
Palapa B1	STS-7	Hughes	HS-376	Perumtel, Indonesia
Insat 1B	STS-8	Ford	–	ISRO, India
Westar 6	STS-41B	Hughes	HS-376	WU, USA
Palapa B2	STS-41B	Hughes	HS-376	Perumtel, Indonesia
SBS 4	STS-41D	Hughes	HS-376	SBS, USA
Telstar 3C	STS-41D	Hughes	HS-376	AT&T, USA
Anik D2	STS-51A	Hughes	HS-376	Telesat, Canada
Anik C1	STS-51D	Hughes	HS-376	Telesat, Canada
Morelos 1	STS-51G	Hughes	HS-376	SCT, Mexico
Arabsat 1B	STS-51G	Aerospatiale	–	ASCO, Arab League
Telstar 3D	STS-51G	Hughes	HS-376	AT&T, USA
Aussat A1	STS-51I	Hughes	HS-376	Aussat, Australia
ASC 1	STS-51I	RCA	S-3000	ASC, USA
Morelos 2	STS-61B	Hughes	HS-376	SCT, Mexico
Aussat A2	STS-61B	Hughes	HS-376	Aussat, Australia
Satcom K2	STS-61B	RCA	S-4000	RCA, USA
Satcom K1	STS-61C	RCA	S-4000	RCA, USA

Challenger, the White House ordered the National Space Transportation System to cease commercial operations. Furthermore, safety concerns prompted the cancellation of the wide-bodied Centaur that was to have enabled the Shuttle to dispatch the new generation of heavyweight planetary probes.[22]

A close call for Columbia

On 23 July 1999, STS-93 was launched to deploy the Chandra X-Ray Observatory. Five seconds after lift-off, Eileen Collins, the first female Shuttle commander, reported warning lights indicating that Columbia had suffered a major electrical failure. A sudden voltage drop on an electrical bus had disabled the primary controllers on two of the three SSMEs. The controllers were to run the engines in response to commands issued by the general-purpose computers. Although the ascent was able to continue using the backup controllers on another power bus, there was now no redundancy in those engines. Just as it appeared that the scare was over, it was found that SSME #3 (one of the two affected by the electrical problem) was leaking hydrogen, and therefore was operating about 100 degrees hotter than the others. If it became too hot, it would have to be shut down, which would, depending on when this occurred, prompt an abort in which Columbia would have either to return to the launch site or try to reach Spain or North Africa. Collins's luck held. The leak was only 2 kilograms per second, but with the 'hot' engine drawing more oxygen than usual, it was a question of whether Columbia could achieve low orbit before the oxygen ran out. In the event, SSME #3 leaked 1,500 kilograms of hydrogen and drew 1,800 kilograms more oxygen than its companions. Nevertheless, when the emptying of the ET's oxygen tank forced the engines to shut down, Columbia was only a few kilometres short of the intended orbit.[23]

The investigation traced the hydrogen leak in SSME #3 to the thin pipes within the engine's interior wall. Pumping hydrogen through these pipes prior to injecting it into the combustion chamber served to warm the cryogenic fluid and to cool the engine bell, and so protect it from the hot exhaust plume. Three of the 1,000 pipes that lined the bell had burst, and the fact that these were adjacent indicated that some object had struck the engine. The injector that fed oxygen into the combustion chamber via 600 tiny holes had previously been repaired. The procedure for dealing with a damaged injector line was to plug the appropriate hole with a 2-centimetre-long pin. One of the two pins on that particular injector was found to be absent. Evidently, as the engine was started, the pressure had ejected the pin, which had struck the wall of the bell. Tests indicated that the impact of the pin had weakened the pipes sufficiently to enable the pressure of the circulating hydrogen to leak out, and a study of the launch video confirmed that hydrogen was being dumped into the exhaust plume. If the impact had damaged a dozen or more of the pipes, the leak may well have been sufficient to prompt the engine to be shut down, resulting in an abort to the Kennedy Space Center as soon as the spent SRBs had been jettisoned. When investigating the electrical problem that had disabled two of the primary controllers, an examination of the cable trays running beneath the floor of the payload bay revealed damaged insulation on several sections of cable, and there was evidence of arcing between a bare wire and an adjacent screw. The other orbiters were examined and found to have similar faults, with Endeavour being the worst affected with 20 degraded wires, in some cases exposing bare wire. Some of the 350 kilometres of cable in an orbiter pass through sections that are routinely accessed by technicians undertaking maintenance between missions, and this essential access had caused the wear and tear. In addition to repairing this damage, NASA revised the maintenance procedures.

Losing Columbia

The reason for delaying STS-51L's launch had been to allow time for the icicles on the service structure to melt, lest these be shaken loose by the sonic shock of the vehicle lifting off and strike the delicate tiles of the thermal protection system on the belly of the orbiter. Although the ascent of Columbia for STS-107 on 16 January 2003 seemed to be uneventful, a routine examination the next day of the film from the long-range tracking cameras revealed that at $T+82$ seconds a piece of foam detached from the left leg of the bipod mount on the intertank of the ET that supported the orbiter's nose, and struck the orbiter. The leading edge of the bipod mount was protected from aerodynamic heating during the ascent by a 'ramp' of foam that had been applied by hand. However, if new foam was not firmly bonded to older foam this left a 'flaw'. If rain water seeped into a void, it would freeze when the cryogenic propellants were loaded into the ET and the expanding ice would open the flaw and enable the vibrations and aerodynamic stress of launch to detach a segment of foam. Although foam strikes had occurred previously without serious consequences, a study was ordered to determine the likely extent of the damage to the thermal protection tiles. In fact, foam had been seen to fall from the bipod mount on *four* previous flights. The point of reference for the study was STS-50 in 1992, when a piece of foam had grazed Columbia's underside at an angle of 3 degrees and etched a furrow in its tiles 23 centimetres in length, 10 centimetres in width and 4 centimetres in depth.[24] The analysis in the second week of STS-107's flight found that the energy of its impact would have been 5 to 16 times greater owing to the angle of the strike being steeper, and would have caused more damage. Nevertheless, it was concluded that this should not pose a serious threat. However, some engineers at NASA's Langley Research Center criticised this analysis, and Robert Daugherty went so far as to suggest in e-mails to colleagues in Houston that flight controllers should be alert to the possibility that they might see off-nominal readings from certain sensors during re-entry and they should not hesitate to recommend that the crew fly offshore and bale out, leaving Columbia to ditch in the Atlantic.[25,26] Meanwhile, the astronauts were informed of the foam strike and reassured that there was no danger. They died when Columbia was destroyed while re-entering the atmosphere on 1 February.

In executing the Shuttle Contingency Action Plan that had been formulated after the loss of Challenger in 1986, a Mishap Investigation Team was activated to coordinate the recovery of the débris, which was widely scattered across Texas, New Mexico, Arizona, Utah, Nevada and California. Subsequently, an Investigation Board was set up under the chairmanship of retired admiral Harold Gehman. The problem was the lack of evidence in the telemetry. As late as mid-March, the foam strike was under consideration together with the possibility of a strike by either a micrometeoroid or a piece of orbital débris.[27] This situation changed with the recovery on 19 March in Texas of the 'black box' that had recorded data from a large number of sensors whose output was not transmitted to the ground.[28] This Modular Auxiliary Data System had been fitted in Columbia many years previously for the 'Orbiter Experiment' to document the vehicle's characteristics.[29] Its data showed that hot plasma had penetrated the left wing, progressively melting its internal structure until it suddenly broke off.

When the film of the ascent was computer-enhanced in early May, the trajectory of the foam was refined and it became evident that it had hit the leading edge of the wing. The crucial realisation was that a piece of foam was *more* dangerous than a chunk of ice, because the lightweight foam would *decelerate* rapidly in the airstream, which would in turn increase the energy of the strike when the vehicle slammed into it.[30] On 29 May, the Southwest Research Institute in San Antonio, Texas, began a series of tests to recreate the foam strike. A 10-metre-long gas-powered gun fired a 0.76-kilogram piece of foam – similar to the piece that hit Columbia – into a fibreglass mock-up of the leading edge of a wing at the calculated speed of 850 kilometres per hour. When it was realised that the shock was sufficient to break a reinforced carbon–carbon panel of the type on the leading edge of an orbiter, the mock-up was fitted with real panels, and a test on 7 July produced the proverbial 'smoking gun'.[31] The foam made a hole in the panel some 40 centimetres wide, and even damaged some of the measuring equipment. Columbia had re-entered the atmosphere with a gaping hole in the leading edge of its left wing.[32]

The three remaining Shuttles, Atlantis, Discovery and Endeavour, were grounded for over two years while means of detecting and repairing damage to the thermal protection system were investigated. One consequence of the loss of Columbia was that the plan to continue to upgrade the vehicles to extend their service life beyond 2010 was cancelled, and it was decided to phase them out at that time.

NOTES

1.   http://www.museum.tv/archives/etv/C/htmlC/communication/communication.htm
2.   http://www.boeing.com/defense-space/space/bss/factsheets/376/syncom/syncom.html
3.   http://kevinforsyth.net/delta/delta2.htm
4.   *Aviation Week & Space Technology*, 12 May 1986, p. 20.
5.   *Aviation Week & Space Technology*, 23 June 1986, p. 18.
6.   *Aviation Week & Space Technology*, 2 June 1986, p. 20.
7.   http://www76.pair.com/tjohnson/thrfail.txt
8.   *Aviation Week & Space Technology*, 15 September 1986, p. 21.
9.   *Aviation Week & Space Technology*, 7 July 1986, p. 22.
10.   *Aviation Week & Space Technology*, 9 March 1987, p. 266.
11.   Presidential Directive no. 37, 11 May 1978.
12.   *Aviation Week & Space Technology*, 9 January 1984, p. 21.
13.   *Aviation Week & Space Technology*, 13 February 1984, p. 23.
14.   *Spaceflight*, July–August 1980, p. 282.

15. *Aviation Week & Space Technology*, 16 April 1984, p. 18.
16. 'The dynamics of the Solar Maximum Mission spacecraft capture and redeployment on STS-41C', K.J. Grady, AAS 85-060 Guidance and Control 1985, Advances in the Astronautical Sciences, vol.57, p. 495.
17. *Aviation Week & Space Technology*, 5 November 1984, p. 66.
18. *Aviation Week & Space Technology*, 22 April 1985, p. 18.
19. *Aviation Week & Space Technology*, 2 September 1985, p. 22.
20. *Aviation Week & Space Technology*, 9 September 1985, p. 21.
21. *Aviation Week & Space Technology*, 23 September 1985, p. 21.
22. *Aviation Week & Space Technology*, 23 June 1986, p. 16.
23. http://www.house.gov/science/readdy_092399.htm
24. http://www.floridatoday.com/columbia/columbiastory2A44772A.htm
25. http://edition.cnn.com/2003/TECH/space/02/21/sprj.colu.investigation.ap/
26. http://www.pbs.org/newshour/updates/columbia_02-21-03.html
27. http://www.space.com/missionlaunches/sts107_theories_030219.html
28. http://www.space.com/missionlaunches/sts107_oex_030320.html
29. http://Spaceflightnow.com/shuttle/sts107/030324tape/
30. http://www.caib.us/news/documents/impact_velocity.pdf
31. http://news.bbc.co.uk/go/pr/fr/-/1/hi/world/americas/3053336.stm
32. Columbia Accident Investigation Report, NASA, August 2003.

3

Back to expendables

TITAN SHOCK

After its mishap in 1985, the Titan 34D remained grounded until 18 April 1986 when one lifted off carrying the final KH-9 film-return reconnaissance satellite. Unfortunately, at T + 8 seconds it exploded at an altitude of 800 feet. The fragmental débris and burning propellant caused severe damage to SLC-4E and the nearby facilities. The explosion also created a toxic cloud that rose to an altitude of 8,000 feet before being blown out to sea.[1] Each strap-on of a Titan 34D comprised five full segments and a half segment, a forward enclosure and an aft enclosure. Over the years, some 940 segments of this kind had been used successfully. No Titan had been lost so catastrophically in its 30-year history.[2] The preliminary report on 9 June ruled out an O-ring problem like the one that had caused the loss of Challenger several months previously, and offered a number of failure modes involving 'debonding' in one of the solid rocket motors, but insufficient material was recovered to reach a definitive conclusion.[3] An examination of other segments from the same batch (which had been made five years earlier, and held in storage) showed that the flaw was not age related.[4] A number of motor segments were cut into 2-foot-high slices and dissected into pie shapes for inspection. The flaws discovered in these segments strongly implied that a debonding of the rubber insulation had allowed the hot gas in the motor to make contact with the steel casing, weakening it sufficiently for the 700-psi pressure to open a hole 7 inches in diameter, which caused the explosion.[5,6] The Air Force initiated an intensive recovery that included improving the handling procedures for the solid motors, inspecting their segments using non-destructive methods, and adding instrumentation to monitor their operation.[7,8]

This second consecutive failure for the Titan 34D was particularly ill timed because, with the Shuttle also out of service, the Air Force had no means of dispatching its heavy satellites. With a KH-11 lost in 1985 and now a KH-9, the USA had just one operational imaging reconnaissance satellite. This had been launched on 4 December 1984 and was almost out of propellant.[9] The Titan 34D resumed

An artist's impression of a DSP satellite.

service on 26 October 1987, with one from Vandenberg deploying a KH-11.[10] On 29 November 1987 the first mission from Canaveral deployed a DSP early warning satellite. Some 30 minutes after the launch of the third on 2 September 1988, Air Force Secretary Edward C. Aldridge announced that the nation was "well on [its] way to assuring access to space through a robust, flexible launch capability". Unfortunately, 5 hours later when the Transtage reached the apogee of the geosynchronous transfer orbit and attempted the 110-second circularisation burn, a fractured pressurisation pipe caused its engine to misfire, stranding a Chalet–Vortex electronic intelligence-gathering satellite.[11] Fortunately, the Shuttle resumed operations later in the month.

SPARE PART DELTAS

At the time of the loss of Challenger on 28 January 1986, only six Deltas remained, and on 3 May the first to be launched was lost as a result of an electrical failure that shut down its first stage at T + 71 seconds. As the Atlas and Delta both used Rocketdyne engines, commonalities in the main electronics relay boxes and wiring harnesses meant that both had to be grounded during the investigation. *All* of the US launch vehicles were now out of service! The Delta resumed flying on 5 September 1986. In order to meet the sudden demand, McDonnell Douglas was able to build three from spare parts. Two were made using old MB-3 Thor engines, with the more powerful Castor IVA (TX-780) strap-ons raising its performance to match that of the 3920/PAM. These hybrids (which were given the code 4925, with the '5' indicating that the Star 48 of the PAM was now considered to be a stage of the launch vehicle rather than part of the payload) dispatched Marcopolo 1 for BSB on 27 August 1989 and Insat 1D for India on 12 June 1990. The third improvised Delta was a one-off two-stage 5920 that augmented an RS-27 with the new strap-ons, and was utilised to launch the COBE satellite for NASA on 18 November 1989.

ARIANESPACE

When the European Space Agency was established in 1975, it set out to develop a three-stage launch vehicle. To exploit the rate at which the Earth rotates on its axis to maximise the ability of the vehicle to put a payload into geosynchronous transfer orbit, a launch facility was built near the equator, at Kourou in French Guiana in

South America. The development of the cryogenic third stage (the first time that Europe used this technology) was protracted and the first Ariane was not launched until 24 December 1979. This was an 'all up' test with all 'live' stages, and it succeeded.[12] The second test on 23 May 1980 was a failure.[13] The four SEP Viking 5 engines of the first stage fired up normally, and the vehicle lifted off at T + 3.3 seconds, but 1.1 seconds later the chamber pressure in one engine fluctuated at a frequency of about 1 kilohertz. This ceased after 1.6 seconds, then momentarily reappeared for a fraction of a second at T + 28 seconds, at which time the temperature in the chamber rapidly rose, the pressure fell, and the vehicle began to be subjected to a powerful rolling torque. By T + 104 seconds the roll had increased to 60 degrees per second, which had the effect of interrupting the flow of propellant to the other three engines, and a few seconds later the vehicle broke up under the stress.[14] Since the wreckage fell close offshore, the engine that suffered the failure was salvaged intact.[15] The investigation found that there was an issue with its propellant injector. The vehicle was grounded until June 1981 while the engine was redesigned and certified.[16] The two ensuing tests were successful, but on the first operational flight on 9 September 1982 the turbopump gearing in the third stage suffered damage, shutting off the engine midway through the 720-second burn to enter geosynchronous transfer orbit, losing a pair of satellites and grounding the vehicle until June 1983.[17,18,19] Arianespace's policy was to offer the owner of a lost satellite priority for the launch of a replacement as soon as that became available.

The Ariane 1 configuration was designed to insert 1,500 kilograms into geosynchronous transfer orbit, but it was possible to increase this to 1,850 kilograms. Further improvements raised this capacity to 2,000 kilograms for the Ariane 2, and to 2,600 kilograms for the Ariane 3. The Ariane 3 launch on 12 September 1985 got off to an excellent start, with the second stage shutting down as planned, but some 8 seconds later the third stage failed to start, with the loss of another two satellites. The telemetry confirmed that the firing command sequence had been issued but the ignition was 0.4 second late and failed to achieve the required thrust. It was concluded that an injector valve leak had inhibited proper ignition. The vehicle was grounded once more, this time until February 1986.[20,21,22]

In the aftermath of the loss of Challenger there were simply not enough American launch vehicles to

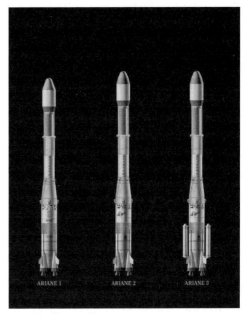

The Ariane 1, 2 and 3 configurations.

meet the demand. Arianespace, however, was in a position to expand its operations in the short term, and promptly won three commercial satellites withdrawn from the Shuttle: Aussat A3, Insat 1C and SBS 5. During the 32 months that the Shuttle was grounded, nine Ariane launches sent 16 satellites successfully on their way, but lost another one. This failure occurred on 31 May 1986, when the third stage of an Ariane 2 had a partial, unsustained ignition, followed by another ignition 0.12 second later that was above nominal pressure, after which it finally shut down. The ignition system of the HM-7B engine used one pyrotechnic cartridge to force hot gas into the injector of the combustion chamber, and another to force gas to spin-up the turbine of the turbopump. The investigation concluded that the thermal performance of the system had been weak. Since there was no evidence of a manufacturing fault, it was decided to install a more powerful igniter, to redefine the ignition sequence, and to undertake a series of high-fidelity test firings in a high-altitude chamber to recertify the engine.[23],[24],[25] As the third stage was common, all flight operations had to be suspended, and the next launch did not occur until 16 September 1987.

REINVIGORATED ATLAS

The first six Block-I NAVSTAR satellites were launched between 1978 and 1980 on converted Atlas-E/F missiles to assess the feasibility of the Global Positioning System.[26] The operational system was to have six sets of four satellites in 20,000-kilometre orbits inclined at 55 degrees, with the six planes evenly distributed in longitude. Establishing this system was to be a major contract for the Shuttle. By ferrying satellites up in batches of four, each flight would be able to populate one plane of the network. Even although they were not destined for geostationary orbit, such satellites required a PAM-D2 kick motor. Their deployment was to start in 1987, and the system was to be in operation within two years. The grounding of the Shuttle by the loss of Challenger in January 1986 reinforced the need for a 'mixed fleet' in which a variety of expendable rockets would complement the reusable vehicle. As a result of the 'Shuttle only' policy, McDonnell Douglas had ceased production of the Delta, and General Dynamics and Martin Marietta were about to close down the Atlas and Titan production lines. There was therefore an imperative to restart production across the range. The Air Force withdrew some government satellites from the Shuttle (including the Block-II NAVSTARs). It requested bids for expendable launchers, and on 8 August 1986 chose four contenders.[27],[28] In addition to an upgraded Delta from McDonnell Douglas, a variant of the Atlas–Centaur from General Dynamics, and a Titan 34D from Martin Marietta, Boeing and Hughes proposed a new vehicle that would utilise the F-1 engine that had been developed for the Saturn V.[29] In January 1987 the Air Force issued McDonnell Douglas with the contract to launch the NAVSTAR satellites one at a time.

Meanwhile, the new mixed-fleet strategy had opened the door for commercial launches.[30] Martin Marietta set out to compete with Arianespace, estimating that it should be able to secure 15 commercial launches between 1989 and 1994 for satellites offloaded from the Shuttle.[31] In June 1987 the company announced that it intended to invest $100 million in the production of 18 Atlas–Centaurs (which it later renamed the Atlas I), three of which were booked in October by the National Oceanic and Atmospheric Administration for its geostationary weather satellites.

The first Atlas I was launched successfully on 25 June 1990, but the second, on 18 April 1991, was frustrated at $T + 360$ seconds when one of the two engines of the Centaur did not start. The asymmetric thrust made the vehicle tumble, and the range safety officer destroyed it. The telemetry readings were normal until one of the turbopumps refused to spin up, preventing the engine from starting. Since there was no physical evidence, the cause could not be conclusively established. The hypothesis was that a 'foreign object' – "perhaps a nut or a bolt" – had been left inside the turbopump.[32] This loss of a Japanese communications satellite was written off as one of those mishaps that occur from time to time.[33] Having flown over 200 missions without a failure, the Centaur had a remarkable record. Nevertheless, it was modified to yield extra torque in order to overcome a 'slow start' in a turbopump, and the software was rewritten to shut down and rerun the startup cycle in the event of an engine failing to start.[34] The Atlas I resumed flying on 14 March 1992. But on 22 August 1992 the Centaur failed to start.[35] It was a replay of the previous failure and since the 'foreign object' argument could not be accepted twice there was clearly a systems failure.[36] The hypothesis was that the pre-launch chill down procedure had resulted in moisture from the ambient air freezing inside the turbopumps. The increased torque *should* have produced a successful start, but had not. The new software *had* ordered a second startup cycle, but without result.[37] It was decided to add a solenoid valve to prevent air from entering the turbopump and freezing onto its blades.[38]

With the Centaur grounded again, urgently required satellites were offloaded to other launch vehicles. On the next flight on 25 March 1993, the Atlas I stranded the first of the UHF Follow-On (UFO) communications satellites for the US Navy.[39] At $T + 24$ seconds the Atlas's two auxiliary engines started to lose thrust, levelling off at 65 per cent, which delayed the 5.5-*g* acceleration that triggered the jettisoning of the auxiliaries and left the sustainer with insufficient propellant to complete its burn. The Centaur extended its first firing by 24 seconds in order to attain the required trajectory, but then ran dry during the burn to enter geosynchronous transfer orbit.[40,41] The investigation identified the problem as the pressure regulator for the oxygen turbopump that fed both of the auxiliaries.[42] Further analysis established that an inadequately tightened screw in the regulator allowed a calibration shaft to rotate out of adjustment. The shaft was redesigned, and quality control improved.[43] The good news for Pratt & Whitney was that the Centaur had ignited properly. The fault in the Atlas was readily rectified, and on 3 September 1993 a replacement UFO satellite was successfully deployed.

ARIANE 4

With a stretched first stage augmented by a mix of strap-ons, the Ariane 4 could put payloads of between 2,000 and 4,900 kilograms into geosynchronous transfer orbit.

The launch of an Ariane 40, the basic variant of this launch vehicle, on 7 July 1995 with a number of payloads including the Cerise satellite.

The basic three-stage vehicle was the Ariane 40; the 42P had two solid strap-ons; the 44P had four solid strap-ons; the 42L had two liquid strap-ons; the 44LP had four strap-ons (two solid and two liquid); and the 44L (the most powerful variant) had four liquid strap-ons. Within a year of its introduction on 15 June 1988, the previous models had been retired.

The first failure was on 22 February 1990 when an Ariane 44L lost Superbird B and BS-2X, which were American-built communications satellites that were being launched for Japanese companies.[44] In this configuration, the Viking 6 liquid engines of the strap-ons ignite for lift-off and burn for 144 seconds; they are shed 10 seconds later, 1 minute before the first stage shuts down. In this case, the telemetry indicated that one of the four engines in the core stage had had a problem very early on. It was clear, however, that this fault was upstream of the engine.[45]

The Ariane 4 drew water from a tank located at the top of the stage to pressurise the propellant tanks and to regulate the combustion chamber pressure via the gas generator. In the case of the 44L, water was also fed through pipes to the strap-on engines. In this case the eight engines ran up to full power and at T + 3.3 seconds the vehicle was released. Within the space of half a second, at T + 6 seconds, the chamber pressure of one of the main engines dropped from the nominal 58 bars to about half of this value, and never recovered. The flight control system promptly gimballed the other three engines to compensate for the asymmetric thrust. Even so, as the vehicle rose from the pad it side-slipped towards the 77-metre-tall umbilical tower, clearing it by a mere 2 metres. At T + 90 seconds, the gimballed engines reached their maximum angle of 5 degrees and the vehicle began to tip over, and at T + 101 seconds the aerodynamic stress began to tear it apart and it finally exploded at an altitude of 9 kilometres. On a previous mission, air trapped in a water feed pipe had caused a momentary dip in combustion chamber pressure in one of the strap-ons. The design had been changed, but it appeared that once again the flow of water had been impeded. With the Atlas–Centaur now operating in competition, Arianespace could ill afford a lengthy grounding.

A record insurance loss

On 24 January 1994 the third stage of an Ariane 44LP shut down 80 seconds into its 750-second burn.[46] This was only the second failure in 35 launches. At $T+60$ seconds the bearing of the oxygen turbopump, which spun at 13,000 revolutions per minute and was cooled by being immersed in the liquid oxygen, began to rapidly over-heat. Nineteen seconds later, the speed of the bearing and the outlet pressure both fell, and the engine shut down one second later as the bearing rup-tured.[47] The investigation conducted 100 tests on 30 bearings (in some cases in a complete engine) to further characterise the heat dissipation para-meters of the bearing. In order to preclude a recurrence, the bearing would henceforth be coated with molybdenum disulphide to make it self-lubricating, and a purge line would be added to the bearing cavity to ensure that the cooling and helium flushing systems would be more reli-able.[48,49,50] The loss of Eutelsat 2F5 and Türksat 1, both of which were of the Aerospatiale Spacebus 2000 series, resulted in a total insurance claim of $350 million – one of the greatest to date. By mid-1994 Arianespace had hoped to ramp up the pace of Ariane 4 launches to one every three weeks, but this failure had undermined that as-piration.[51] The Ariane 44LP resumed operations on 17 June 1994 with Intelsat 702.

The launch of an Ariane 44L, the most powerful variant of this launch vehicle, on 29 March 2002 with Astra 3A and JCSAT 8.

Telstar 402

Within 10 minutes of being released by its Ariane 42L on 8 September 1994, Telstar 402 suffered a leak of the helium gas that was to pressurise the propellant for its attitude control thrusters, which prevented it from orienting itself, and the following day AT&T Skynet Satellite Services declared it a write-off.[52,53] The fact that the satellite had fallen silent in the act of pressurising its propellant suggested an explosion.[54] The investigation focused on the explosively actuated valves used to pressurise the propulsion system.[55] At first, such valves had been made of stainless steel, but over recent years there had been a trend towards titanium in order to save mass. It was concluded that when the second of the redundant pair of pyrovalves opened the hydrazine line, this set off an explosion.[56] As such valves were believed to have been responsible for the losses of Mars Observer and Landsat 6, AT&T sued Martin Marietta, whose Astro Space had supplied all of these spacecraft, citing incompetence and deception for failing to correct "known defects" in the propulsion system. The case was settled out of court in May 1995.[57] Meanwhile, in October 1994 NASA had organised an industry-wide teleconference to review the use of such valves and, in view of concern about how titanium reacted with hydrazine, it was concluded that the performance margins were ill understood. It was therefore decided to revert to stainless steel in some cases – notably for Mars Global Surveyor and Cassini.[58] AT&T had put Telstar 402 on an Ariane in an effort to jump the queue on the repeated grounding of the Atlas. Although the loss was not a launch vehicle failure, Arianespace promised to launch a replacement as a matter of priority, and the third satellite, known as Telstar 402R, went up on an Ariane 42L on 23 September 1995.

PanAmSat 3

President Ronald Reagan terminated Intelsat's monopoly in 1984 to enable private companies to offer telecommunications services on a commercial basis. In 1985 the Pan American Satellite Company (later PanAmSat) was created. It promptly bought a spare GE Astro Space 3000 series satellite, named it PanAmSat 1 and booked it on the maiden launch of the Ariane 4 for a bargain fee. Upon deciding to expand and become the first company to provide a global service, it ordered a trio of HS-601s, the first of which was launched on 8 July 1994.[59] Unfortunately, PanAmSat 3 (which held the distinction of being Hughes's 100th commercial comsat) was lost on 1 December 1994 when an Ariane 42P failed.[60,61,62] Telemetry indicated that the problem developed at the ignition of the third stage, with the combustion chamber failing to achieve its nominal pressure. The engine used a gas generator cycle in which the engine's turbine was driven by hydrogen burned in a generator upstream of the main chamber. The pressure in the gas generator should have been about 338 psi but it was only 220 psi; so instead of being 530 psi, the pressure in the main chamber was only 400 psi, and produced only 70 per cent of the required thrust. The engine ran for 740 seconds of the planned 780-second burn, then the control system shut it down with 700 kilograms of propellant remaining, leaving the vehicle to follow a ballistic arc that ended with re-entry over the Atlantic. This fault was clearly of a different nature to the one earlier in the year.[63] That this was the first use of an upgraded version of the engine was also a likely factor. When the investigation attributed the loss to contaminants in the propulsion system, the control measures were made more rigorous.[64]

The Ariane 4 resumed flying on 28 March 1995, successfully deploying Brasilsat B2 and Hot Bird 1.[65,66] With two state of the art communications satellites riding the first rocket after a launch vehicle failure, the insurance premium set a record.[67,68] In an effort to clear its backlog, Arianespace set a fast pace. PanAmSat 4 went up on an

Ariane 42L on 3 August 1995, but PanAmSat 5 (a Space Systems/Loral 1300 series) was offloaded to International Launch Services for launch on a Proton.[69,70,71] When Hughes delivered PanAmSat 3R, this was launched by an Ariane 42L on 12 January 1996.

Since the introduction of the Ariane 1 in 1979, a total of seven Ariane launches had failed, of which three involved the Ariane 4. Including this most recent loss, five failures were of the third stage, and two of those occurred in 1994.

DELTA II

When the Air Force awarded McDonnell Douglas the contract to launch the NAVSTAR Block-II satellites, the company set out to develop the more powerful Delta II as a two-phase programme. The first stage of the 6000 series introduced the Extra-Extended Long Tank and the upgraded RS-27A engine augmented by Castor IVA strap-ons, the second stage had the AJ-10-118K engine, and the third stage was a Star 48B kick motor. The vehicle had a capacity to geosynchronous transfer orbit of 1,400 kilograms, and a shroud adapted from the Titan IIIC could accommodate large modern payloads. All 17 fired between 14 February 1989 and 24 July 1992 were successful. The second phase of this upgrade introduced the much larger Hercules (later Alliant Techsystems) GEM-40 Graphite–Epoxy Motors. The first of the 7000 series was successfully launched on 26 November 1990.

On 5 August 1995 a Delta II from Canaveral had a strap-on that failed to jettison. This was the first failure of a Delta since 3 May 1986. At T + 66 seconds the three air-started strap-ons ignited, and the six that had augmented the lift-off were shed. Two of the air-started motors detached at T + 130 seconds, but one remained in place. The 'dead weight' of the 1,360-kilogram casing led to a significant velocity shortfall when the first stage shut down and was released. The second stage extended its burn by 35 seconds in order to recover the trajectory, but when it attempted its follow-on burn its tanks ran dry 10 seconds into the 44-second manoeuvre, and the Star 48B was subsequently able to achieve an apogee of only 34,780 kilometres.[72] Since the release point was well outside the expected 3-sigma dispersion range (a measure of accuracy), this was classified as a launch vehicle failure.[73] The satellite, a 3000 series supplied by Lockheed Martin Astro Space to Korea Telecom, was later able to achieve geostationary orbit, but at the cost of a significant fraction of its station-keeping propellant.[74]

A Delta II firework display
On 10 January 1997 a Delta II was to have carried the first batch of satellites for the Iridium constellation, but for four consecutive days the countdown was scrubbed before finally being postponed to 19 January.[75,76] In the meantime, a Delta II was lost seconds after lifting off from Canaveral on 17 January. This was the first catastrophic failure of a Delta II in 55 launches, and it caused the vehicle to be grounded for an investigation.[77,78] Only 200,000 of the 782,000 pounds of thrust for launch was provided by the RS-27A core engine, the remainder being from six of the

The launch of a Delta II on 20 March 2004 with the 11th Block-IIR NAVSTAR satellite.

The explosion of the Delta II on 17 January 1997 carrying the 1st Block-IIR
NAVSTAR satellite.

nine strap-ons.[79] At T + 13 seconds, at a height of 1,600 feet, the vehicle exploded.
"Barely had the crackling noise of the solids reached the press site when onlookers
saw, to their bewilderment, the rising rocket suddenly erupt into a massive fireball,"
as one eye-witness put it.[80] For the next 10 minutes, 250 tonnes of hardware and
propellant rained over a radius of 3,000 feet around Pad 17A, extending about half
way to the press site, near the old 'Mercury Control' building.[81] The damage to the
umbilical structure was minimal, but since the second stage had been loaded with
aerozine and nitrogen tetroxide, which are both toxic, the 73-member launch team in
the pad blockhouse donned respirators in case fumes penetrated the building. This
proved to be a wise precaution as a piece of débris struck a cable track and provided
a route for smoke to seep into the building. Worse off, however, were the 20 cars
parked adjacent to the blockhouse. Owing to the trend towards larger, more

powerful strap-ons for the Delta II, it was decided to abandon the blockhouse in favour of McDonnell Douglas's Delta Launch Control Center located in the 1st Space Launch Squadron's Operations Building, which was already under construction some 4 kilometres off. The equivalent facility for SLC-2 at Vandenberg was 13 kilometres from the pad, and the blockhouse was occupied only for non-hazardous activities. In April 1993 the Air Force had awarded the company the contract to launch the NAVSTAR Block-IIR series. As it happened, this Delta II had been carrying the first of these satellites.[82] However, its loss did not impair the GPS service because the satellite was to have replaced an older one that was still functioning.

A series of photographs by Carleton Bailie for *Aviation Week & Space Technology* showed black smoke venting from the side of one of the solid strap-ons about six seconds after it ignited, implying a flaw in its casing.[83],[84] The Air Force investigation, led by Colonel Ronald J. Haeckel, found from the telemetry that the self-destruct system had intervened at $T+13$ seconds, when it detected a vehicle distortion a fraction of a second after what seemed to be the explosion of a strap-on.[85] In view of the fact that more than 300 such strap-ons had flown successfully, this failure was unexpected.[86] Although the investigation did not identify the definitive cause, it found damage to fibres in the outer five composite layers of the motor in question, and concluded that a fracture had indeed developed in its casing six seconds after lift-off.[87] In an effort to preclude a recurrence, it was decided to conduct ultrasonic inspections of the new motors and to revise their handling procedures. The 14-tonne motor had been transported in a horizontal configuration on a 'travelling trunnion' system on which one of the supports was close to that part of the casing that had failed. This was considered significant as it was only the second time that this device had been used. The procedure was revised to ensure that the casings of the 36-foot-long, 3.3-foot-diameter graphite–epoxy structures were not subjected to anomalous stresses.[88],[89]

An artist's impression of a Block-IIR NAVSTAR satellite.

A close run thing

With the Delta II reinstated on 5 May 1997, a vehicle from Vandenberg put the first five Iridium satellites into low polar orbit.[90] In fact, this deployment was threatened by a fault in the attitude control system of the second stage.[91] About an hour into the mission – having maintained its attitude to release the first three satellites one at a time – the pressure in its gaseous nitrogen tank started to decrease markedly, and was nearly depleted by the time the fourth satellite was released. Although the vehicle was no longer under positive control and was slowly tumbling, the final satellite was successfully deployed. The Delta II returned to service at Canaveral on 20 May with Thor 2 for Norway.[92],[93] It was hoped to make one launch per month for the next 24 months to re-establish the schedule drawn up prior to the grounding.

ATLAS II

In May 1988 the Air Force issued Martin Marietta a contract to launch 10 DSCS-III satellites, starting in 1992. However, as the Atlas–Centaur could put only 2.3 tonnes into geosynchronous transfer orbit this required the development of an upgraded vehicle. This Atlas II had the uprated MA-5A power plant on which the auxiliary chambers were replaced by RS-27 engines, a stretched tankage, and a longer Centaur to increase the capacity to 2.8 tonnes, sufficient for a DSCS-III with an Integrated Apogee Boost Subsystem. The inaugural launch of the Atlas II on 7 December 1991 deployed Eutelsat 2F3, and the first DSCS-III went up on 11 February 1992.

The Atlas IIA had a Centaur with an uprated engine that increased the capacity to geosynchronous transfer orbit to 3 tonnes. Martin Marietta had expected that this would be able to launch the Intelsat 7 series, but these satellites put on so much mass in development that the first stage had to be reinforced to support four Castor IVA strap-ons in order to increase the capacity of 3.7 tonnes. The first Atlas IIA was launched on 10 June 1992 with Intelsat K. After its introduction on 16 December 1993 with Telstar 401,[94],[95] the more powerful Atlas IIAS began launching the Intelsat 7 series.

COMMERCIAL TITAN

In September 1986, after it had been decided to prohibit the Shuttle from deploying commercial satellites, Martin Marietta announced that in 1989 it would make available a new version of the Titan 34D with a lengthened second stage and an enlarged shroud to carry satellites that had been developed for the Shuttle.[96] This Commercial Titan could accommodate various third stages depending on the payload requirements, including the PAM, an IUS, a Centaur G-Prime (developed for Shuttle payloads) and the new Transfer Orbit Stage (TOS). The Titan was expensive to operate, but the fact that it could accommodate a heavy load meant that if it was fitted with a dispenser it would be able to deploy a pair of satellites and the cost shared between the customers. In addition, it was offered to the Air Force as a Medium Launch Vehicle, but this contract (which was issued in January 1987) went to McDonnell Douglas. Martin Marietta soon secured 10 launch options.[97] Commercial Titan Incorporated was formed in May 1987, and negotiated an agreement with the Air Force for launch services from Pad 40 at Canaveral.

The introduction of the Commercial Titan was good news for the owners of oversized satellites. As with the HS-381, the Hughes HS-393 (at that time the largest commercial communications satellite) had been designed to exploit the width of the

The launch of an Atlas IIA on 30 June 2000 with TDRS 8.

The launch of an Atlas IIAS on 18 September 2002 with Hispasat 1D.

Shuttle's bay, but it was considerably taller than the HS-381, and had upgraded HS-376-type transponders that could relay up to 120,000 telephone calls simultaneously. Intelsat had ordered six and booked them on the Shuttle, but after the loss of Challenger it had to await the introduction of either the Ariane 4 or the Commercial Titan.[98] Japan was in the same predicament, with two satellites. Japan's JCSAT 1 was launched on an Ariane 4 on 6 March 1989, and the first of the Intelsat 6 series followed on 27 October 1989. The first Commercial Titan on 1 January 1990 deployed Skynet 4A for the British military and JCSAT 2. For the second Commercial Titan, however, the dispenser's control system was set as if it had two satellites, whereas it had only one, and the firing command was sent to one pyro cable with the separation system connected to the other! The result was that on 14 March 1990 it failed to release Intelsat 603 (also known as 6F3).[99] To prevent the satellite being dragged back into the atmosphere along with the spent stage, it was ordered to separate from the trapped Orbus perigee kick motor and fire its own thrusters to raise its orbit. Even although the valuable satellite was stranded in low orbit, it was at least in space, and if a Shuttle crew could attach another motor it should be able to attain geostationary orbit. NASA was therefore invited to rescue a satellite that it was originally to have dispatched. Meanwhile, on 23 June 1990 the third Commercial Titan successfully deployed Intelsat 6F4.

Rescuing Intelsat 603

Several motors were considered for this mission. As Intelsat 603 had used a substantial fraction of its propellant to stave off orbital decay (which would reduce its operating life from 15 years to 12 years if it attained its assigned station) an IUS was an attractive option because a two-stage vehicle would be able to make the circularisation manoeuvre. However, bolting an HS-393 onto an IUS would not be easy. The second option was the TOS that had been developed specifically to boost heavy Shuttle payloads into geosynchronous transfer orbit. This had the advantage that it, too, was cleared for flight on the Shuttle, but had yet to be used. The third option was to fit the same type of motor as that used originally. In fact, the TOS and the first stage of the IUS used variants of this same motor. It was eventually decided to carry a new Orbus in a modified TOS cradle and mate it with the satellite. The challenge was to find a means of capturing the satellite, which was much larger than any that astronauts had worked on previously. A stinger was impractical because the support ring at the base of the satellite was 3 metres wide. The only option was to develop a capture bar incorporating a trunnion pin to enable the remote manipulator to manoeuvre the captive satellite into position directly above the motor. If the motor had a suitable adapter, the satellite could be connected to the motor using the bar, the arm could be withdrawn, the trunnion pin removed, and the payload released. The tricky part would be capturing the satellite. The astronaut would position the bar diametrically across the base of the satellite and activate the mechanism that would engage both ends simultaneously. One complication was that the target would be rotating, but as the rate was only half a revolution per minute underwater trials indicated that this would not pose a problem.

On its début mission in May 1992 as STS-49, Endeavour was launched to rescue

The launch of a Commercial Titan on 23 June 1990 with Intelsat 6F4.

Intelsat 603. Bruce Melnick swung Pierre Thuot on the arm up to the slowly rotating satellite while Rick Hieb remained in the payload bay. After Thuot had carefully aligned the bar across the ring on the bottom of the enormous cylinder, he eased it forward to establish contact. An automatic mechanism was supposed to trigger latches to secure both ends simultaneously, but Thuot had not contacted the target with sufficient force to trigger the mechanism. On a second attempt, he made a firmer contact and the latches fired. Unfortunately, when Thuot tried to use the bar to cancel the residual spin, it slipped off and the satellite resumed its spin. Furthermore, Thuot's actions had caused it to precess, which made positioning the bar for a third capture much more difficult. Not only did his attempts to attach the bar fail, but each time he touched the satellite its precessional motion became more pronounced. After three hours he admitted defeat and Endeavour withdrew, leaving Intelsat 603 in a slow 50-degree nutation. It was belatedly realised that the simulations in the water tank had not accurately replicated the degree to which the satellite would be disturbed by a light contact. The clue was the fact that it had reacquired its spin after the bar had slipped off. This meant that the liquid propellant had retained sufficient momentum to spin up the satellite again, and the precession was due to the liquid sloshing around. If he ever managed to capture the satellite, Thuot would need to hold it steady until the liquid had shed its kinetic energy, and then take care each time he moved it. Despite having trained in a water tank, they had been caught unawares by the physical properties of fluids in weightlessness. Overnight, the satellite's handlers eliminated both the precession and the residual spin. When Endeavour returned the next day Thuot ventured out again, confident that the satellite would be captured. This time he eased the bar against the ring and activated the latches manually. However, they failed to engage. After five hours he retreated, this time leaving the satellite in a 'flat' spin. When Endeavour had withdrawn, the controllers once again set out to stabilise their charge, but it was beginning to look as if it was beyond rescue.[100]

While the crew slept, their colleagues on the ground tried to find an alternative way of capturing the rogue satellite. It was all too evident that the capture bar would have to be abandoned. How could an astronaut retrieve the 4-metre-diameter, 6-metre-tall drum without using the bar? And, in any case, the bar would have to be in place to enable the arm to fit the satellite on the motor. Could Thuot capture the satellite by hand? Could he hold it in place while Hieb added the bar? If Thuot stood on the arm to hold the satellite, Hieb would not be able to reach it unless the orbiter manoeuvred really close in. If Thuot lost his grip, the satellite might strike and damage the payload bay doors or the vertical stabiliser. Meanwhile, the crew of Endeavour had also reached the conclusion that the ideal tool for grappling a satellite was the human hand, and they added a twist of their own. The Shuttle carried a spare suit, and if Tom Akers, who was scheduled for a later excursion, was to lend assistance, Thuot could then grab the satellite and slowly swing it down into the bay, where Hieb and Akers would be waiting for it. They could then hold it steady while Thuot attached the bar, free of any concern about disturbing the satellite's motion. As soon as the bar was in place, they could revert to the original plan. When this idea was put to Houston, a test was ordered to verify that the airlock could accommodate three spacesuited astronauts. It was a tight fit, but was feasible, so in the absence of a better idea the plan was approved.

Endeavour had propellant for just one more rendezvous. If Intelsat 603 eluded them this time it would have to be written off, even though it was perfectly healthy. If they failed, it would not be for lack of experience. Dan Brandenstein, chief of the Astronaut Office, manoeuvred the orbiter to position the satellite a mere 2 metres above the bay, with its base facing down. Hieb had taken up station on a foot restraint on the starboard sill; Akers was on a truss that spanned the mid-bay, and the satellite was just above their heads. Melnick, further demonstrating his prowess with the arm, stationed Thuot around the far side. For 10 minutes they simply observed the satellite's motion. It had a slight

Pierre Thuot, riding the Shuttle's arm attempts to engage the ring fixture on the base of Intelsat 603 using the 'capture bar'.

Pierre Thuot (on the arm), Richard Hieb (on the sill) and Tom Akers (in the centre of the payload bay) manually manipulate Intelsat 603.

nutation with a 3-minute rate. They waited until it was vertical with respect to the bay, and then six hands simultaneously grasped the base ring and held the satellite still for several minutes to allow its fluids to settle. Although they had captured the drum in a vertical attitude, it was not conveniently oriented for fitting the capture bar, which Hieb had stowed on the side wall below his station. Working in concert, they rotated the massive satellite, a little at a time, through 120 degrees. When it was restabilised, Hieb let go, retrieved the bar, and positioned it just beneath the ring. From his position on the sill he could not reach the controller in the centre of the bar to engage the latches, so he held it in position with one hand and the ring with the other. Akers then acted as an anchor while Thuot manoeuvred beneath the satellite, indicating his instructions to Melnick by way of hand movements. With the latches of the bar engaged, Thuot tightened a number of bolts to complete the attachment, then dismounted the arm. Once the foot restraint had been deleted from the arm, Melnick grasped the satellite via the trunnion pin on the bar, and Akers, who had held the satellite in a vise-like grip for 90 minutes, was finally able to let go. After a clamp had been attached to each end of the bar, the satellite was mated with the motor. There was a moment of concern when the command to eject the stack from its cradle was issued and nothing happened, but the backup circuit functioned and Intelsat 603 finally set off on the first stage of its journey to geostationary orbit. NASA set up a review of the rescue by a panel chaired by Eugene Covert of the Massachusetts Institute of Technology, and this reported that the operation had not been cost-effective, and was not worth the risk of the astronauts' lives.[101]

Uncompetitive

On 25 September 1992 the fourth Commercial Titan launched Mars Observer, and the first TOS accelerated it to escape velocity. Unfortunately, on 22 August 1993, three days prior to the burn that was to put it into orbit around the planet Mars, the spacecraft fell silent while pressurising its propulsion system.[102]

In 1987, it had been expected that there would be five launches per year for at least five years, but the Commercial Titan was uncompetitive, and after this brief operational history it was withdrawn.[103]

NOTES

1. *Aviation Week & Space Technology*, 28 April 1986, p. 16.
2. *Aviation Week & Space Technology*, 5 December 1994, p. 47.
3. *Aviation Week & Space Technology*, 9 June 1986, p. 20.
4. *Aviation Week & Space Technology*, 9 March 1987, p. 266.
5. *Aviation Week & Space Technology*, 7 July 1986, p. 22.
6. *Aviation Week & Space Technology*, 8 August 1988, p. 60.
7. *Aviation Week & Space Technology*, 22 June 1987, p. 37.
8. *Aviation Week & Space Technology*, 20 July 1987, p. 26.
9. *Aviation Week & Space Technology*, 2 November 1987, p. 24.
10. *Aviation Week & Space Technology*, 2 November 1987, p. 24.

11. *Aviation Week & Space Technology*, 12 September 1988, p. 26.
12. *Spaceflight*, January 1980, p. 32.
13. *Spaceflight*, September–October 1980, p. 321.
14. *Aviation Week & Space Technology*, 2 June 1980, p. 17.
15. *Aviation Week & Space Technology*, 23 June 1980, p. 16.
16. *Aviation Week & Space Technology*, 5 March 1990, p. 20.
17. *Aviation Week & Space Technology*, 27 September 1982, p. 20.
18. *Aviation Week & Space Technology*, 18 October 1982, p. 68.
19. *Aviation Week & Space Technology*, 1 November 1982, p. 24.
20. *Aviation Week & Space Technology*, 16 September 1985, p. 26.
21. *Aviation Week & Space Technology*, 23 September 1985, p. 22.
22. *Aviation Week & Space Technology*, 7 October 1985, p. 25.
23. *Aviation Week & Space Technology*, 9 June 1986, p. 21.
24. *Aviation Week & Space Technology*, 7 July 1986, p. 22.
25. *Aviation Week & Space Technology*, 14 July 1986, p. 31.
26. *Spaceflight*, January 1980, p. 2.
27. *Aviation Week & Space Technology*, 30 June 1986, p. 21.
28. *Aviation Week & Space Technology*, 18 August 1986, p. 22.
29. *Aviation Week & Space Technology*, 4 August 1986, p. 34.
30. *Aviation Week & Space Technology*, 23 June 1986, p. 18.
31. *Aviation Week & Space Technology*, 23 March 1987, p. 25.
32. *Flight International*, 10–16 July 1991, p. 20.
33. *Aviation Week & Space Technology*, 29 May 1991.
34. *Aviation Week & Space Technology*, 7 September 1992.
35. *Aviation Week & Space Technology*, 31 August 1992.
36. *Aviation Week & Space Technology*, 28 September 1992.
37. *Aviation Week & Space Technology*, 7 September 1992.
38. *Aviation Week & Space Technology*, 15 January 1993.
39. http://leonardo.jpl.nasa.gov/msl/QuickLooks/ufoQL.html
40. *Aviation Week & Space Technology*, 29 March 1993.
41. *Aviation Week & Space Technology*, 5 April 1993.
42. *Aviation Week & Space Technology*, 10 May 1993.
43. *Aviation Week & Space Technology*, 14 June 1993.
44. *Aviation Week & Space Technology*, 26 February 1990, p. 32.
45. *Aviation Week & Space Technology*, 5 March 1990, p. 18.
46. *Aviation Week & Space Technology*, 31 January 1994, p. 27.
47. *Flight International*, 2–8 February 1994, p. 17.
48. *Flight International*, 2–8 March 1994, p. 30.
49. *Aviation Week & Space Technology*, 27 June 1994, p. 81.
50. *Aviation Week & Space Technology*, 18 July 1994, p. 83.
51. *Aviation Week & Space Technology*, 18 July 1994, p. 83.
52. *Flight International*, 21–27 September 1994, p. 26.
53. *Aviation Week & Space Technology*, 19 September 1994, p. 70.
54. *Flight International*, 11–17 January 1995, p. 18.
55. *Aviation Week & Space Technology*, 9 January 1995, p. 27.
56. *Aviation Week & Space Technology*, 17 April 1995, p. 24.
57. *Aviation Week & Space Technology*, 27 January 1997, p. 62.
58. *Aviation Week & Space Technology*, 9 January 1995, p. 27.
59. *Flight International*, 31 May–6 June 1995, p. 54.

60. *Aviation Week & Space Technology*, 5 December 1994, p. 15.
61. *Flight International*, 14–20 December 1994.
62. *Flight International*, 4–10 January 1995.
63. *Aviation Week & Space Technology*, 12/19 December 1994, p. 25.
64. *Aviation Week & Space Technology*, 2 January 1995, p. 58.
65. *Aviation Week & Space Technology*, 3 April 1995, p. 27.
66. *Flight International*, 5–11 April 1995, p. 24.
67. *Aviation Week & Space Technology*, 6 February 1995, p. 62.
68. *Aviation Week & Space Technology*, 13 February 1995, p. 17.
69. *Flight International*, 5–11 October 1994, p. 19.
70. *Flight International*, 21 December 1994–3 January 1995, p. 22.
71. *Flight International*, 7–13 June 1995, p. 29
72. *Aviation Week & Space Technology*, 14 August 1995, p. 63.
73. http://kevinforsyth.net/delta/delta2.htm
74. *Flight International*, 30 August–5 September 1995, p. 24.
75. *Flight International*, 15–21 January 1997.
76. *Aviation Week & Space Technology*, 20 January 1997, p. 64.
77. *Flight International*, 29 January–4 February 1997, p. 26.
78. This was only the 14th failure in 241 Thor and Delta launches over a span of 36 years, of which 10 had failed to achieve orbit and the others had placed their payloads in useless orbits. On 5 August 1995 one of the three air-started solids of a Delta II failed to release, and Koreasat 1 was placed into an undesired (but usuable) orbit. The previous total failure was #178 on 3 May 1996 when the first stage veered off course and was destroyed by the range safety officer at T + 90 seconds carrying the GOES G metsat. The only other 'low' incident was #134 on 13 September 1977 when a faulty Castor IV solid caused an explosion at T + 54 secods, destroying Europe's OTS experimental comsat. The Delta II's most recent launches had carried Mars Global Surveyor and Mars Pathfinder in November and December 1996, respectively.
79. It was the 7925 'standard' contemporary configuration of the Delta II.
80. *Spaceflight*, April 1997, p. 112.
81. *Spaceflight*, May 1997, p. 156.
82. The Block-IIR built by Lockheed Martin was significantly more powerful than the earlier models built by Rockwell International. The first Block-I rode an Atlas from Vandenberg on 22 February 1978. The initial 10 satellites demonstrated the operation of the system. The first Block-II rode a Delta II on 14 February 1989, and the Block-IIA began on 26 November 1990. These were used to create an operational constellation of 24 satellites in 12-hour orbits at 20,200 kilometres in six orbital planes inclined to the equator. The Block-IIR satellites are being launched to replace the Block-II/IIA as they wear out. A new series (Block-IIF, built by Rockwell - see *Aviation Week & Space Technology*, 29 April 1996, p. 64) will be introduced in 2005. To clarify - only the Block-IIR were supplied by Lockheed Martin.
83. *Aviation Week & Space Technology*, 27 January 1997, p. 30.
84. *Aviation Week & Space Technology*, 3 February 1997, p. 29.
85. *Aviation Week & Space Technology*, 7 April 1997, p. 28.
86. *Aviation Week & Space Technology*, 5 May 1997, p. 27.
87. *Aviation Week & Space Technology*, 12, May 1997, p. 24.
88. *Aviation Week & Space Technology*, 5 May 1997, p. 27.
89. *Aviation Week & Space Technology*, 26 May 1997, p. 36.
90. *Aviation Week & Space Technology*, 25 May 1998, p. 26.

91. *Aviation Week & Space Technology*, 12 May 1997, p. 25.
92. *Aviation Week & Space Technology*, 26 May, 1997, p. 36.
93. Thor 2 was a 3,235-pound HS-376HP for Telenor in Norway.
94. *Flight International*, 12–18 January 1994, p. 19.
95. *Flight International*, 21–27 September 1994, p. 26.
96. *Aviation Week & Space Technology*, 20 April 1987, p. 66.
97. *Aviation Week & Space Technology*, 9 March 1987, p. 266.
98. *Aviation Week & Space Technology*, 23 March 1987, p. 30.
99. *Flight International*, 4–10 April 1990, p. 14.
100. *Aviation Week & Space Technology*, 18 May 1992, p. 22.
101. *Aviation Week & Space Technology*, 16 November 1992, p. 22.
102. *Aviation Week & Space Technology*, 30 August 1993, p. 20.
103. *Aviation Week & Space Technology*, 20 April 1987, p. 66.

4

Heavyweights

TITAN IV

When the Shuttle was declared operational in 1982 the Air Force became concerned that it took so long to 'turn around' that it would never achieve the advertised flight rate, and that without considerable further development it would not meet its target of being able to insert a 14.5-tonne payload into polar orbit. On 7 February 1984 the Department of Defense issued a Space Launch Strategy calling for the development of an expendable launch vehicle capable of placing a Shuttle-class payload into geostationary orbit.[1] In March 1984 the Air Force announced its Request for Proposals from industry for such a Complementary Expendable Launch Vehicle, with the rate of its development being the crucial factor because the existing production lines were in the process of closing down. The two leading entries were for upgraded forms of the Atlas and the Titan III, with the Titan III winning.[2] In the 1960s, the Air Force had developed a version of the Titan IIIC with a seven-segment strap-on for its Manned Orbiting Laboratory, which would be launched with a two-man Gemini spacecraft on its nose.[3] Although this Titan IIIM was cancelled in 1969, the new vehicle would be a stretched Titan 34D with these seven-segment strap-ons in a configuration designated the Titan 34D7. It would also have a much larger shroud in order to accommodate the satellites which had been designed for the Shuttle's cavernous payload bay. The Air Force received development approval in October 1984, and in June 1985 it awarded the production contract for 10 vehicles. The initial plan was that they would all employ the wide-bodied Centaur G-Prime that was under development for the Shuttle, but the contract was revised to include the option of using the IUS.[4] It had been expected that these vehicles would be placed in storage and used as conditions warranted, perhaps being fired at the rate of two per year. However, after the loss of Challenger the Air Force urgently needed a reliable replacement vehicle to launch its heavy satellites.

In August 1986, the Air Force approved a major restructuring of the programme that modified the design to include a two-stage variant – that is, with no upper stage

Titan IVA models (left to right): Titan IVA–IUS, the 1st vehicle of this type, which was launched on 14 June 1989 with a DSP satellite; Titan IVA on 8 June 1990 with four satellites for the Naval Ocean Surveillance System; Titan IVA–Centaur awaits launch on 7 February 1994 with the Milstar DFS 1 satellite.

– able to achieve polar orbit from Vandenberg. In the expectation of sustaining a rate of six launches per year, the contract was increased to 23 vehicles. As part of this review, the Complementary Expendable Launch Vehicle was renamed the Titan IV. The Air Force's withdrawal from the Shuttle programme made this the main launcher for Department of Defense satellites.[5,6] Although the plan had been to mount the inaugural flight by October 1998, the schedule slipped. On 14 June 1989 a Titan IVA lifted off from Pad 41 with an IUS, and this inserted a DSP satellite into geostationary orbit. On 8 March 1991 the first from SLC-4E at Vandenberg carried the Lacrosse 2 radar-imaging reconnaissance satellite.[7] After successfully flying three from each launch site, the seventh failed.[8]

First failure

On 2 August 1993 a Titan IVA without an upper stage lifted off from SLC-4E and exploded at T + 101 seconds, just a few seconds before the core stage was to ignite.[9] The video showed a normal ascent until a light-coloured ring of smoke puffed outward from the vehicle moments before it was transformed into a fireball. The investigation blamed a faulty repair to one of the solid rocket motors that had permitted a burn-through of the casing.[10,11] The telemetry indicated pitch and yaw perturbations immediately prior to the explosion due to gas venting from this hole. After the slurry mix of propellant had been poured into its casing, each segment was capped by a rubbery material designed to retard erosion of the field joint. In cases where a void formed between this 'restrictor' and the propellant (a flaw referred to as debonding) this was repaired. After a motor containing a repaired segment on a Titan 34D exploded in 1986, the repair procedure was revised and had since given no cause for concern. In this case, however, the repair had been by far the most extensive – involving an area of some 5,000 square inches of the restrictor's surface compared to the more typical 350 square inches. As there had been multiple voids, it had been decided not to repair each void individually but instead to apply a large pie-shaped patch. It had been believed that when the ignition system fired a jet of flame down the bore of the stacked motor the pressure would *seal* this cut, but a test undertaken for the investigation showed that the transient pressures would actually *open* it.[12] This had evidently allowed a fast flame front to burn through the 10-centimetre-thick steel casing. It was decided to remain within known parameters, and not attempt to repair voids larger than those that had been shown to be safe. As an expedient to resume flying as soon as possible, all repaired segments (a total of 14) were removed from already stacked motors and placed in storage to await inspection.

Initially, Titan IVA launches from Canaveral had utilised Pad 41, but following the launch of the final Commercial Titan from Pad 40 in 1992 with Mars Observer this pad was refurbished. Its first use, on 7 February 1994, marked the return to service of the Titan IVA after the loss in August 1993.

A Titan IVA–Centaur on Pad 40 at Canaveral.

Programme pressures

In late 1993 – following the catastrophic failure of a Titan IVA and the losses of Mars Observer, NOAA 13 and Landsat 6 – Martin Marietta ordered a 'blue riband' review of systems engineering right across its space programmes.[13,14] When the panel reported in December, it blamed "too much employee turnover, an unclear organisational structure, and a breakdown in accountability".[15]

Slow pace

The Titan IVA–Centaur that left Pad 41 on 3 May 1994 had an unenviable record on the ground. It was rolled out from the Vertical Processing Facility on 14 June 1991, only to be rolled back a year later when inspections found corrosion in the joints of its solid rocket motors. After the investigation into a Centaur failure in August 1992 it was rolled out again, only to be held up by a repeat of the Centaur problem in March 1993. In July 1993, in order not to exceed the 18-month limit for solid rocket motors to sit in the open, it was rolled back due to the late delivery of its payload. Following the loss of a Titan in August 1993 due to the explosion of one of its solid rocket motors, a segment in the grounded vehicle containing one of the suspect repairs was replaced.[16,17,18]

In contrast, the Titan IVA launched from Canaveral on 27 August 1994 set a record for its rapid preparation, having spent "only 91 days" on the pad – which the Air Force said was a sign that the vehicle was "acquiring maturity".[19,20,21] Indeed, on 22 December 1994 the Titan IVA that lifted off from Pad 40 marked the fourth launch in the 12 months since the loss in 1993, all of which had been from the Cape. Another launch on 6 November 1995 deployed a Milstar, the most advanced communications satellite yet developed by the Department of Defense.[22,23]

The first Titan IVA to leave from SLC-4E since the loss in August 1993 lifted off on 5 December 1995 with a classified payload.[24] This hiatus had been to update the pads for the new solid rocket motors that were to be used on the Titan IVB.[25]

A Titan IVA–Centaur awaits launch on 6 November 1995 with the Milstar DFS 2 satellite.

A Milstar satellite stowed in its shroud and (artist's illustration) in its deployed configuration in orbit.

The final Titan IVA

When the final Titan IVA–Centaur lifted off from Canaveral on 12 August 1998, it became the second one to be lost.[26,27] The telemetry established that the power from the battery to the guidance system had faltered at T+39.4 seconds.[28] The outage was only for a fraction of a second but it was long enough to cause the guidance system to lose its horizon and, when the power was restored, the system had issued an erroneous attitude correction that pitched the vehicle over at an altitude of 20,000 feet, which in turn had triggered the self-destruct system at T+41.3 seconds.[29] The investigation traced the fault to a defect in an electrical harness in the second stage of the core, in which eroded insulation caused a momentary short circuit.[30,31,32] This $1 billion mishap lost the National Reconnaissance Office a Mercury signals-intelligence satellite.[33]

The Titan IVB

In view of the problems with the solid rocket boosters of the Titan 34D and Shuttle, the Air Force had given the contract for a Solid Rocket Motor Upgrade to Hercules (later Alliant Techsystems).[34,35] Unlike the seven-segment steel-cased motors of the Titan IVA, the new ones had filament-wound composite casings and had just three segments to reduce the number of joints. In addition, they were of greater diameter in order to accommodate more propellant and the motor had a hydraulic actuator for thrust-vectoring.[36,37] The core stage of the Titan IVB ignited at T+130 seconds, some 15 seconds before the solids were jettisoned.[38] It could exceed its predecessor's payload by 25 per cent, increasing the low-orbit capacity from 17.7 to 22.3 tonnes, and the capacity to polar orbit from 13.6 to 17.3 tonnes.[39] The optional upper stage could be selected to suit the payload. The single-stage Centaur could put 5.75 tonnes into geosynchronous transfer orbit, and the two-stage IUS could place a 2.5-tonne satellite directly into geostationary orbit. The Titan IVB was intended to be cheaper to operate and more reliable than its predecessor. Whereas each Titan IVA was built to accommodate its payload, the mechanical and electrical interfaces for the payload were standardised on the Titan IVB. Launch vehicles were to be built for stock, with mission-specific kits being fitted when its

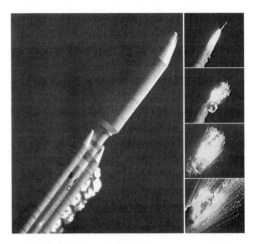

The destruction of the Titan IVA–Centaur at T + 41 seconds on 12 August 1998.

payload was assigned, and operational improvements were intended to reduce the time to prepare a vehicle for launch.[40] This $1 billion programme marked the most significant upgrade in the Titan's 35-year history.[41]

It had been planned to introduce the Titan IVB in 1996, but its début mission slipped to 23 February 1997, when it lifted off from Pad 40 with an IUS that put a DSP satellite into geostationary orbit.[42,43] On 15 October 1997 a Titan IVB with a Centaur dispatched the Cassini spacecraft for its mission to Saturn. The first Titan IVB launch from Pad 41 was on 7 November, when a Centaur deployed a classified satellite. On 8 May 1998, a Titan IVB from Pad 40 with a Centaur deployed a payload for the National Reconnaissance Office.

Two down
On 9 April 1999 the second (and last) Titan IVB from Pad 41 put an IUS with a DSP satellite into low parking orbit.[44,45] The Orbus 21 first stage of the IUS accomplished the burn for geosynchronous transfer orbit, but at apogee the Orbus 6E second stage shut down prematurely while attempting to circularise. Since its début in 1982, the IUS had suffered only two operational failures and was considered reliable.[46] A NASA representative joined the Air Force investigation because an IUS was to deploy the Chandra X-ray Observatory from a Shuttle in July 1999 (a mission that had later to be postponed).[47] The nozzle of the second-stage motor of an IUS forms three nested sections while stacked. After the first stage has been shed, the two extendable sections of the nozzle are 'cranked' into position. In considering the possibility that the nozzle had failed to deploy, the investigation found that the first stage had not separated properly. The thermal wrap and tape that were applied to a harness and connector had inhibited the disconnection of a plug linking the two stages, with the result that they remained attached at one point, which formed a 'hinge'.[48] The dangling first stage damaged the nozzle of the second stage, and when

that motor ignited it set the vehicle tumbling in a way that the attitude control system could not counter.[49] Ironically, the technicians had followed the procedure that had been in force since the introduction of the IUS, which in retrospect was flawed by its failure to take account of the potential for disabling the internal mechanism of the separation connector. The 2.5-tonne DSP did not have the propellant to undertake its own circularisation burn. Despite the loss, the Air Force insisted that its missile-launch warning system still had "complete world-wide coverage".[50]

Twenty-one days later, on 30 April, the Centaur of a vehicle launched from Pad 40 stranded a Milstar satellite![51,52] The investigation found a software error in the guidance system.[53,54,55] The 'constant' for the roll rate had been entered with the decimal point one place to the left, making it one-tenth of the value. In attempting to cancel an anomalous roll during its first burn, the Centaur had consumed 85 per cent of its hydrazine attitude control propellant, and thereafter ran dry when trying to do so during a later manoeuvre, with the result that it released its payload in an orbit with an apogee that fell far short of geosynchronous altitude. Of course, the great mystery was that the software verification process at Lockheed Martin Astronautics had missed this slip. In fact, data that would have shown something amiss in a test a week prior to launch had not been monitored, and anomalous indications in the final hours of the countdown were misinterpreted.[56] At $1.23 billion ($880 million for the satellite and $433 million for the launch vehicle) this was the most costly satellite loss to date for the Department of Defense.[57]

Although the Titan IVB had suffered two upper stage failures in succession – one an IUS and the other a Centaur – it was possible to proceed with the next mission as it did not require an upper stage. At T + 12 minutes after lifing off from Vandenberg on 22 May 1999, the National Reconnaissance Office payload was released to make its own way to its operating station.[58,59,60]

The recent problems involving the Titan IV were investigated by an Independent Assessment Team on Mission Success chaired by Thomas Young, former president and chief operating officer of Martin Marietta. This highlighted a number of quality control issues derived from the shedding by Lockheed Martin of experienced personnel and an overemphasis on cost-cutting – in short, the improper application of the 'faster-better-cheaper' mantra.[61,62,63] This report prompted major management and quality control changes.

The hiatus ended on 8 May 2000, when a Titan IVB was launched from Pad 40 with an IUS that deployed a DSP satellite to replace the one that had been lost.[64,65,66]

PROTON

The Proton was developed in the early 1960s by Vladimir Chelomei, who was one of Sergei Korolev's rivals. The first stage of the Proton had an oxidiser tank in the centre of six narrower fuel tanks, and an engine at the base of each of the peripheral assemblies. Unlike Korolev, Chelomei used storable hydrazine and nitrogen

A Titan IVB–Centaur with a Milstar II satellite lifts off.

A pair of N-1 moonrockets at the Baikonur Cosmodrome.

tetroxide as propellants. The two-stage version tested in 1965 carried a scientific payload named Proton, which gave the vehicle its popular moniker. With a third stage, it could insert '20 tonnes' into low orbit, and was used in this form to launch heavy satellites and space station modules. With a fourth stage derived from the fifth stage of Korolev's giant N-1 moonrocket, known as the Block-D,[67] it could deliver satellites into geostationary orbit and dispatch heavy probes to the Moon and into deep space.

The Proton initially had a high proportion of failures, with 16 out of 25 launches in the first five years suffering a mishap, but went on to become a reliable vehicle with just 10 further losses by the time of the 100th launch in 1982.[68] After the loss of a Raduga satellite on 24 December 1982 there was a long run of successes until an Almaz radar-imaging satellite was lost on 29 December 1986.[69] A commmunications satellite was stranded in geosynchronous transfer orbit on the next launch, on 30 January 1987, when the Block-D failed to make the circularisation burn.[70] A Raduga satellite was inserted into geostationary orbit on 19 March 1987. It was initially denied that the launch on 24 April 1987 had been a failure,[71] but it was obvious that it had been carrying three satellites for the GLONASS system, the Soviet version of GPS. This system was at an altitude of 19,100 kilometres, with satellites in three orbital planes inclined at 65 degrees, with a 45-degree spacing between the satellites in a plane and the planes separated by 120 degrees in longitude.[72] In this case, the Block-D's engine had shut down prematurely, falling 1,350 kilometres short of the requisite apogee. A restart command from the ground was ineffective, and it became apparent that the engine was damaged when the circularisation burn failed to occur.

The timer, oblivious to the situation, then proceeded to eject the satellites.[73],[74] The investigation concluded that the failures of 30 January and 24 April were due to a system upgrade, and that there was no reason to ground the older Block-D. Flights resumed on 11 May 1987 with a Block-D successfully inserting a Gorizont communications satellite into geostationary orbit. A Proton suffered a third-stage failure on 18 January 1988, a Block-D failed on 17 February 1988, and on 9 August 1990 a vehicle was unfortunately lost as a result of a worker leaving a rag in the second stage's propulsion system.[75],[76]

Nevertheless, as the Proton was a heavy-lifter with a good record on 200 launches, in early 1992 a Lockheed delegation made an impromptu visit to Khrunichev to propose a joint venture, and one year later, with approval of both the US and Russian governments, Lockheed Khrunichev Energiya Incorporated (LKEI) was created to market the Proton. Although a Gorizont satellite was lost on 27 May 1993 when an engine in the second stage exploded due to a fuel contaminant,[77],[78],[79] several months later LKEI signed up its first customer.[80]

International Launch Services
In April 1993 Martin Marietta acquired Astro Space from General Electric, giving it a high-profile presence in the commercial satellite business. On 23 December 1993 it announced its decision to buy the Space Systems Division from General Dynamics, and this acquisition on 2 May 1994 not only gave the company two of the main US launch vehicles, but also the powerful Centaur upper stage.[81] In March 1994 it announced an agreement in which the Russian Energomash company would develop and supply RD-180 engines. On 29 August 1994 Lockheed and Martin Marietta announced that they were to merge, and on 15 March 1995 Lockheed Martin was created.[82],[83] This brought the Atlas, Titan and Proton vehicles, with their complementary capabilities, into a single stable. The Atlas and the Proton could jointly challenge Arianespace, the leader in the commercial launch market, and so Lockheed Martin set out to develop a strategy that would increase its share of this market.[84],[85] One issue was the US–Russian Launch Trade Agreement of September 1993 limiting LKEI to nine Proton commercial geostationary launches through to 2000, whereas Khrunichev's manufacturing capacity would allow 20 launches.[86],[87] In addition, a floor had been imposed on the fee to ensure that the Russians did not 'soak up' the market.[88] The first launch for Lockheed Martin was the Atlas IIAS on 22 March 1995, on which the Martin Marietta logo on the shroud that enclosed Intelsat 705 had been painted over.[89] At the Paris Air Show in June 1995 the company announced that it was to form a joint venture with LKEI. This new company, International Launch Services (ILS), would jointly market the Proton and Atlas.[90],[91] The objective was to capture 50 per cent of the commercial launcher market.[92] If a problem grounded one vehicle, a satellite would be transferred to the other vehicle.[93] Since the start of the decade, the Proton-K had achieved a 96 per cent success, so getting insurance for a satellite was not a problem.[94] In fact, LKEI had bookings worth $1 billion for the Proton.[95] In January 1996 Russia requested that the quota limiting commercial launches be relaxed, and it was increased to allow a total of 16 launches by the end of 2000.[96],[97]

The Proton-K made its début with ILS on 9 April 1996 with Astra 1F for the Société Européenne des Satellites (SES) of Luxembourg.[98],[99] Following two Russian launches in September, on 17 November 1996 a modified Block-D that was to have dispatched the Mars '96 spacecraft malfunctioned during the escape manoeuvre and dived back into the atmosphere.[100],[101] The second ILS launch was slipped to 24 May 1997 and carried Telstar 5 for the Loral Skynet system.[102] The first three stages were used to insert the Block-DM and its payload (a total of 23 tonnes) into an initial 200-kilometre parking orbit inclined at 51.6 degrees, and after a 65-minute coast this stage fired its engine for the first time to enter geosynchronous transfer orbit. Five hours later, it made a burn that simultaneously raised the perigee to 6,650 kilometres and reduced the inclination to 17.6 degrees. It then released the satellite which, 6 hours and 41 minutes after launch, used its own engine to circularise above the equator. This burden-sharing enabled the launcher to carry heavier satellites than would otherwise have been practicable. With two successes and firm bookings through 2000, ILS ramped up its launch rate, and with the Atlas also in its portfolio, it expected to give Arianespace a real run for its money. On 18 June 1997 an ILS Proton-K deployed seven satellites for Iridium.[103] This employed a new version of the Block-DM on which the payload adapter was strengthened and widened to 132 inches (from 78 inches), the propulsion system was upgraded with a new pneumatic and hydraulic subsystem, and the oxidiser tankage was modified to compensate for the loads involved in carrying the dispenser for the cluster of satellites. Three such launches had been booked. The first failure of an ILS Proton-K was on 25 December 1997. The first firing of the Block-DM stage successfully achieved geosynchronous transfer orbit, but when it attempted to circularise and eliminate the inclination 6 hours later it shut off just 1 second into the 110-second burn. Because the payload, AsiaSat 3, was to have been placed directly into geostationary orbit, it carried propellant only for station-keeping and hence had to be written off. After the Asia Satellite Telecommunications Company of Hong Kong had claimed the insurance, the underwriters sold the HS-601HP back to Hughes, which renamed it HGS 1.[104]

The Proton-M
A satellite that is placed into geosynchronous transfer orbit must perform its own circularisation. An HS-601 consumed 1,175 kilograms of propellant doing this, and hence by the time that it reached its operating station its mass was just 1,800 kilograms.[105] While the Block-DM stage of the Proton-K could put 2,600 kilograms directly into geostationary orbit, constraints (such as the delayed shedding of the aerodynamic shroud) requested by the suppliers of western satellites effectively reduced its capacity to 2,100 kilograms. Nevertheless, the fact that the Block-DM could make the circularisation burn meant that the Proton-K could readily deliver an HS-601. In the competitive launch vehicle market, however, this was not good enough. In September 1995, ILS announced that it would introduce the Proton-M in 1998.[106],[107] This would be capable of either inserting 7,600 kilograms into geosynchronous transfer orbit – which would be a 50 per cent increase over the Proton-K – or placing 4,500 kilograms into geostationary orbit. By being able to carry *two* HS-601 satellites, it would be able compete with the (yet to fly) Ariane V.

A Proton-K lifts off on 6 September 1996 with Intelsat 3F2 for ILS.

A Proton-M is rolled out to the pad.

Although the first stage was to be improved, the key would be the Briz-M multiple-start hypergolic upper stage that was to supersede the Energiya Block-DM. This was a modified Briz-K stage of the Rokot – a launcher based on the SS-19 intercontinental-range ballistic missile that was was being marketed by Khrunichev and Daimler–Benz Aerospace to put payloads of up to 1 tonne into low orbit with launches from Plesetsk.[108] The Briz-M comprised a cylindrical motor section in the middle of a toroidal tankage that could be jettisoned. It was controlled by a closed-loop, triple-redundant guidance system that could be commanded in flight.[109,110] An advantage of this stage was that it used the same propellants as the Proton, which would simplify preparations.[111,112] The enhancements to the main stages were to be the introduction of a digital control system (adapted from that of the Zenit), the uprating of the performance of the RD-253 first-stage engines by 7 per cent, the modification of the feed system to reduce the residual propellant wastage, and the reinforcement of the second and third stages to carry a heavier payload.[113] A future option was to establish a facility at a near-equatorial latitude to enable the Proton to compete on equal terms with the Ariane V. However, because this would involve the construction of a wholly new site, such an option, if it were to be pursued, would more likely be in Florida in order to exploit the existing infrastructure.[114]

Frustrations

A Proton-K with a Briz-M carrying a Raduga communications satellite intended for geostationary orbit to serve the Russian military was lost on 5 July 1999.[115,116] At T+280 seconds, two minutes into the second stage's burn, telemetry indicated a significant rise in temperature in the combustion chamber of one of the engines, and at T+330 seconds radar tracking noted a trajectory deviation.[117,118] After the débris fell on a village 1,000 kilometres downrange, the Kazakhs banned Proton launches pending the investigation. This concluded that defective welding of the seam holding a cover in place on an engine turbine had failed, causing flow changes, and allowing a particle of aluminium to ignite in engine number three, which exploded.[119] It was also speculated that the incident may have been "exacerbated by the presence of microparticles introduced during the fuelling process". It had been the first launch from Area 81 at Baikonur, a pad that had been refurbished for the Proton, and had been delayed by problems in loading the propellants, so the contamination theory seemed likely.[120] The Briz-M stage, making its début, never got the chance to fire. A Proton-K lifted off with two Yamal satellites on 6 September, marking the lifting of the ban on launches. Although Protons had deployed multiple Iridium and GLONASS satellites into low and medium orbits, this was the first time that one carried a pair of geostationary satellites.[121] Loral had joined with Energiya, Gasprom (the Russian natural gas production company), and Gascom (its communications subsidiary) to develop, launch and operate satellites by mounting a Loral communications system on an Energiya multi-purpose bus for the 300,000 petroleum workers in the remote sites of Siberia.[122] This was the first type of communications satellite to be built by Energiya for a decade. Unfortunately, one of the satellites suffered an electrical fault that prevented it deploying its solar panels.[123,124] On 26 September an ILS Proton-K deployed LMI-1, the first Lockheed Martin–Intersputnik satellite, which was to provide communications for the Commonwealth of Independent States, Eastern Europe, Asia and Africa.[125,126,127]

A Proton-K was lost on 27 October 1999 when the second stage shut down at T+222 seconds, some 90 seconds into its 4-minute burn. This was a replay of the 5 July failure, except that this time the débris did not cause much damage. The payload, Express A1, was to have replaced an old Gorizont satellite in order to update Russia's Intersputnik communications network.[128,129] Although neither loss was an ILS mission, the Proton was grounded since this was the second failure in only four launches. The investigation concluded that in both cases a turbopump had caught fire and exploded due to metallic or mineral contaminants.[130,131,132] An engine test on 5 November demonstrated ignition in the turbine exhaust duct. Examination of the engine found metallic and non-metallic contaminants in the internal chambers, including particles of sand. A piece of asbestos fabric found near a valve head came from the fuel line downstream of the engine starter valve, and hence must have been introduced during manufacture. It turned out that all of the failed engines were from a batch made in 1993 by the Voronezh Mechanical Engine Plant, which had resumed operations after nine months of inactivity without recertifying its tools for production.[133] Seven second stages in stock that also contained such engines were stripped and rebuilt. As it had been planned to introduce the 'phase 2' upgrades to the engines of the second and third stages in 2000, it was decided to bring forward these changes, which included fitting screens in the oxidiser line and using more nickel alloys that would be less likely to catch fire if contaminated.[134]

While Khrunichev took solace from the fact that, excluding these two failures, there had been only one Proton loss in the past decade (during which 97 vehicles had lofted 142 satellites), the grounding was bad news for ILS because the premature

shutdown of the RL-10 engine in the upper stage of Boeing's Delta III on 4 May 1999 had also grounded the Atlas. The Proton-K resumed flying on 12 February 2000 with an ILS launch that deployed Garuda 1 for the Asia Cellular Satellite Company. At 4.5 tonnes, this was the largest commercial satellite yet to be carried.[135,136,137] On station, the Lockheed Martin A2100 bus unfurled a pair of 12-metre-diameter umbrella mesh antennas supplied by the Harris Corporation to provide regional cov-

An artist's impression of the Garuda 1 satellite.

erage for 10,000 simultaneous L-Band and C-Band telephone calls, to do single-handedly from geostationary orbit what Iridium and GlobalStar required constellations of satellites at lower altitudes to do. Unfortunately, in September 2000 a problem denied the satellite one of the antennas, restricting its mobile communications service.[138,139,140] After being launched on 12 March 2000, Express A2 took the slot assigned to its predecessor in order to relieve the ageing Gorizont 12 of the Stasionar network.[141] The insurance claim funded the construction of Express A1R as a replacement for the lost satellite.[142] These satellites were built by Prikladnoy Mekhaniki of Krasnoyarsk employing C-Band and Ku-Band systems supplied by

Alcatel of France, and were for the Russian Satellite Communications Company.[143,144,145] The Briz-M stage was successfully tested on a Proton-K on 6 June.[146] With the upgraded engines certified, the Zvezda 'service module' for the International Space Station was finally launched on 12 July. On its maiden launch on 7 April 2001, the Proton-M put the 2-tonne Ekran-M 24 satellite into geostationary orbit.[147,148] This was the final member of the 1970s-era series, each of which provided a single television channel.

An artist's impression of a satellite in the Ekran-M series.

When the Block-DM stage of a Proton-K stranded Astra 1K in a low orbit on 26 November 2002, ILS offered its Proton clients a transfer either to the Proton-M (which did not use the Block-DM) or to the

new Atlas V.[149] The investigation found that
the anomaly occurred at the start of the
second burn (as did three other failures of this
stage since 1996, when ILS began Proton
operations). Although it was not possible to
pinpoint the fault, there was an excess of fuel
in the engine when it was reignited, producing
such a high temperature that the engine was
destroyed.[150] The fuel build-up was attributed
to "stray particles" that clogged the engine
components, and the remedy was to improve
the preparation of the propellants. According
to the *Moscow Times*, the ILS venture came

An artist's impression of a satellite in
the Nimiq series.

"close to collapse" after this incident because (it argued) the commercial launcher
market was dwindling, and because the Proton was in *competition* with the Atlas V.
And indeed, whereas ILS had launched Nimiq 1 (a high-powered DBS satellite for
Telesat Canada) on a Proton-K in 1999, it earmarked Nimiq 2 for the Atlas V.[151] But
with the Atlas V falling behind schedule, ILS had to reassign this satellite to the
Proton-M, and it was successfully deployed by the Briz-M on the inaugural
commercial flight on 30 December 2002.[152] In early 2003, as part of a restructuring
effort, Intelsat cancelled its 10-01 satellite, which was to have ridden a Proton-M, but
reassigned 10-02 from Sea Launch to ILS as compensation.[153] The Proton-M proved
itself again when on 16 March 2004 it deployed Eutelsat W3A, an Astrium Eurostar
E3000 bus that, at 4.2 tonnes, was the most complex of Eutelsat's satellites to
date.[154,155,156]

JAPAN'S H-SERIES

In 1986 Japan introduced its H-1 launch vehicle, which was a license-built version of
the McDonnell Douglas 1000 series Delta augmented by nine Castor II strap-ons
and a second stage powered by the cyrogenic LE-5 engine built by Mitsubishi.[157]
After nine successful flights, the H-1 was superseded by the H-2, which had an
indigenous first stage with the LE-7 cryogenic engine and a second stage using the
uprated LE-5A.[158] The début of the H-2 on 1 February 1994 from the Tanegashima
Space Centre on an island south of Kyushu, marked Japan's first step towards
assuring independent access to space. On 28 August 1994 the second H-2 inserted
ETS 6 into geosynchronous transfer orbit but the satellite's liquid-propellant apogee
rocket failed.[159,160] On 18 March 1995 the third vehicle first dropped off the Space
Flyer Unit (a package of microgravity experiments to be retrieved by the NASA
Shuttle) in low parking orbit and then put the Hughes-built GMS 5 meteorological
satellite into geosynchronous transfer orbit.[161,162] There was then a prolonged hiatus,
awaiting the delivery of the next payload, the first Advanced Earth Observing
Satellite (ADEOS).[163] This was launched on 17 August 1996 but fell silent on 30 June
1997. The investigation concluded that the malfunction was triggered by the

breakdown of a soldered part at the base of the solar panel.[164] This prompted the postponement of the Communications and Broadcasting Engineering Test Satellite (COMETS), set for launch later in 1997, since it had systems in common with the lost satellite.[165] Because COMETS was to function as a geostationary relay for experiments on other satellites, these also had to be delayed.[166] In the event, when COMETS was launched on 21 February 1998, the second firing of the LE-5A engine was cut short and the satellite was stranded in a low orbit.[167] The investigation concluded that the nozzle had suffered a burn-through.[168] Nevertheless, 20 per cent of the experiments were able to be undertaken.[169]

Too costly

When the development of the H-2 started in the mid-1980s, operating cost had been secondary to the technical challenge. As they were now in a recession, the Japanese realised that their heavy launcher was so costly to operate that it was uncompetitive commercially, which meant that there was no prospect of generating revenues to offset the investment.[170,171] This led to the development of the H-2A, which was to be cheaper to manufacture and to operate. The first stage was to have the LE-7A with a thrust of 110,000 kilograms rather than 100,000 kilograms, and use shorter, fatter Nissan solid strap-ons giving a thrust of 175,000 kilograms.[172] The LE-5B of

the second stage would have a thrust of 14,000 kilograms rather than 12,500 kilograms.[173,174,175] Different combinations of strap-on motors would cater for a range of payloads.[176] The basic vehicle with two solid rockets would be able to insert 4 tonnes into geosynchronous transfer orbit, or 2.2 tonnes directly into geostationary orbit – an improvement of 200 kilograms over the H-2. The auxiliary strap-ons were bought by Mitsubishi from the Thiokol Corporation, which had stretched its Castor IVA to improve the performance by 30 per cent.[177] Adding two auxiliaries would increase the capacity to geosynchronous transfer orbit to 4.25 tonnes, and four would raise it to 4.5 tonnes.[178] With a fee of $80 million, the H-2A would be able to compete against the Ariane 4 and Atlas II for satellites in the HS-601 class. The Mitsubishi-led Rocket Systems Corporation readily sold 10 launches to Hughes Space and Com-

A cutaway of the H-2, the second such vehicle being prepared with the ETS-6 satellite, and an artist's impression of the COMETS satellite.

munications which, acting on behalf of its customers, was spreading its satellites around the different launch vehicles.

The demise of the H-2

On 15 November 1999 the first stage of an H-2 shut down 107 seconds short of what was to have been a 5-minute 46-second powered ascent. After coasting ballistically, the vehicle staged 30 seconds early. The second stage – which was marking the début of the LE-5B that had been built for the H-2A – ignited 8 seconds later, but was unable to make up the inherited velocity shortfall and was destroyed by the range safety officer at an altitude of 45 kilometres. The débris fell into the Pacific near the Bonin Islands.[179,180] A review of the launch video showed a white trail, which suggested that hydrogen had been leaking from the first stage.[181] As there had been several drops in the temperature and chamber pressure of the engine prior to its shut down, it appeared that it had been starved of hydrogen, but it was impossible to say whether this was due to a leak from the hydrogen tank, a crack in the nickel alloy

The liquid hydrogen turbopump of an LE-7 engine.

plumbing running between the hydrogen pre-burner and the turbopump, or a failure of the turbopump. In January 2000 an underwater imaging probe operated by the Japanese Marine Science and Technology Centre located the wreckage of the first stage on the sea floor at a depth of almost 3 kilometres. The 12-tonne LE-7 engine was raised and sent to the National Aerospace Laboratory in Tokyo, where it was confirmed that it had been starved of hydrogen, and the cause came as a shock.[182,183,184] An 8-inch blade in a 3-bladed inducer, which spun at 42,000 revolutions per minute to feed liquid hydrogen into the turbopump, was broken.[185] This was not a manufacturing problem; it was a fatigue fracture resulting from repeated exposure to excessive force. When a bladed-wheel rotates in a liquid, the pressure decrease near the blades can generate bubbles of gas. In this case, as the blades and bubbles swept round at different speeds, pressure fluctuations on the blades as they hit the bubbles caused a resonance that resulted in fatigue failure. This

The LE-7A engine of the H-2A.

phenomenon of 'rotating cavitation' had been observed during the development of the LE-7, and although it had been considered that the blades were strong enough to resist it, this had proved to be false.[186] The vibrations also cracked a weld in the inlet to the turbopump, and caused the engine to be starved of fuel. The lost payload was the 1,600-kilogram Multi-function Transport Satellite (MTSAT 1) that had been built by Space Systems/Loral for the Japanese Ministry of Transport to supplement air traffic control communications in the Asia Pacific region. The replacement (MTSAT 1R) was not shipped to Japan until March 2004 for launch in early 2005.[187,188,189] Unfortunately for Japan, Hughes exercised a clause in its contract that allowed for a painless cancellation in the event of two consecutive failures of the H-2.

Since the H-2 had suffered a number of failures for *different* reasons, the final mission in 2000 was cancelled and its payload reassigned to the H-2A. However, because the development of the LE-7A engine was proving more difficult than expected, the first launch of the H-2A was slipped to early 2001.[190,191,192,193]

The H-2A
By early 2001 the inaugural launch of the H-2A was still scheduled for 22 July, but it slipped to 29 August. It had a pair of Nissan's solid rockets augmenting the first stage, and carried a Vehicle Evaluation Package to document the stresses that a payload would have to endure.[194,195] This launch marked the first time since July 1998 that a Japanese vehicle achieved orbit! If it had failed, it may well have resulted in cancellation of the programme. The second test was slipped to early 2002 and the first operational mission was postponed from mid to late 2002 to provide more time to evaluate the data from the two tests.[196,197] The second H-2A lifted off on 3 February 2002 using four large auxiliary strap-ons. The second stage entered the planned 500-kilometre parking orbit, coasted for 12 minutes, and reignited for insertion into geosynchronous transfer orbit. However, while the 450-kilogram Mission Demonstration Satellite to test new technologies for use on communications satellites was successfully deployed, the 86-kilogram Demonstrator of Atmospheric re-entry System with Hyper-velocity (DASH) was not released.[198] The investigation found that the signal to initiate the separation sequence had not been issued by the launch vehicle. The interface between the launch vehicle and the Data Processing Unit of the piggyback payload was minimal, preventing a comprehensive test in the launch configuration, and the special non-flight cable and connector used in ground tests did not reveal a pin-assignment error on the drawing of the Data Processing Unit.[199] In a revision of the plan, the Data Relay Test Satellite (DRTS) and Unmanned Space Experiment Recovery System (USERS) were assigned to the first operational mission, and ADEOS 2, which was to replace the satellite that fell silent in 1997, slipped one place.[200] In order to recover from the protracted development, some payloads were offloaded.[201] MBSAT, which had been built by Space Sytems/Loral for Japan's Mobile Broadcasting System, was reassigned to the Atlas III for launch in 2003. On 10 September the H-2A made its operational début. After releasing USERS into the 450-kilometre parking orbit with a payload of Japanese microgravity experiments, the second stage reignited and put the DRTS into

geosynchronous transfer orbit, and on reaching apogee this satellite circularised in order to stand in for COMETS as a real-time communications link between the ground and other satellites.[202] On 14 December 2002 the fourth H-2A successfully placed ADEOS 2 into Sun-synchronous orbit.[203] At 3,700 kilograms, this was the largest satellite launched by a Japanese rocket. Unfortunately, contact was lost on 25 October 2003, possibly due to increased solar activity, and it was declared a total loss.[204,205,206]

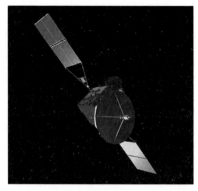

An artist's impression of the Data Relay Test Satellite.

In mid-2003, Japan announced that it was to upgrade the H-2A to put 10 tonnes into geosynchronous transfer orbit to enable the H-2A to keep pace with the upgrading of the Ariane V. This was to be done by increasing the diameter of the first stage from 3 metres to 4 metres, and adding a second engine to the second stage.[207] Then, at the Paris Air Show in June, Boeing, Arianespace and Mitsubishi Heavy Industries announced a plan to provide "a powerful new commercial launch service ... to ensure on-time launches for customers around the world". The objective was to enable customers to shift their payloads from one launch provider to another in the group, in the event that the Ariane V, H-2A or Zenit 3SL was grounded by failure.[208]

On 29 November 2003, one of the two Nissan strap-ons on the sixth H-2A suffered a nozzle burn-through which disabled the pyrotechnic package that was to have jettisoned it.[209,210,211] By T + 11 minutes, the 'dead weight' of this appendage had caused the vehicle to veer so far off course that the second stage would not be able to recover the trajectory. It was therefore destroyed.[212] The second pair of Information Gathering Satellites were to have augmented the first pair (launched by the previous H-2A on 28 March) for a scheduled constellation of eight satellites intended to enable Japan to monitor North Korean missile tests. (The programme was in trouble anyway, because the performance of the satellites was reportedly disappointing.)

The 6th H-2A lifts off on 29 November 2003.

Although the Japanese had designed the H-2A to make it attractive as a commercial launcher, and had formed a strategic alliance with its competitors, all of its early flights carried indigenous payloads and it has yet to prove itself in this wider context.

NOTES

1. Memorandum from Secretary of Defense to Secretaries of the Military Departments, *et al.*, *Defense Space Launch Strategy*, 7 February 1984.
2. *Aviation Week & Space Technology*, 25 June 1984, p. 159.
3. *Spaceflight*, April 1980, p. 155.
4. *Aviation Week & Space Technology*, 11 June 1984, p. 23.
5. Assured Access: 'The Bureaucratic Space War', *Dr. Robert H. Goddard Historical Essay* by E.C. Aldridge.
6. http://web.mit.edu/org/s/spacearchitects/Archive/space wars.doc
7. *Aviation Week & Space Technology*, 18 March 1991.
8. http://www.globalsecurity.org/space/systems/t4table.htm
9. *Aviation Week & Space Technology*, 9 August 1993, p. 22.
10. *Aviation Week & Space Technology*, 23 August 1993, p. 26.
11. *Aviation Week & Space Technology*, 5 December 1994, p. 47.
12. *Aviation Week & Space Technology*, 20 September 1993, p. 92.
13. *Aviation Week & Space Technology*, 10 January 1994, p. 26.
14. *Aviation Week & Space Technology*, 5 September 1994, p. 41.
15. *Flight International*, 19–25 January 1994, p. 23.
16. *Aviation Week & Space Technology*, 11 October 1993.
17. *Aviation Week & Space Technology*, 8 November 1993, p. 73.
18. *Aviation Week & Space Technology*, 9 May 1994, p. 24.
19. *Flight International*, 21–27 September 1994, p. 26.
20. *Aviation Week & Space Technology* 5 September 1994, p. 46.
21. *Aviation Week & Space Technology*, 12 September 1994, p. 53.
22. *Flight International*, 15–21 November 1995, p. 20.
23. This was the third Titan IVA of the year, and the 14th since the type was introduced in 1989, of which six had used the Centaur upper stage and four had used the IUS (the others had not needed an upper stage because the payload had been able to make its own set up manoeuvres).
24. *Aviation Week & Space Technology*, 11 December 1995, p. 20.
25. *Aviation Week & Space Technology*, 12 September 1994, p. 53.
26. http://www.globalsecurity.org/space/library/news/1998/ns9899.htm
27. http://www.globalsecurity.org/space/systems/t4table.htm
28. http://www.globalsecurity.org/space/library/news/1998/nr98-09-02.htm
29. *Spaceflight*, November 1998, p. 440.
30. *Aviation Week & Space Technology*, 24 January 1999, p. 24.
31. *Aviation Week & Space Technology*, 13 September 1999, p. 24.
32. http://www.globalsecurity.org/space/library/news/1999/n19990119_990066.htm
33. *Aviation Week & Space Technology*, 19 April 1999, p. 40.
34. *Aviation Week & Space Technology*, 26 October 1987, p. 3.
35. *Aviation Week & Space Technology*, 5 December 1994, p. 47.
36. *Flight International*, 22–28 September 1993, p. 22.
37. *Aviation Week & Space Technology*, 20 September 1993, p. 92.

38. *Aviation Week & Space Technology*, 3 March 1997, p. 29.
39. *Aviation Week & Space Technology*, 6 November 1995, p. 61.
40. *Aviation Week & Space Technology*, 6 November 1995, p. 61.
41. *Aviation Week & Space Technology*, 10 February 1997, p. 90.
42. *Aviation Week & Space Technology* 3 March 1997, p. 29.
43. *Aviation Week & Space Technology*, 24 March 1997, p. 60.
44. *Aviation Week & Space Technology*, 19 April 1999, p. 40.
45. *Aviation Week & Space Technology*, 3 May 1999, p. 46.
46. In 1983 a hydraulic problem had prevented the second stage of an IUS from circularising, and in 1985 the second stage had recovered from the under performance of the first stage.
47. In the case of Chandra, the first stage of the IUS was to produce an interim orbit with an apogee of 60,000 kilometres, which the second stage was to extend to 140,000 kilometres, so it would not be the usual flight profile.
48. *Aviation Week & Space Technology*, 23 August 1999, p. 34.
49. *Spaceflight*, July 1999, p. 269.
50. *Spaceflight*, June 1999, p. 224.
51. *Aviation Week & Space Technology*, 10 May 1999, p. 28.
52. *Spaceflight*, August 2000, p. 312.
53. *Aviation Week & Space Technology*, 21 June 1999, p. 21.
54. *Aviation Week & Space Technology*, 26 July 1999, p. 27.
55. *Aviation Week & Space Technology*, 2 August 1999, p. 31.
56. *Aviation Week & Space Technology*, 13 September 1999, p. 24
57. *Aviation Week & Space Technology*, 10 May 1999, p. 28.
58. *Aviation Week & Space Technology*, 31 May 1999, p. 34.
59. *Spaceflight*, August 1999, p. 310.
60. This was the sixth Titan IVB.
61. *Aviation Week & Space Technology*, 3 May 1999, p. 31.
62. *Aviation Week & Space Technology*, 10 May 1999, p. 28.
63. *Aviation Week & Space Technology*, 13 September 1999, p. 24
64. *Aviation Week & Space Technology*, 15 May 2000, p. 24.
65. *Spaceflight*, August 2000, p. 312.
66. The Air Force intended to launch three more DSP satellites, and then switch to the follow-on programme in 2004.
67. Sergei Korolev designated the stages of his launchers alphabetically, and in cyrillic the sequence is A, B, V, G, D and so when he modified the fifth stage of the N-1 to serve as the fourth stage of the Proton it was out of sequence - even although to a Westerner it seemed more appropriately named!
68. http://www.tbs-satellite.com/tse/online/lanc_proton.html
69. *Jane's Space Directory*, vol. 10, 1994–1995, p. 211.
70. *Aviation Week & Space Technology*, 9 February 1987, p. 26.
71. *Aviation Week & Space Technology*, 11 May 1987, p. 34.
72. http://www.spaceandtech.com/spacedata/constellations/glonass_consum.shtml
73. *Aviation Week & Space Technology*, 4 May 1987, p. 24.
74. *Aviation Week & Space Technology*, 18 May 1987, p. 22.
75. http://www.tbs-satellite.com/tse/online/lanc_proton.html
76. *Jane's Space Directory*, vol. 10, 1994–1995, p. 211.
77. *Aviation Week & Space Technology*, 7 June 1993.
78. *Aviation Week & Space Technology*, 28 June 1993.

79. *Jane's Space Directory*, vol. 10, 1994–1995, p. 211.
80. *Aviation Week & Space Technology*, 20 September 1993, p. 90.
81. http://www.globalsecurity.org/military/industry/general_dynamics.htm
82. http://www.fas.org/man/company/lock-mart.htm
83. *Aviation Week & Space Technology*, 5 September 1994, p. 41.
84. *Flight International*, 22–28 March 1995, p. 6.
85. *Aviation Week & Space Technology*, 24 April 1995, p. 40.
86. *Aviation Week & Space Technology*, 6 September 1993, p. 22.
87. *Aviation Week & Space Technology*, 22 January 1996, p. 57.
88. *Aviation Week & Space Technology*, 15 April 1996, p. 22.
89. *Flight International*, 29 March–4 April 1995, p. 28.
90. *Aviation Week & Space Technology*, 19 June 1995, p. 27.
91. *Aviation Week & Space Technology*, 14 August 1995, p. 51.
92. *Flight International*, 21–27 June 1995.
93. *Flight International*, 2–8 August 1995, p. 25.
94. *Aviation Week & Space Technology*, 24 April 1995, p. 40.
95. *Aviation Week & Space Technology*, 14 August 1995, p. 51.
96. *Aviation Week & Space Technology*, 22 January 1996, p. 57.
97. http://www.telecomweb.com/reports/satellite2001/news/article3.htm
98. http://www.ilslaunch.com/launches/
99. http://www.ilslaunch.com/newsarchives/newsreleases/rec135/
100. *Aviation Week & Space Technology*, 25 November 1996, p. 71.
101. http://nssdc.gsfc.nasa.gov/planetary/text/mars96_timeline.txt
102. *Aviation Week & Space Technology*, 2 June 1997.
103. *Aviation Week & Space Technology*, 23 June 1997, p. 66.
104. *Aviation Week & Space Technology*, 5 January 1998, p. 28.
105. *Flight International*, 20–26 September 1995, p. 26.
106. *Aviation Week & Space Technology*, 31 July 1995, p25.
107. *Flight International*, 20–26 September 1995, p. 26.
108. *Flight International*, 19–25 April 1995, p. 25.
109. http://www.ilslaunch.com/proton/1protonm/
110. http://www.astronautix.com/lvs/probrizm.htm
111. *Aviation Week & Space Technology*, 24 April 1995, p. 43.
112. *Spaceflight*, January 2003, p. 11.
113. *Flight International*, 30 October–5 November 1996, p. 28.
114. *Aviation Week & Space Technology*, 15 April 1996, p. 22.
115. *Spaceflight*, December 1999, p. 491.
116. *Spaceflight*, January 2000, p. 6.
117. *Spaceflight*, October 1999, p. 401.
118. *Aviation Week & Space Technology*, 19 July 1999, p. 27.
119. *Aviation Week & Space Technology*, 13 September 1999, p. 28.
120. *Aviation Week & Space Technology*, 12 July 1999, p. 32.
121. *Spaceflight*, November 1999, p. 447.
122. *Flight International*, 21–27 February 1996, p. 21.
123. *Spaceflight*, January 2000, p. 7.
124. *Spaceflight*, December 2000, p. 489.
125. *Spaceflight*, November 1999, p. 447.
126. *Spaceflight*, December 1999, p. 491.
127. *Spaceflight*, January 2000, p. 6.

128. *Aviation Week & Space Technology*, 22 November 1999, p. 58.
129. *Flight International*, 26 October–1 November 1994, p. 26.
130. *Aviation Week & Space Technology*, 13 September 1999, p. 28.
131. *Spaceflight*, February 2000, p. 47.
132. *Spaceflight*, April 2000, p. 136.
133. *Aviation Week & Space Technology*, 29 November 1999, p. 25.
134. *Aviation Week & Space Technology* 21 February 2000, p. 47.
135. *Flight International*, 24–30 May 1995, p. 31.
136. *Aviation Week & Space Technology* 21 February 2000, p. 47.
137. *Spaceflight*, May 2000, p. 184.
138. http://www.tbs-satellite.com/tse/online/sat_garuda_1.html
139. *Spaceflight*, August 2001, p. 316.
140. http://www.spaceandtech.com/digest/sd2000-26/sd2000-26-001.shtml
141. *Spaceflight*, June 2000, p. 226.
142. http://www.spaceandtech.com/digest/sd2000-25/sd2000-25-006.shtml
143. Prikladnoy Mekhaniki was previously known as NPO PM, which stood for Research and Production Association for Applied Mechanics.
144. http://www.rosaviakosmos.ru/ORG/npopm_e.html
145. *Aviation Week & Space Technology*, 21 February 2000, p. 50.
146. http://www.ilslaunch.com/proton/1protonm/
147. http://www.ilslaunch.com/proton/1protonm/
148. *Spaceflight*, July 2001, p. 270.
149. *Spaceflight*, February 2003, p. 50.
150. *Spaceflight*, March 2003, p. 94.
151. *Spaceflight*, October 2001, p. 403.
152. *Spaceflight*, March 2003, p. 98.
153. *Spaceflight*, April 2003, p. 143.
154. *Spaceflight*, May 2004, p. 182.
155. http://www.ilslaunch.com/newsarchives/newsreleases/rec256/
156. http://www.skyrocket.de/space/doc_sdat/eutelsat-w3a.htm
157. http://www.astronautix.com/lvs/h1.htm
158. *Aviation Week & Space Technology*, 19 July 1999, p. 57.
159. *Flight International*, 25–31 January 1995, p. 24.
160. *Spaceflight*, April 1999, p. 149.
161. *Aviation Week & Space Technology*, 13 March 1995, p. 110.
162. *Aviation Week & Space Technology*, 27 March 1995, p. 24.
163. *Aviation Week & Space Technology*, 26 August 1996, p. 68.
164. *Spaceflight*, April 1999, p. 149.
165. *Aviation Week & Space Technology*, 7 July 1997, p. 31.
166. *Aviation Week & Space Technology*, 14 April 1997, p. 61.
167. *Spaceflight*, April 1999, p. 149.
168. *Aviation Week & Space Technology*, 16 March 1998, p. 61.
169. *Aviation Week & Space Technology*, 9 March 1998, p. 17.
170. *Aviation Week & Space Technology*, 31 January 1994, p. 52.
171. *Aviation Week & Space Technology*, 4 April 1994, p. 56.
172. *Aviation Week & Space Technology*, 24 June 1996, p. 52.
173. *Aviation Week & Space Technology*, 15 July, 1996, p. 25.
174. http://www.astronautix.com/lvs/h2.htm
175. http://www.astronautix.com/lvs/h2a.htm

176. *Aviation Week & Space Technology*, 21 October 1996, p. 44.
177. *Aviation Week & Space Technology*, 17 March 1997, p. 40.
178. *Spaceflight*, November 2001, p. 445.
179. *Aviation Week & Space Technology*, 22 November 1999, p. 38.
180. *Spaceflight*, February 2000, p. 47.
181. *Aviation Week & Space Technology*, 29 November 1999, p. 25.
182. *Spaceflight*, April 2000, p. 135.
183. *Aviation Week & Space Technology*, 31 January 2000.
184. *Spaceflight*, March 2000, p. 92.
185. *Aviation Week & Space Technology*, 13 March 2000, p. 19.
186. http://www.nasda.go.jp/projects/rockets/h2a/documents/f4/sheet/h2af4_27_e.html
187. *Spaceflight*, June 2000, p. 226.
188. http://www.loral.com/inthenews/040319.html
189. http://www.bom.gov.au/sat/MTSAT/MTSAT.shtml
190. *Spaceflight*, September 2000, p. 355.
191. *Aviation Week & Space Technology*, 13 December 1999, p. 38.
192. *Spaceflight*, March 2001, p. 94.
193. After five successful flights the H-2 had suffered two consecutive failures, so it was cancelled in order to switch the effort to its 'commercial' successor.
194. *Spaceflight*, June 2001, p. 228.
195. *Spaceflight*, October 2001, p. 408.
196. *Spaceflight*, November 2001, p. 445.
197. *Spaceflight*, February 2002, p. 52.
198. *Spaceflight*, April 2002, p. 140.
199. www.isas.ac.jp/dtc/dash-e/dash-e.html
200. *Spaceflight*, May 2002, p. 185.
201. *Spaceflight*, September 2002, p. 360.
202. *Spaceflight*, November 2002, p. 447.
203. *Spaceflight*, March 2003, p. 94.
204. *Spaceflight*, January 2004, p. 10.
205. http://www.daviddarling.info/encyclopedia/A/ADEOS.html
206. http://www.*Spaceflight*now.com/h2a/f4/
207. *Spaceflight*, October 2003, p. 405.
208. *Spaceflight*, October 2003, p. 402.
209. http://www.astronautix.com/lvs/h2a.htm
210. With the exception of the first flight (and the one that carried ADEOS 2), all of the H-2A launches had used the 2024 configuration.
211. *Spaceflight*, March 2004, p. 101.
212. *Spaceflight*, February 2004, p. 47.

5

Lightweights

BLACK ARROW

In 1955 the Royal Air Force awarded a contract to de Havilland Aircraft to develop the Blue Streak single-stage missile powered by a pair of two-chamber Rolls Royce engines using kerosene and oxygen. Despite performing well in tests, and on schedule for service by 1965, it was cancelled in April 1960 when the British government decided to buy the Skybolt air-launched ballistic missile that was under development for the US Air Force. When the Skybolt was cancelled, the British transferred their interest to the submarine-launched Polaris. The European Space Research Organisation was established in 1962 to pursue collaborative projects. When the European Launcher Development Organisation was formed in 1964 to develop the Europa launch vehicle, the Blue Streak was selected as the first stage because it was by far the most powerful liquid rocket available, and work began to develop a French second stage and a German third stage. Unfortunately, this project was plagued with difficulties, and eventually cancelled. In parallel, the Royal Aircraft Establishment was given the go-ahead to develop a small launcher named Black Arrow as a derivative of the Black Knight, the two-stage rocket built to test materials for the re-entry vehicle of the Blue Streak, which had flown 22 times without a failure.[1] The rationale was that this project would keep Britain in the launcher business at a low level. The budget was so tight – only £9 million – that the engines were refined in ground trials rather than by development flights.[2]

The first test from the Woomera rocket range in Australia was on 28 June 1968, and was to be a sub-orbital test of the first two stages. Within seconds it started to twist and corkscrew. Although the dummy third stage sheared off, the vehicle continued its tortuous climb. At an altitude of 8 kilometres it toppled over and started to tumble, so it was destroyed by the range safety officer. The telemetry indicated that one of the four engine pairs had been repeatedly slewing from one end of its movement range to the other. The investigators ran a computer simulation that indicated that this violent movement had almost certainly been caused by a loss of signal, which suggested a broken wire.[3] On 4 March 1969 another two-stage vehicle flew this sub-orbital test flawlessly, clearing the way for a three-stage vehicle to attempt to place a demonstration satellite in low orbit on 2 September 1970. Unfortunately, telemetry and tracking indicated that the second stage's engine was losing thrust. It shut down almost 30 seconds ahead of schedule. The Waxwing motor of the third stage ignited successfully, but was unable to compensate for the inherited shortfall in velocity, and fell into the ocean. The enquiry determined that a fractured pipe in the second stage had allowed the nitrogen pressurant to vent.

On 28 October 1971 a Black Arrow lifts off on its transparent plume carrying the Prospero satellite.

Starved of oxidiser, the combustion had declined, and eventually shut down the engine. In the spring of 1971 a Parliamentary Select Committee met to assess the future of British space efforts. On 29 July 1971, before the Select Committee could report, the government cancelled the Black Arrow.[4] Nevertheless, the next launch was allowed to proceed, and on 28 October this put into a near-polar orbit the 66-kilogram Prospero satellite built by the British Aircraft Corporation.[5] That same year, following the advice of the Select Committee, Britain withdrew from the European Launcher Development Organisation (which in 1975 merged with the European Space Research Organisation to become the European Space Agency). This, then, is an example of a purely political failure of a launch vehicle. It was arguably one of the most aesthetically pleasing launchers – tiny, proportioned like a rifle-bullet and rising on a crystal clear oxygen-rich exhaust. The last remaining Black Arrow was put on display in the Science Museum in London.

SCOUT

The Scout four-stage solid-propellant launcher was designed by the Langley Research Center as a relatively inexpensive multi-stage solid-propellant rocket using off-the-shelf hardware to insert a 60-kilogram satellite into a circular orbit at an altitude of some 500 kilometres. The Algol first stage motor was developed from

Navy's Polaris; the Castor second stage was derived from the Army's Sergeant; and the Antares and Altair upper stages were adapted by Langley from the Navy's Vanguard. Chance Vought Aircraft of Dallas, Texas, was made prime contractor in March 1959. The vehicle was assembled in a horizontal configuration and hoisted upright for launch. The first Scout was fired from Wallops Island in Virgina on 1 July 1960 on a vertical trajectory as a sounding rocket.[6] The first attempt to launch a satellite was the third mission. To the eye, it appeared to be a good launch, but ground radar indicated a problem after 136 seconds.[7,8,9] The Antares developed a rolling moment which, although it soon disappeared, confused the tracking radar, which indicated on the plot board that the vehicle was some 50 degrees off course. The range safety officer was obliged to activate the auto destruct system (an innovation at the time). In reality, having recovered from the roll, the vehicle was performing well. The next flight on 16 February 1961 was successful. On 30 June, on the fifth launch, the third stage malfunctioned. The first launch from Vandenberg Air Force Base in California on 26 April 1962, the 10th of the series, failed. These early failures prompted a recertification of the launcher. Significantly, the in-depth investigations of the rocket's subsystems made during this review revealed that each Scout failure had been caused by a different problem. Institutional factors also played a role – the Langley engineers, the contractors, and their Air Force partners had not been acting as an integrated team. However, the principal cause of the mishaps was the need to make everything happen so fast – parts had been 'borrowed' from vehicles in stock to fix problems on those about to launch. However, after this recertification, the Scout suffered only occasional failures of a 'random' nature. The 22nd mission caused spectacular damage at the Wallops facility. A flame appeared above the fins at the base of the first stage 2.5 seconds after lift-off. Two seconds later, this stage was engulfed by fire. The burn-through caused the vehicle to execute wild gyrations. "It got about 300 feet high and broke into three parts: the first stage went in one direction; the second stage went in another; and the third and fourth stages fell more or less back on the launch pad and burned. It was a disaster."[10]

The Scout was repeatedly upgraded to increase the payloads to 200 kilograms, with the variants designated alphabetically. Nevertheless, the final version was very similar in appearance to the original. In 1967 launches began from Italy's San Marco platform in the sea off the coast of Kenya in Africa, this equatorial launch site maximally exploiting the rotation rate of the Earth to further increase the payload to 270 kilograms. In January 1991 management of the Scout programme was reassigned to the Goddard Space Flight Center, but by then the flight rate had fallen to about one per year. The final launch was on 8 May 1994, with a satellite for the Ballistic Missile Defense Organisation. By then, a number of new lightweight vehicles had either been introduced or were in development to launch commercial satellites. Over a total of 118 launches, the Scout had achieved the enviable success rate of 96 per cent, with its success rate since 1976 being 100 per cent. This stemmed from its simple technology and from standardised procedures and configuration control.[11]

The launch of the 5th Scout on 30 June 1961.

AIR-LAUNCHED PEGASUS

In 1987 the Orbital Sciences Corporation (OSC) set out to develop an
air-launched three-stage rocket to
place small satellites into low orbit,
and the Department of Defense made
an advance booking for six vehicles.
On 5 April 1990, the inaugural
vehicle with two small satellites was
released at 40,000 feet by a B-52 from
NASA's Dryden Flight Research

A Pegasus climbs away under first-stage thrust.

Center in California, thereby initiating operations for the first privately owned
launcher.[12,13] With a 1.27-metre-diameter shroud, it could place up to 212 kilograms
into a circular orbit at an altitude of 740 kilometres. This was a niche market, but the
prospects for commercial operations were promising. The flight on 17 July 1991
introduced OSC's Hydrazine Auxiliary Propulsion System, effectively a fourth stage,
which executed two manoeuvres to circularise the orbit prior to dispatching the
payload, which in this case was a collection of seven microsatellites.

The first problem
The Hydrazine Auxiliary Propulsion System was meant to refine the accuracy of the
final orbit, but on 19 May 1994 it underperformed by 12 per cent, placing a Space
Test Experiment Platform for the Air Force in an unwanted elliptical orbit.[14,15] On
its maiden flight the orbital manoeuvres had been monitored in real-time by an
ARIA aircraft. This time, the telemetry was recorded on board for later replay,
which proved fortunate for the investigators. As the stage was not scheduled to be
used again for 18 months, there was no need to hold up launches pending the
investigation.

The XL's début
OSC's more powerful Pegasus XL was longer than the original, to accommodate 24
per cent more propellant in the first stage and 30 per cent in the second stage. The
third stage was unchanged. The 60
per cent overall increase in performance was to enable it to insert 340
kilograms into a circular orbit at an
altitude of 740 kilometres. Its introduction on 27 June 1994 also marked
the first use of the Lockheed L-1011
Tristar, 'Stargazer', which had been
bought from Air Canada and modified to carry the rocket on its belly.[16]
In fact, the B-52 could not carry the
stretched form of the launch vehi-

The L-1011 'Stargazer' with a Pegasus mounted
on its belly.

cle.[17] However, at T + 75 seconds, several seconds after the second stage ignited, the rocket veered off course and lost speed, prompting the safety officer to destroy it.[18] The investigation revealed that the vehicle suffered an anomalous roll due to sideslip, also referred to as 'phantom yaw', derived from inadequacies in the aerodynamic model used to design the new configuration, which relied on computer simulations without verification by wind tunnel tests.[19] The flight control software was revised to counter this disturbance.

Payload glitches
When it was confirmed that the original Pegasus was not susceptible to sideslipping, the Advanced Photovoltaic and Electronic Experiments payload was launched for the Air Force.[20] This introduced OSC's 'standard platform', which,

The Orbcomm satellites were designed to be stacked on the nose of a Pegasus.

by remaining attached to the vehicle to use its communications system, eliminated the requirement for the payload to include such hardware – maximising the *effective* payload. Minor software revisions had to be uplinked to the Sun-sensors to enable the vehicle to maintain stability in orbit. In 1993 OSC teamed with Teleglobe Incorporated of Canada, and created Orbcomm to provide a global messaging service using a constellation of 36 satellites.[21,22,23,24] The small disk-shaped satellites were designed by OSC to be stacked inside the shroud of a Pegasus.[25,26] The first two satellites were launched on 3 April 1995 in order to evaluate the system.[27] In checking them out, Orbcomm 1 was found to have a fault in its subscriber communications subsystem that prevented it from transmitting to clients, and Orbcomm 2 had a malfunction in its gateway receiver prohibiting its response to commands from the ground.[28,29,30] However, both spacecraft were coaxed to operational status.[31,32]

Another XL loss
Launched on 22 June 1995, the second Pegasus XL suffered a mechanical problem.[33] The first set of pyrotechnics on the annular interstage fired to jettison the first stage,

and 0.8 second later other charges separated the interstage from the aluminium ring at the rear of the second stage, but the interstage snagged.[34] The investigation found that one of the three skid-guides had been improperly installed, and prevented the interstage from sliding off. The continued presence of the interstage limited the motion of the gimbal on the engine nozzle, and the vehicle began to pitch soon after the shroud was shed. After two and an half loops, during which the interstage was eventually shaken off, the vehicle restabilised itself, but the dramatic excursion had overwhelmed the inertial measurement unit and the vehicle veered off course, which prompted the safety officer to destroy it.[35] The fact that the root cause was a straightforward case of human error was good news, as it meant that there was nothing wrong with the vehicle.[36] As previously, the lost payload was a Space Test Experiment Platform for the Air Force. The first Pegasus XL success came on 8 March 1996. The payload, REX II, was a GPS attitude determination and control experiment for the Air Force. The last of the original variant was launched on 16 May 1996, and successfully deployed a Miniature Seeker Technology Integration experiment for the Ballistic Missile Defense Organisation.

Mixed fortunes

Having launched the six packages for the Department of Defense (and unfortunately lost half of them), OSC turned its attention to two long-delayed satellites for NASA. A Pegasus XL successfully released the Total Ozone Mapping Spectrometer Earth Probe on 1 July 1996. Another launched the Fast Auroral Snapshot Explorer for the Small Explorer series on 21 August. On 4 November, another vehicle attained the desired 550-kilometre circular orbit, but the system in the third stage that was to provide power for 'transient events' failed, inhibiting the pyrotechnics from releasing the two satellites.[37,38] The fault was detected towards the end of the first orbit, when radar tracking 'saw' only one object. Telemetry from Argentina's SAC B satellite confirmed that it was still in the forward position of the dual-payload dispenser. A ground command was able to open the satellite's solar panels, but its attitude control system was overwhelmed by the mass of the spent stage, and as this tumbled it cast a shadow over the solar arrays and caused the satellite to drain its battery. The High-Energy Transient Explorer built for NASA by the Massachusetts Institute of Technology was worse off. It was inside a container further aft, and its transmission

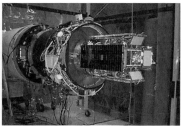

The HETE and SAC B satellites are stacked on the nose of a Pegasus, with HETE (aft) being completely enclosed within the deployment system.

was masked. Unable to deploy its solar panels, it soon exhausted its battery.[39] After a string of successes, on 4 March 1999 the WIRE satellite for astronomical studies failed soon after being released into the desired orbit, when it jettisoned a cover three days ahead of schedule, allowing heat to vent the coolant for its telescope and causing it to spin out of control.[40] The next flight on 18 May successfully inserted the TERRIERS satellite for ionospheric studies into orbit, but an attitude control fault prevented the satellite from maintaining its solar panels facing the Sun, and within hours it was dead, having exhausted its battery.[41]

Losing 'Hyper X'

The X-43A 'Hyper X' was a sub-scale prototype that NASA had developed to test a supersonic-combustion (scramjet) engine.[42,43] As the aircraft had to be accelerated prior to starting its engine, it was to be boosted by a hybrid version of the Pegasus (designated the HXLV) using only the first stage. The avionics normally carried on the second stage were relocated to the booster. There would be no aerodynamic shroud, the 12-foot-long aircraft would ride on the exposed tip. The rocket was to climb to an altitude of 100,000 feet and accelerate to Mach 7, whereupon a gas-driven piston would release the payload. The X-43A was to run its engine for 10 seconds to sustain its speed.[44] The fact that the scramjet required air meant that rather

An artist's impression of the X-43A 'Hyper-X' on the nose of its Pegasus booster.

than making its usual zooming climb to 200,000 feet, the Pegasus would require to pursue a 'depressed' trajectory, and its control system had to be considerably reinforced because flying so fast within the atmosphere would impose aerodynamic loads 50 times greater than usual. In addition, because there were no upper stages and the payload was so light (1,275 kilograms), some 2,500 kilograms of ballast had to be carried to prevent the booster exceeding the desired speed. Although a hybrid, this was not to be a one-off test, as the MicroCraft Company had built four copies of the X-43A.[45] The first vehicle was released by NASA's B-52 off the coast of California on 2 June 2001. About 8 seconds later, the booster began to shed débris, and it was destroyed by command when it veered off course. A replay of the video showed what appeared to be a piece of the reinforced tail fin snapping off.[46] The investigation concluded that the mishap was probably due to multiple causes related to the *ad hoc* requirements imposed on the hybrid booster.[47] When the programme resumed on 27 March 2004, both vehicles performed flawlessly.[48] A follow-up flight on 16 November attained a speed just short of Mach 10.[49,50]

ATHENA

In 1993 the Lockheed Company decided to apply its expertise in solid-propellants – gained from the Polaris, Poseidon and Trident missiles – to develop the Lockheed Launch Vehicle (LLV) to place small satellites into low orbit, or into geosynchronous transfer orbit.[51] When the company merged with Martin Marietta in 1995 this became the Lockheed Martin Launch Vehicle (LMLV). Following the first successful flight, the programme was renamed Athena. The Athena I was a two-stage vehicle capable of inserting 800 kilograms into low orbit. Its first stage was a Castor 120 (which had been derived by Thiokol from the motors of the Trident and Peacekeeper missiles) and its second stage was an Orbus 12D supplied by United Technologies. The Athena II would have two Castor 120 stages in tandem and a payload of 1,975 kilograms. An Athena III would have strap-on solids to further increase the capacity to 3,600 kilograms. In each case, the payload was to be manoeuvred into the desired orbit by the hydrazine-burning Orbit Adjust Module that had been developed by Primex Technologies.[52]

Disastrous début

As the Athena I lifted off for its inaugural launch on 15 August 1995 from SLC-6 at Vandenberg, looking diminutive among the pad facilities built for much larger rockets, the programme was some nine months behind schedule.[53] The vehicle was carrying the 150-kilogram GEMstar data-relay satellite for CTA Space Systems. It was to put the satellite into a polar orbit at an altitude of 680 kilometres, but at T + 160 seconds it veered off course and the range safety officer destroyed it some 466 kilometres downrange, at an altitude of 148 kilometres.[54,55] The tracking camera footage indicated that an anomalous pitch excursion started at T + 79 seconds. The investigation focused on the 500 telemetry channels from T + 65 seconds to the shutdown of the first stage at T + 82 seconds. After the first stage had been jettisoned, the vehicle coasted (as planned) until the second stage ignited at T + 150 seconds, but at T + 121 seconds the 'coning' motion it had

The inaugural launch of the Athena I on 15 August 1995.

inherited dislodged the shroud 24 seconds before it was to have been shed. Although the second stage rapidly stabilised itself, the range safety officer destroyed it because it was not on the desired trajectory.[56],[57] In fact, *two* independent failures were found, either of which would have led to the loss of the vehicle.[58] The thrust-vectoring system had induced the attitude excursion shortly prior to first stage shutdown. The inertial measurement unit had failed while coasting, at T + 127 seconds. The Castor 120 was a modified missile booster with a carbon–epoxy casing and a 'submerged' flexible nozzle bearing that was driven through 5.5 degrees of deflection by cold-gas-pressurised hydraulic actuators to vector the thrust to steer the vehicle. The spent hydraulic fluid was vented, but it caused a fire that eroded the insulation of the cable with the signal from the sensors that measured the orientation of the nozzle, causing a short circuit that made the control system deflect the nozzle – which induced the attitude excursion. The thrust-vectoring system was similar to that of the Peacekeeper, except that the hydraulic fluid in that case had not been vented. The solution was to install a reservoir to collect the spent hydraulic fluid, thereby forming a closed system. The subsequent inertial measurement unit fault resulted from corona arcing in its power supply. Litton, the manufacturer, had tested the unsealed package for this phenomenon, but only to the 70,000-foot limit of its vacuum chamber. A test by Lockheed Martin for the investigation established that it began to arc at 86,000 feet! The solution was to apply a compound to protect the components that were susceptible to arcing, and to seal the package. It was an object lesson in favour of high-fidelity testing. The Castor 120 had been adopted for several other 'lightweight' launch vehicles, but this was its first flight. In light of this failure, the development of the Athena III was put on hold.[59] Despite this setback, Lockheed Martin was confident that in June and July 1996 it would be able to launch the Lewis and Clark satellites that NASA had initially booked with OSC's Pegasus and later reassigned to the Athena, but this optimism was to prove unfounded.[60],[61],[62] After two years of redevelopment, it was hoped to launch an Athena I in May 1997, but this was slipped to provide more time to analyse a test of the revised thrust-vectoring system.[63]

An Athena I from Vandenberg finally placed Lewis into orbit on 22 August 1997, but the satellite immediately malfunctioned, leading NASA initially to postpone and later to cancel the second satellite.

Mixed results

In 1989 the Spaceport Florida Authority was formed to support the activities of the commercial launch companies using the Castor 120 motor. It refurbished Pad 46, right on the tip of the Cape Canaveral promontory, which had previously been used to test the Trident missile for the Navy.[64],[65]

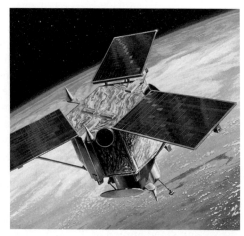

An artist's impression of an Ikonos satellite.

On 7 January 1998 an Athena II from Pad 46 placed NASA's Lunar Prospector on course for its survey of the Moon. In January 1999 an Athena I from Canaveral launched ROCSat 1, the first of three remote-sensing satellites built for Taiwan by TRW.[66] In early 1999 Raymond S. Colladay, the president of Lockheed Martin Astronautics, which made the Athena, observed that the market for small satellites was developing very slowly.[67] On 27 April 1999 an Athena II was dispatched from Vandenberg to insert the Ikonos 1 high-resolution commercial imaging satellite into Sun-sunchronous polar orbit at an altitude of 675 kilometres, but the shroud snagged.[68] The electrical failure that was suspected was confirmed by the investigation.[69,70] In fact, it was a design flaw that could have struck previously. The aluminium–lithium shroud was in two pieces, each with a pyrotechnic charge at its base. A few milliseconds after these charges fired, others were to split the shroud lengthwise to enable its segments to fall away. However, the shock from detonating the charges at the base momentarily dislodged the connectors that were to carry the signal to the charges inside the shroud. Although the structure was free, since its halves had not separated it was held in place by the payload. Unable to compensate for the 635-kilogram 'dead weight', the third stage fell into the atmosphere over the South Pacific. An Athena II with a revised shroud successfully launched the replacement satellite on 24 September 1999.[71]

Despite a run of seven commercial launch successes, the prospects for Athena I were uncertain due to the shrinking market. However, the future for Athena II brightened in 2001 when NASA added the rocket to its launch services contract, along with the Delta and the Atlas. On 29 September 2001 an Athena I made the first orbital launch from the newly built facility on Kodiak Island in the Aleutians. When the Clark satellite was cancelled, its launcher had been reassigned to NASA's Vegetation Canopy Lidar, but this was grounded in 2000 by technical

An Athena II ready to launch on 24 September 1999 with Ikonos 2.

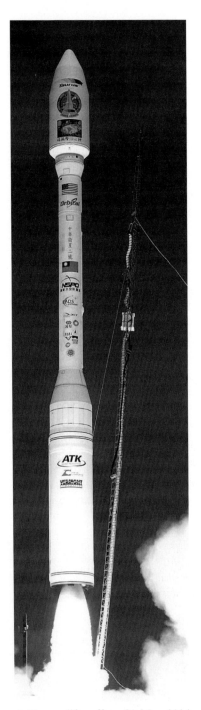

A Taurus lifts off on 20 May 2004 with ROCSat 2.

problems. With no prospect of a commercial launch, the vehicle was offered to the Air Force, which assigned Picosat for the Department of Defense's Space Test Program. Picosat was to have ridden a Russian Dniepr rocket in 1998 but the prohibition on flying military payloads on non-US rockets had prevented this. It was accompanied by a trio of student-built satellites that had been sponsored by the Air Force.[72] In view of the dearth of commercial payloads, Lockheed Martin announced that it was placing the Athena in "stand by".[73]

TAURUS

The Taurus was a Pegasus stripped of its wings and mounted on a Castor 120 for vertical launch.[74,75] Introduced on 13 March 1994, it started well with five straight successes. Then on 21 September 2001 the drive shaft of the actuator for the thrust-vectoring system seized momentarily, causing the vehicle to veer off course. Even although the system recovered, the resulting velocity shortfall made it re-enter over the Indian Ocean. In addition to QuickTOMS for NASA and Orbview 4 for Orbimage (a company that was part owned by OSC), it had a commercial package for Celestis containing a residue of the ashes from 50 cremations. Since the Alliant Techsystems Orion 50S engine of the first stage of the Pegasus formed the second stage of the Taurus, the problem with the thrust-vectoring system raised the possibility that this had contributed to the loss of the X-43A.[76]

MED-LITE DELTA

In December 1994 NASA invited bids for the 'Med-Lite' contract to supply a launch vehicle for payloads between the mass ranges served by the Pegasus and the Delta II.[77] A requirement was that the vehicle be able to orbit a payload with a mass half of that of the Delta II at the same cost per unit of mass. On 1 February 1995 OSC submitted its Taurus but McDonnell Douglas won the contract with the 7300 and 7400 forms of the Delta II, which employed only three and four strap-ons respectively.[78,79,80,81] The first launch on 24 June 1999 deployed the Far Ultraviolet Spectroscopic Explorer for the Origins programme.[82] NASA hoped to launch up to 14 such payloads over the next decade, including many of the Discovery-class deep space missions.[83,84,85]

The 1st 'Med-Lite' Delta II model 7320 lifts off on 24 June 1999 with the FUSE satellite.

NOTES

1. http://www.fathom.com/course/21701717/session4.html
2. *Backroom Boys*, Francis Spufford, Faber and Faber, 2003.
3. http://www.fathom.com/course/21701717/session5.html
4. http://www.geocities.com/CapeCanaveral/Launchpad/6133/blackarrow.html
5. http://www.fathom.com/course/21701717/session6.html
6. http://www.vought.com/heritage/special/html/sscout8.html
7. *Spaceflight Revolution*, J.R. Hansen, SP-4308, NASA, 1995.
8. http://history.nasa.gov/SP-4308/ch7.htm
9. 'Get 'em Up, Scout!', J.A. Harriss, *Air & Space*, March 1989.
10. *Spaceflight Revolution*, J.R. Hansen, SP-4308, NASA, 1995.
11. http://www.nasa.gov/centers/langley/news/factsheets/Scout.html
12. OSC cheekily dubbed the carrier aircraft an air-breathing first stage.
13. http://www.orbital.com/SpaceLaunch/Pegasus/index.html
14. *Aviation Week & Space Technology*, 30 May 1994, p. 30.
15. *Flight International*, 1–7 June 1994, p. 17.
16. *Spaceflight*, March 1997, p. 78.
17. *Aviation Week & Space Technology*, 30 May 1994, p. 31.
18. *Aviation Week & Space Technology*, 4 July 1994, p. 30.
19. *Aviation Week & Space Technology*, 1 August 1994, p. 28.
20. *Aviation Week & Space Technology*, 8 August 1994, p. 69.
21. *Aviation Week & Space Technology*, 26 April 1993.

22. http://www.orbcomm.com/
23. *Aviation Week & Space Technology*, 17 August 1993.
24. *Aviation Week & Space Technology*, 10 April 1995, p. 67.
25. *Flight International*, 15–21 March 1995, p. 24.
26. *Aviation Week & Space Technology*, 27 September 1993, p. 49.
27. *Flight International*, 12–18 April 1995, p. 17.
28. *Aviation Week & Space Technology*, 24 April 1995, p. 17.
29. *Flight International*, 12–18 April 1995.
30. *Flight International*, 3–9 May 1995, p. 21.
31. http://leonardo.jpl.nasa.gov/msl/QuickLooks/orbcommQL.html
32. http://centaur.sstl.co.uk/SSHP/data/data_orbcomm.html
33. *Spaceflight*, March 1997, p. 78.
34. *Aviation Week & Space Technology*, 3 July 1995, p. 21.
35. *Flight International*, 18–24 October 1995, p. 32.
36. *Aviation Week & Space Technology*, 4 September 1995, p. 60.
37. *Spaceflight*, March 1997, p. 78.
38. *Aviation Week & Space Technology*, 11 November 1996, p. 28.
39. NASA Press Release 96-231, 8 November 1996.
40. http://www.news.cornell.edu/Chronicle/99/3.11.99/WIRE.html
41. *Aviation Week & Space Technology*, 24 May 1999, p. 61.
42. *Aviation Week & Space Technology*, 9 June 1997, p. 32.
43. *Aviation Week & Space Technology*, 22 November 1999, p. 18.
44. In doing so, the X-43A would claim the Mach 6.7 record set by the rocket-powered X-15 piloted by Pete Knight in 1967.
45. *Aviation Week & Space Technology*, 31 March 1997, p. 22.
46. *Spaceflight*, September 2001, p. 361.
47. *Spaceflight*, January 2002, p. 14.
48. http://www.dfrc.nasa.gov/Newsroom/ResearchUpdate/X43A/
49. http://www.Spaceflightnow.com/news/n0410/16x43a/
50. http://news.bbc.co.uk/1/hi/sci/tech/4018117.stm
51. *Flight International*, 21–27 September 1994, p. 39.
52. *Aviation Week & Space Technology*, 14 August 1995, p. 64.
53. *Flight International*, 30 August–5 September 1995, p. 24.
54. *Aviation Week & Space Technology*, 21 August 1995, p. 18.
55. *Spaceflight*, June 1998, p. 202.
56. *Aviation Week & Space Technology*, 28 August 1995, p. 67.
57. *Aviation Week & Space Technology*, 4 September 1995, p. 53.
58. *Aviation Week & Space Technology*, 18/25 December 1995, p. 96.
59. *Aviation Week & Space Technology*, 7 April 1997, p. 42.
60. *Flight International*, 30 August–5 September 1995, p. 24.
61. *Flight International*, 22–28 March 1995, p. 20.
62. *Flight International*, 5–11 July 1995, p. 20.
63. *Aviation Week & Space Technology*, 19 May 1997, p. 17.
64. *Flight International*, 11–17 January 1995, p. 43.
65. *Flight International*, 3–9 May 1995, p. 21.
66. *Flight International*, 21–27 June 1995, p. 39.
67. *Aviation Week & Space Technology*, 8 February 1999, p. 69.
68. *Aviation Week & Space Technology*, 3 May 1999, p. 45.
69. *Aviation Week & Space Technology*, 24 May 1999, p. 61.

70. *Aviation Week & Space Technology*, 14 June 1999, p. 82.
71. *Aviation Week & Space Technology*, 4 October 1999, p. 41.
72. *Spaceflight*, January 2002, p. 18.
73. *Spaceflight*, February 2002, p. 50.
74. *Aviation Week & Space Technology*, 21 April 1997, p. 65.
75. *Aviation Week & Space Technology*, 13 March 2000, p. 57.
76. *Spaceflight*, December 2001, p. 494.
77. *Aviation Week & Space Technology*, 12–19 December 1994, p. 27.
78. *Aviation Week & Space Technology*, 4 March 1996, p. 61.
79. *Aviation Week & Space Technology*, 21 April 1997, p. 65.
80. In line with the four-digit designation scheme, the 7300 series had three GEMs and the 7400 had four.
81. http://www.spaceandtech.com/spacedata/elvs/delta2_sum.shtml
82. http://fuse.pha.jhu.edu/
83. *Aviation Week & Space Technology*, 7 November 1994, p. 67.
84. *Aviation Week & Space Technology*, 5 December 1994, p. 40.
85. Whereas the Delta 7900 series uses nine GEM strap-ons and can place approximately 5 tonnnes into an easterly low orbit, the Delta 'Med-Lite' would have a capacity of 2 tonnes.

6

Boom and bust

BOOMING PROSPECTS

After NASA demonstrated the advantages of the high-capacity Ka-Band with its Advanced Communications Technology Satellite (ACTS) in 1993, companies sought licences from the US Federal Communications Commission to operate satellites offering global communications to mobile systems and to fixed sites in developing countries that lacked a conventional communications infrastructure.[1,2,3] The Commission required that any company granted a licence must start the construction of its first two satellites within one year of the licence being granted, finish them within four years, and have the entire constellation operating within six years. In January 1995 the Commission issued licences to Motorola-led Iridium, Loral-led GlobalStar and TRW-led Odyssey.[4,5] In contrast to the established practice of operating geostationary communications relays, all of these proposals were for constellations in *non*-geostationary orbits.

An artist's impression of the Advanced Communications Technology Satellite.

A rosey forecast
In early 1995 the Teal Group, a market analysis company in Fairfax, Virginia, issued a forecast that over the next 10 years almost 1,000 satellites would require launching, 68 per cent of which would be for communications, with most of these placed in *non*-geostationary orbits.[6] The increasingly competitive market for global anywhere-to-anywhere mobile

telephone systems was confidently expected to generate $26 billion by 2005, and to have in excess of 33 million subscribers by 2012.[7] In fact, the great fear was that industry would not be able to manufacture sufficient handsets to satisfy the demand when the systems entered service in 1998! As a result of competition for the limited operating frequencies, and the high cost of establishing the requisite satellite constellations, this market was expected to be dominated by a small number of well-financed companies. With so many satellites to be deployed, the future for the launch vehicle companies looked good.

Sea Launch

In April 1995 Boeing established the Sea Launch company as a joint venture with Russia's Energiya, the Ukraine's Yuzhnoye and Yuzhmash, and Norway's Kvaerner.[8] Although based in Long Beach, California, launches were to be from a platform in the Pacific. The Russians and Ukrainians were to supply the rocketry and Kvaerner was to build the launch platform (named Odyssey) as a conversion of an oil rig, together with the service and command ship.[9] The Zenit 3SL launch vehicle was to be the two-stage Ukrainian Zenit 2 with its first stage strengthened to support the extra mass of the third stage. With a thrust of 1.6 million pounds, the four-nozzle RD-170 of the first stage was slightly more powerful than the Rocketdyne F-1,[10] five of which had been clustered for Wernher von Braun's Saturn V. The Energiya-built Block-DM that had for many years been used with the Khrunichev-built Proton-K was to be given upgraded avionics and employed as the third stage.[11] The fact that all three stages burned kerosene and oxygen would simplify operations.[12] The advantage of the Sea Launch system was that by locating the platform on the equator the vehicle would be able to take full advantage of the rate at which the Earth rotates on its axis, and be ideally placed to put satellites into geosynchronous transfer orbit. The Zenit 3SL would be able to place a payload of 5 tonnes into geosynchronous transfer orbit.[13] And to accommodate the most modern satellites, the shroud would be 4 metres in diameter. It was hoped to start flying in 1998.

GlobalStar opts for Zenit 2

The GlobalStar constellation called for 48 satellites (and eight in-orbit spares) at an altitude of 1,000 kilometres in eight orbital planes inclined at 52 degrees to the equator. The design was for "repeaters in the sky" with switching on the ground. Each of the 450-kilogram satellites would have 16 channels with a simultaneous capacity of 64,000 calls. Whereas Iridium sought the top 1 per cent of the global paging and cellular telephone markets, GlobalStar was aiming to route calls within nations, so the two companies were in different segments of the market.[14] GlobalStar confidently predicted revenues of $1.6 billion by 2002 from 2.7 million subscribers, and to increase its subscriptions to 16 million over the next decade.[15] In May 1995 Space Systems/Loral, which was building the satellites for GlobalStar, booked 36 of them with Yuzhnoye for launch from Baikonur by Zenit 2.[16] They were to be carried in batches of 12, in three tiers of four, and deployed in sequence by a dispenser that had originally been developed to enable intercontinental-range ballistic missiles to deliver multiple independently targeted warheads.

A Ukrainian Zenit 2 launch vehicle is prepared and dispatched.

Inmarsat forms ICO

The International Maritime Satellite Organization (Inmarsat), collectively owned by 73 countries or their agents, believed that it would have an advantage over individual corporations in raising funds. On 3 March 1994 Inmarsat's Council discussed deploying a constellation of satellites to provide a global service to handheld telephones (using a fourth-generation system that it dubbed 'Inmarsat-P'), but "uncertainties in the business potential for global handheld services" resulted in the postponement of any decision.[17,18] Nevertheless, the Council authorised continuing study.[19] It had earlier funded external studies of three possible operating strategies, with constellations utilising low, medium, and geostationary orbits,[20,21] and had selected medium orbit.[22,23] The favoured proposal was for a dozen satellites at an altitude of 10,600 kilometres, in two planes inclined at 45 degrees to the equator. One reason for this orbit was Inmarsat's aspiration to offer a service to rival the Global Positioning System provided by the similarly deployed NAVSTAR satellites run by the US Department of Defense.[24,25,26] In December 1994 Inmarsat decided to go ahead, and in January 1995 created ICO Global Communications to manage it.[27,28] In 1996, the Comsat Corporation of America (which owned 22 per cent of Inmarsat) blocked the proposal to provide a service to rival the GPS.[29]

Lawsuit

In July and August 1995 TRW was granted US patents related to the operation of a constellation of communications satellites in medium orbit which, it argued, would pose "serious challenges to competitors who wish to provide similar services".[30] This was a reference to ICO. However, while TRW planned 12 satellites in three orbital planes, ICO intended to use 10 satellites in two planes, and while a TRW satellite would have 37 beams, an ICO satellite would have 160 beams. ICO said that the only feature the two systems had in common was the operating altitude. In essence, TRW claimed the rights to medium orbit, which it defined to be altitudes between 10,000 and 18,000 kilometres.[31] Motorola's Iridium and Loral's GlobalStar were not affected, as they were intended to operate below this zone. The root of the dispute was a contract issued by Inmarsat to TRW in 1993, as one of several independent

An artist's impression of an ICO satellite, with (in the background) a depiction of the entire constellation.

studies to investigate how it might provide global mobile communications. TRW reported on the 'concept' of medium orbit, for which an architecture had never really been developed previously.[32] Of course, when Inmarsat set out to develop such a system, TRW had expected to receive the production contract, and when this was given to Hughes Space and Communications, it was argued that Inmarsat had given TRW's idea to a competitor.[33,34] Undeterred by the dispute, on 3 October, ICO finalised the $1.4 billion contract with Hughes to supply 10 satellites based on its HS-601 bus rated at 8 kilowatts (the standard configuration provided typically 4.5 kilowatts) with a 10-year lifetime. At this time, Hughes invested $100 million in ICO. A second contract in December that authorised Hughes to arrange launch services for 'on-orbit delivery' increased the total contract value to $2.3 billion.[35,36] With 40 satellites either in-build or on-order, Hughes had recently made bulk bookings with several of the launch providers.[37] It gave three of the ICO satellites to Sea Launch, four to ILS (three for the Proton and one for the Atlas IIAS) and five to McDonnell Douglas for the new Delta III, all of which were to be launched over a 20-month interval starting in late 1998.[38]

Warning forecasts

The Teal Group reported that 1995 was the worst year twice over, first because 112 was the lowest number of payloads since 1964 and second because the seven launch vehicle failures in 83 launches was the worst since 1969.[39] Nevertheless, as Teal had forecast at the beginning of 1995, the market for commercial satellites was evidently on the verge of a dramatic increase. The spurt of launch vehicle development was being driven by the demand for global telecommunications services. However, in December 1996 a report by the market research company Frost and Sullivan warned that demand would likely plummet at the end of the decade.[40] It produced a chart showing 88 launches in 1995 rising to 235 in 1999, falling back to around 93 in 2000. The competing companies would be hit hard once the constellations of small satellites were established in low-to-medium orbits and the demand resumed the more sedate pace of delivering heavyweight satellites into geostationary orbit. The Paris-based Euroconsult group estimated that with 1,100 satellites in need of launch over the coming 10 years, the launch vehicle market would be worth $34 billion.[41] However, its *Launch Services Market Survey* echoed Frost and Sullivan in warning that, as a result of the surge in development to meet "a very sharp peak in demand" in 1997–1999, there would be overcapacity when the market took a downturn in 2004–2005. The launch vehicle companies, however, were confident that a variety of yet-to-be-developed satellite-based applications would sustain the high demand after the global communications systems were operational.[42,43]

Starsem

In early 1997, with the booming market for global communications driving up the need for launch services, about half of the orders were going to the Ariane 4.[44] In an effort to keep pace with satellites for low and medium as well as geostationary orbit, Arianespace and Aerospatiale formed Starsem as a joint venture with Samara, the

A Starsem Soyuz–Fregat is prepared for launch from the Baikonur Cosmodrome.

Russian maker of the Soyuz launcher, to market this vehicle with a range of upper stages in a manner that would complement the Ariane series.

Teledesic

On 14 March 1997, the Federal Communications Commission awarded Teledesic a licence to operate a constellation of Ka-Band satellites. The original 1994 design for this constellation had 840 satellites with 40 in each of 21 orbital planes, and was to have cost some $9 billion.[45] After a three-year study, Boeing redesigned the network, was made the prime contractor, and invested $100 million in the company. The redesign (announced in April) reduced the number of satellites to 288, with 24 in each of 12 orbital planes. The original plan was for 800-kilogram satellites that would operate at 350–400 kilometres. In the revised network the satellites were to be 1,300 kilograms, more capable, and fly at 800 kilometres. However, as each satellite would cost four times as much, the scaled down constellation would *not* be cheaper. In addition to linking up terrestrial telephone systems, Teledesic was to provide "bandwidth on demand" to companies that required to move massive amounts of data rapidly.[46] The plan was to begin launching satellites in 2001 in order to enter service by the end of 2002.[47]

Zenit problems

On 20 May 1997 a Zenit 2 exploded 48 seconds after lifting off from Baikonur with an electronic intelligence-gathering satellite of the Tselina 2 series for the Russian military. The débris fell 12 kilometres downrange.[48],[49] The investigation blamed the

An Iridium satellite undergoes test.

RD-170 kerosene and oxygen engine that powered the first stage.[50] This was the fifth failure in 27 launches of the Zenit 2 since its introduction in 1985, in each case losing a Tselina 2 satellite: on 28 December 1985 the second stage had failed to ignite; on 4 October 1990 an engine in the first stage exploded after 3 seconds; when the vehicle resumed flying on 30 August 1991 one of the engines of the second stage exploded; and there was a replay on the next mission on 5 February 1992.[51] This latest failure boded ill for Sea Launch, whose launch vehicle was basically a Zenit 2 with an upper stage.

TRW makes peace with ICO
When the Federal Communications Commission accepted further proposals in September 1997, ICO submitted a letter of intent. The issue faced by the Commission was that as ICO was a subsidiary of Inmarsat, it was a foreign company.[52,53] ICO was therefore viewed as a 'test case' of the Commission's obligations to the World Trade Organisation to provide a free market.[54] Despite resey forecasts for mobile satellite communications, there was growing concern that there would be insufficient investors to fund all of the proposed constellations! For example, TRW had been unable to attract the $3.2 billion for its Odyssey system.[55,56] On 17 December TRW abandoned its Odyssey plan, ceded its Ka-Band operating licence, dropped its lawsuit against ICO and invested in the company.[57,58]

Iridium's constellation
In September 1996 Motorola's plant in Chandler, Arizona, delivered its first Iridium satellite.[59,60] At its peak rate, it would be shipping one every five days. Originally, there were to have been 77 satellites in circular orbits at an altitude of 700 kilometres, in seven orbital planes (and the name was chosen because 77 is the atomic number of the element iridium) but the decision taken in 1992 to increase the power of the transponders had the effect of deleting one orbital plane, reducing the number to 66 – but fortunately without causing the project to be renamed Dysprosium![61] At 700 kilograms, the satellites were lightweights. Each would be rated at 1.2 kilowatts, have onboard processing for 48 spot beams, and be capable of intersatellite linkage. The first launch was in May 1997, and the constellation was completed over a 12-month interval by three Proton, three Long March 2C-SD and nine Delta II rockets. In fact, 72 satellites were launched, but as *five* had already failed, there was only one in-orbit spare. On 19 August a Long March 2C-SD replaced two that had failed in *one* plane, the absence of which was degrading the coverage. (A given launcher could replace satellites only in one orbital plane, so it was not so much the number of failures that mattered as their distribution in space.) Iridium initiated commercial operations in November 1998, a few months later than scheduled, but still ahead of the competition.[62,63] The aim was to provide voice, paging, fax and data transmission from anywhere in the world, but take-up was slow owing to the fact that a handset was initially priced at $2,500 and a pager at $500.

GlobalStar's frustration
GlobalStar was close on Iridium's heels. A Delta II from Canaveral put its first four satellites into orbit on 14 February 1998.[64,65] By the time of Iridium's initiation of

The first five Iridium satellites are loaded onto the dispenser.

A Delta II model 7920 lifts off with a payload of five Iridium satellites.

service, GlobalStar had eight of its 48-satellite constellation in orbit, and was planning to launch a further 36 by the end of the year, sending them on three Zenit 2 launchers. The insurance companies were initially reluctant to risk placing so many satellites on a single vehicle, but the overall reduction in cost – if nothing went wrong – was appealing to the operators. GlobalStar was hoping to initiate its service in April 1999.[66] However, when a Zenit 2 failed on 10 September 1998 it lost a dozen of GlobalStar's satellites.[67] At T + 272 seconds, "a faulty command was issued in the control system, shutting down the engines", and the vehicle fell to Earth in the Gorno–Altaisk region of Russia. This put Global-Star's plan on hold. The Zenit contract was cancelled, and the remaining manifest divided

Table – Iridium launches

1	5 May 1997	Delta II	5
2	18 June 1997	Proton	7
3	9 July 1997	Delta II	5
4	20 August 1997	Delta II	5
5	13 September 1997	Proton	7
6	26 September 1997	Delta II	5
7	8 November 1997	Delta II	5
8	8 December 1997	Long March 2C-2D	2
9	20 December 1997	Delta II	5
10	18 February 1998	Delta II	5
11	25 March 1998	Long March 2C-2D	2
12	29 March 1998	Delta II	5
13	6 April 1998	Proton	7
14	2 May 1998	Long March 2C-2D	2
15	17 May 1998	Delta II	5
16	19 August 1998	Long March 2C-2D	2
17	8 September 1998	Delta II	5
18	6 November 1998	Delta II	5
19	19 December 1998	Long March 2C-2D	2
20	12 June 1998	Long March 2C-2D	2
21	11 February 2002	Delta II	5
22	20 June 2002	Rokot-Briz-KM	2
Total			95

Note that when Iridium Satellites took over the assets, 88 satellites had been launched, of which 16 had failed and one had just re-entered, so of the 71 satellites remaining in the constellation, five were in-orbit spares and 66 were operational.[68]

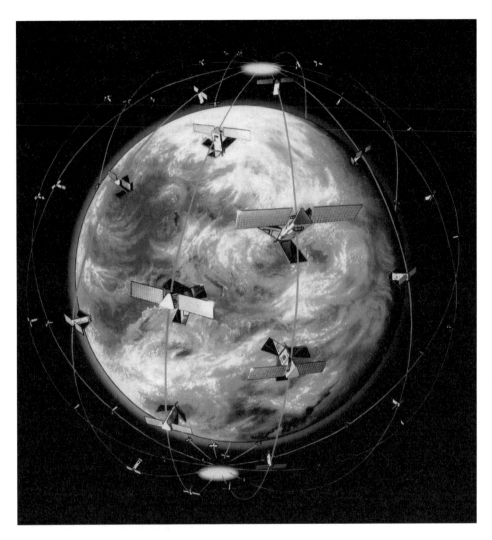

An artist's depiction of the Iridium constellation.

into batches of four for launch by either the Delta II or Starsem's Soyuz. The Department of State's requirement that Starsem sign a technology safeguard agreement delayed the resumption of launches to 1999,[69] but thereafter deployment was rapid and Space Systems/Loral booked the 12 replacement satellites with Starsem.[70]

Sea Launch's début

As a result of the Zenit 2 loss, Sea Launch decided to assign a dummy payload to its inaugural launch and to slip it from December 1998 to March 1999.[71] This vehicle was to have carried Galaxy 11 for PanAmSat on a commercial basis, but this was

Preparing and launching the Sea Launch Zenit 3SL with DirecTV 1R.

reassigned to Arianespace.[72] In the autumn of 1998, Sea Launch had 13 'firm' orders from Hughes Space and Communications and five from Space Systems/Loral. The first commercial mission following the demonstration was to be for ICO, but ICO cancelled two of its three contracts and reassigned the satellites to ILS, which put one on a Proton (thereby increasing its contingent to four) and one on an Atlas IIAS (now two); Boeing's Delta III had five satellites booked.[73,74] Of the dozen satellites to be launched, 10 were to form the operational constellation and the other two were to be in-orbit spares. The optimistic plan was to launch them all over a 15–18-month period starting in May 1999. Sea Launch's Odyssey platform arrived at Long Beach, California, on 4 October 1998, *en route* to its station near Christmas Island on the equator in mid-Pacific.[75,76] On 27 March 1999 the first Zenit 3SL lifted off with 4.7-tonnes of ballast.[77] The first commercial launch was scheduled for August, but had to be postponed.[78] On 9 October the second Zenit 3SL successfully released DirecTV 1R, an HS-601HP for DirecTV Incorporated, which, at 3.5 tonnes, was well within the vehicle's 5.5-tonne capacity.[79,80]

GlobalStar builds up

On Starsem's inaugural launch on 9 February 1999, a Soyuz with the Ikar restartable upper stage lifted off from Baikonur carrying four satellites for GlobalStar.[81] By making six launches in 1999, in the process placing

Preparing a GlobalStar satellite, a depiction of the constellation and an artist's impression of the their deployment.

Table – GlobalStar launches

1	14 February 1998	Delta II	4
2	24 April 1998	Delta II	4
3	9 September 1998	Zenit 2	12
4	9 February 1999	Soyuz–Ikar	4
5	15 March 1999	Soyuz–Ikar	4
6	15 April 1999	Soyuz–Ikar	4
7	10 June 1999	Delta II	4
8	10 July 1999	Delta II	4
9	25 July 1999	Delta II	4
10	17 August 1999	Delta II	4
11	22 September 1999	Soyuz–Ikar	4
12	18 October 1999	Soyuz–Ikar	4
13	22 November 1999	Soyuz–Ikar	4
14	8 February 2000	Delta II	4

Total: 52 (= 48 operational and 4 in-orbit spares)

Note that the Zenit 2 launch on 9 September 1998 failed.[82]

half of the constellation in space, Starsem demonstrated to the market that it could run regular launches.[83,84]

BUST

Bankruptcies!
On 13 August 1999, having defaulted on some $1.5 billion in loans, Iridium filed for Chapter Eleven protection to give Motorola, the majority shareholder, time to restructure the company.[85] Iridium's difficulties raised doubts over the viability of its rivals, and on 27 August 1999, having failed to raise $600 million to clear its debts, ICO followed suit. The $3.1 billion that ICO had previously raised had fallen $1.6 billion short of its overall requirements. Hughes Space and Communications, which was building the satellites for ICO, stood to lose $300 million if it collapsed. The launch vehicle companies were also exposed, but the risk was spread out because the satellites were booked on Delta, Atlas, Proton and Sea Launch. Despite the crisis, the mood was optimistic that ICO would be rescued.[86,87] In contrast, GlobalStar started commercial service on 10 October 1999.[88] It promptly began to market itself aggressively in an effort to turn in a quarterly profit by late 2000 or early 2001.[89] The $216 million loss in the first quarter of 2000 was of little concern, because the company was not as exposed as ICO, and was certainly not in the position of Iridium, which had folded after consuming $5 billion.[90] However, one of the GlobalStar satellites had already failed, so on 8 February 2000 a Delta II added four in-orbit spares, one of which was immediately made operational.[91] The launcher industry was hoping that GlobalStar's operations would inject sufficient confidence

to the market to enable both Iridium and ICO to survive restructuring. However, by the end of 2000 GlobalStar was experiencing difficulty in signing up clients.[92] The basic problem was that the cellular telephone market had failed to expand at the predicted rate, at least not in terms of using satellites to provide global coverage.[93] Meanwhile, on 13 January 2000 Boeing bought Hughes Space and Communications for $3.75 billion, renaming it Boeing Satellite Systems.[94]

Sea Launch loses ICO's satellite

On 13 March 2000, after two successes, Sea Launch had its first loss. The launch got off to a good start, with the first stage of the Zenit 3SL separating at $T+145$ seconds and the shroud jettisoning 32 seconds later, but contact was lost 2 minutes short of the end of the second stage's 6.5-minute burn. The earlier launches had been due east along the equator in order to achieve geosynchronous transfer orbit, but with the ICO F1 satellite heading for a medium orbit at 10,600 kilometres inclined at 45 degrees to the equator the débris fell into the sea several hundred kilometres from the Chilean coast. As there were no tracking stations along this flight path, telemetry was relayed via one of NASA's TDRS satellites. The initial focus of the investigation concerned the verniers that steered

Preparing the ICO F1 satellite.

the second stage. These burned the same propellants as the main RD-120 four-chamber engine, with all four thrusters being fed by a single turbopump.[95,96] It was discovered that the flight control system shut down the engine at $T+450$ seconds after it had sensed that the vehicle was not functioning properly. The theory was that a computer error just after lift-off had allowed a pressure valve to remain open, causing the verniers to either misfire or fail to fire. Sea Launch said that it was more likely to have been a software error than a hardware fault, and later added that it had "strong evidence" that the cause was a software error on the ground that failed to command a valve to close in a pneumatic system that performed several functions, including vectoring the engine.[97] Telemetry indicated that this system had lost 60 per cent of its pressure, leading to a significant deviation in attitude (through the inability to gimbal the engines) which in turn had triggered the self-destruct system. Ironically, this occurred just short of the scheduled end of the second stage's burn, so it had *almost* made it. ICO had booked its second satellite on the Delta III, but after this vehicle suffered two successive failures ICO withdrew its satellite to

A Sea Launch Zenit 3SL lifts off on 28 June 2004 with Telstar 18.

await a successful demonstration flight.[98]

Sea Launch made its return to flight by launching PanAmSat 9 on 28 July 2000, and followed this with a long run of successes.[99] However, the rate of launches was less than expected because the Zenit 3SL had to be stood down whenever the Block-DM stage of a Proton malfunctioned.[100] As a result, the company was not operating profitably – indeed, it had yet to recoup its start-up cost.[101] On 28 June 2004 the early shutdown of the Block-DM of a Zenit 3SL stranded Telstar 18 in a transfer orbit with an apogee that fell short of geosynchronous altitude. However, the Loral 1300 series satellite had been loaded with such a margin of propellant for station-keeping that it was able to reach its operating position without undermining its 13-year nominal operating life.[102] The three-month-long investigation concluded that a wiring problem on the Block-DM triggered a short that induced electrical interference in the circuitry carrying data on the propellant flow rates to the control system, causing this to consume fuel more rapidly than planned, resulting in an early shutdown when the tank ran dry.[103] Sea Launch postponed its next mission to early 2005.

Corporate rescues

Iridium secured $3 billion to keep it afloat during the rescue effort, but the company was wound up in March 2000 after the telecommunications entrepreneur Craig McCaw declined his support.[104,105] Of the 78 satellites that had been launched, 12 were no longer operational, and it was decided to de-orbit those that were functional in order to preclude their becoming orbital débris.[106] However, in December 2000 a venture capital start-up called Iridium Satellites took over the bankrupted Iridium Incorporated,[107] and in March 2001 it purchased the assets for $25 million. The

Table – Sea Launch missions

1	27 March 1999	Demosat
2	9 October 1999	DirecTV 1R
3	12 March 2000	ICO F1
4	28 July 2000	PanAmSat 9
5	20 October 2000	Thuraya 1
6	18 March 2001	XM 2 (Rock)
7	8 May 2001	XM 1 (Roll)
8	15 June 2002	Galaxy 3C
9	10 June 2003	Thuraya 2
10	7 August 2003	EchoStar 9/Telstar 13
11	30 September 2003	Galaxy 13/Horizons 1
12	11 January 2004	Telstar 14/Estrela do Sul 1
13	4 May 2004	DirecTV 7S
14	28 June 2004	Telstar 18

Note that the launches on 12 March 2000 and 28 June 2004 were failures.[108]

Department of Defense then issued a $72 million contract to make use of the satellites over a period of two years, with the option of an extension to 2004 increasing the contract to $252 million.[109] The appeal of the deal was not only that the satellites would ease the pressure on government communications, but also that the government owned most of the expensive handsets. In addition, the new company signed up partners around the world to sell and support point-to-point services in their own territories.[110] On 11 February 2002 five new satellites were launched on a Delta II as spares for the revived constellation.[111] On 20 June 2002 the second commercial flight of the Rokot by Eurockot from Plesetsk added two more.[112,113] With 12 spares complementing the 66 operational satellites (sufficient to sustain operations through to the end of the decade) the new company was beginning to look as if it would make a limited success out of the defunct constellation.

On 17 May 2000, ICO Global Communications was rescued by an investment group headed by Craig McCaw's Eagle River company, which agreed to provide $1.2 billion. ICO–Teledesic Global was created as a holding company to control all the satellites of McCaw's various operations.[114,115] The reborn company, which was named 'New ICO', acquired its predecessor's assets and was restructured to provide digital data and voice services in addition to the satellite equivalent of 3G wireless services, including wireless internet connectivity and other packet-data services.[116] In order to support this new emphasis, and also to supply an additional three satellites, Boeing was instructed to modify the satellites that had been designed for mobile voice telephony. This new company would complement rather than compete with Teledesic: the packet-data transfer offered by New ICO would operate at 144 kilobits per second, which was only twice that of a standard telephone link, whereas the 'Internet in the Sky' that Teledesic was to provide would be hundreds of times faster.[117,118] In 2000 ICO–Teledesic lost $1.2 billion in its attempt to keep the

'Internet in the Sky' afloat.[119] After an Atlas IIAS successfully orbited ICO F2 on 19 June 2001, the Federal Communications Commission issued the company its Ka-Band licence on 17 July,[120],[121] but shortly thereafter the order for the remaining satellites was suspended and later cancelled.

Desert Launch

In 2001 Sea Launch raised the possibility of providing launches from Baikonur. While this would reduce the cost of operating from an ocean platform, it would also significantly reduce the Zenit 3SL's capacity to geosynchronous transfer orbit.[122] In fact, Energiya and Yuzhnoye were both making commercial Zenit 2 launches from Baikonur. Another option was to launch the Zenit 2 from Odyssey, with lightweight payloads for equatorial low orbit. Work began to refurbish a pad at Baikonur to service the Zenit 3SL, but the commitment to use it would depend on demand from the market.[123] The Russians cheekily dubbed the Baikonur option 'Desert Launch'. In fact, Yuzhnoye was proposing a fundamental design change to the Zenit 3SL, to use a first stage comprising a core tank of liquid oxygen mated with two 22-metre-long 2.2-metre-diameter tanks of kerosene. This configuration would increase the geosynchronous transfer orbit capacity by 1,500 kilograms, to 7.5 tonnes, and would enable Sea Launch to compete with the uprated Ariane V.[124] As yet, however, this proposal has not been pursued.

GlobalStar's demise

Although by late 2001 GlobalStar had signed up 60,000 customers, its revenues had declined. Despite cutting operating costs, the company was obliged to revise its business plan.[125] Its operational constellation of satellites was still fully operational, although two spares had been activated to cover failures. On 15 February 2002 the company filed for Chapter Eleven.[126]

Teledesic throws in the towel

GlobalStar's demise prompted Teledesic to reassess its strategy. In February 2002 it announced that a new "advanced network design" would enable it to employ a constellation of 30 satellites in medium orbit rather than 288 in low orbit. The first 12 satellites (costing $1 billion) would provide an initial operating capability with coverage in selected (high revenue-generating) territories. The other 18 satellites would be added later to expand to global coverage.[127] To preclude the Federal Communications Commission rescinding its operating licence, Teledesic issued a contract to Alenia Spazio in Italy to build the first two satellites.[128],[129] On 1 October 2002 it ordered a halt to this construction.[130],[131] Having reviewed the commercial prospects for broadband by satellite, the company had concluded that it was not prudent to continue purely on the speculation that a market would develop. Satellite point-to-point showed "no foreseeable financial market or commercial prospects", it announced.[132] Over the past decade, Teledesic had spent hundreds of millions of dollars on the design and development of satellites for global broadband, but the market had simply failed to develop – just as it had for mobile communications by satellite.

Aftermath
The hyped projections of the mid-1990s for global mobile phone and internet markets had spawned companies which, by their pursuance of constellations of satellites, had stimulated a boom for the satellite makers and the launch providers. However, many of these communications ventures had collapsed due to inappropriate business models, with the result that the prospects for the new launch vehicles proved to be less attractive than expected.

NOTES

1. *Aviation Week & Space Technology*, 4 April 1994, p. 57.
2. *Aviation Week & Space Technology*, 9 October 1995, p. 24.
3. *Aviation Week & Space Technology*, 31 March 1997, p. 48.
4. *Aviation Week & Space Technology*, 21 November 1994, p. 29.
5. *Aviation Week & Space Technology*, 7 July 1997, p. 29.
6. *Aviation Week & Space Technology*, 16 January 1995, p. 55.
7. *Flight International*, 4–10 October 1995, p. 44.
8. *Aviation Week & Space Technology*, 17 April 1995, p. 25.
9. http://www.sea-launch.com
10. *Flight International*, 18–24 January 1995, p. 6.
11. Interestingly, the Block-DM started out as the *fifth* stage of the N-1, was modified to serve as the *fourth* stage of the Proton, and was now adapted to serve as the *third* stage of the Zenit 3SL.
12. *Spaceflight*, January 1999, p. 19.
13. *Aviation Week & Space Technology*, 30 November 1998, p. 56.
14. *Aviation Week & Space Technology*, 25 May 1998, p. 26.
15. *Flight International*, 4–10 October 1995, p. 45.
16. *Flight International*, 31 May–6 June 1995, p. 23.
17. *Flight International*, 16–22 March 1994, p. 15.
18. *Aviation Week & Space Technology*, 4 April 1994, p. 57.
19. *Aviation Week & Space Technology*, 23 May 1994, p. 58.
20. *Aviation Week & Space Technology*, 29 March 1993, p. 58.
21. *Aviation Week & Space Technology*, 11 October 1993, p. 44.
22. *Aviation Week & Space Technology*, 2 August 1993, p. 26.
23. *Aviation Week & Space Technology*, 4 April 1994, p. 57.
24. *Aviation Week & Space Technology*, 29 March 1993.
25. *Aviation Week & Space Technology*, 2 August 1993.
26. *Aviation Week & Space Technology*, 4 April 1994, p. 57.
27. The name ICO was derived from 'Intermediate Circular Orbit'.
28. *Flight International*, 21 December 1994–3 January 1995, p. 22.
29. *Aviation Week & Space Technology*, 22 April 1996, p. 59.
30. *Flight International*, 4–10 October 1995, p. 44.
31. *Aviation Week & Space Technology*, 13 November 1995, p. 68.
32. http://listserv.unb.ca/bin/wa?A2=ind9305&L=canspace&F=&S=&P=9450
33. *Aviation Week & Space Technology*, 13 November 1995, p. 68.
34. *Aviation Week & Space Technology*, 20 May 1996, p. 21.
35. *Aviation Week & Space Technology*, 13 November 1995, p. 68.

36. *Flight International*, 20–26 September 1995, p. 26.
37. *Flight International*, 3–9 January 1996, p. 19.
38. *Aviation Week & Space Technology*, 4 November 1996, p. 92.
39. *Aviation Week & Space Technology*, 12 February 1996, p. 53.
40. *Aviation Week & Space Technology*, 9 December 1996, p. 80.
41. *Aviation Week & Space Technology*, 16 December 1996, p. 86.
42. Of course, the fact that the 'peak' in the late 1990s failed to materialise ('because the bubble burst') meant that there was no subsequent overcapacity of launch vehicles.
43. *Aviation Week & Space Technology*, 7 April 1997, p. 41.
44. *Flight International*, 23–29 April 1997, p. 27.
45. *Aviation Week & Space Technology*, 28 March 1994, p. 26.
46. *Aviation Week & Space Technology*, 5 May 1997, p. 26.
47. *Aviation Week & Space Technology*, 9 March 1998, p. 25.
48. *Aviation Week & Space Technology*, 16 June 1997, p. 216.
49. http://www.russianspaceweb.com/zenit.html
50. *Aviation Week & Space Technology*, 26 May 1997, p. 34.
51. http://www.astronautix.com/lvs/zenit2.htm
52. *Aviation Week & Space Technology*, 9 October 1995, p. 24.
53. http://www.lta.com/res_regulatory/bigleos.htm
54. http://www.findarticles.com/p/articles/mi_m0UKG/is_1998_May_4/ai_56534712
55. *Aviation Week & Space Technology*, 16 October 1995, p. 57.
56. *Aviation Week & Space Technology*, 7 July 1997, p. 29.
57. http://ast.faa.gov/files/pdf/leo-d.pdf
58. http://www.businesswire.com/trw/1997.shtml
59. *Aviation Week & Space Technology*, 3 April 1995, p. 56.
60. *Aviation Week & Space Technology*, 23 September 1996, p. 57.
61. *Aviation Week & Space Technology*, 21 September 1992.
62. *Aviation Week & Space Technology*, 25 May 1998, p. 26.
63. *Aviation Week & Space Technology*, 14 December 1998, p. 21.
64. *Spaceflight*, April 1999, p. 134.
65. http://www.decodesystems.com/globalstar.html
66. *Aviation Week & Space Technology*, 25 May 1998, p. 26.
67. *Spaceflight*, November 1998, p. 439.
68. *Aviation Week & Space Technology*, 14 December 1998, p. 21.
69. *Spaceflight*, December 1998, p. 456.
70. *Spaceflight*, November 1998, p. 439.
71. *Spaceflight*, January 1999, p. 8.
72. http://www.decodesystems.com/iridium.html
73. *Spaceflight*, January 1999, p. 8.
74. *Spaceflight*, May 1999, p. 181.
75. *Spaceflight*, December 1998, p. 457.
76. *Aviation Week & Space Technology*, 30 November 1998, p. 56.
77. *Aviation Week & Space Technology*, 5 April 1999, p. 65.
78. *Spaceflight*, June 1999, p. 224.
79. *Aviation Week & Space Technology*, 27 September 1999, p. 39.
80. *Aviation Week & Space Technology*, 18 October 1999, p. 32.
81. *Spaceflight*, April 1999, p. 134.
82. *Spaceflight*, June 1999, p. 227.
83. *Spaceflight*, April 2000, p. 138.

84. *Spaceflight*, November 1999, p. 444.
85. http://www.decodesystems.com/globalstar.html
86. *Aviation Week & Space Technology*, 6 September 1999, p. 29.
87. *Aviation Week & Space Technology*, 8 November 1999, p. 53.
88. *Spaceflight*, October 1999, p. 407.
89. *Spaceflight*, January 2000, p. 4.
90. *Spaceflight*, August 2000, p. 315.
91. *Spaceflight*, May 2000, p. 181.
92. Although Iridium was no longer in competition, GlobalStar was still encountering difficulties.
93. Another problem specific to Iridium was that its portable user sets were bulky and costly, and it had signed up only 10,000 customers, just one-tenth of the predicted take-up.
94. *Spaceflight*, December 2000, p. 494.
95. *Aviation Week & Space Technology*, 20 March 2000, p. 36.
96. *Spaceflight*, June 2000, p. 224.
97. *Spaceflight*, July 2000, p. 269.
98. *Spaceflight* July 2000, p. 269.
99. *Spaceflight*, November 2000, p. 445.
100. *Spaceflight*, September 2002, p. 359.
101. *Spaceflight*, February 2003, p. 50.
102. *Spaceflight*, September 2004, p. 345.
103. http://www.Spaceflightnow.com/sealaunch/t18/041007report.html
104. *Aviation Week & Space Technology*, 13 March 2000, p. 37.
105. *Spaceflight*, June 2000, p. 227.
106. *Aviation Week & Space Technology*, 27 March 2000, p. 39.
107. *Spaceflight*, March 2001, p. 96.
108. *Spaceflight*, September 2002, p. 358.
109. *Spaceflight*, May 2002, p. 189.
110. *Spaceflight*, May 2002, p. 189.
111. *Spaceflight*, September 2002, p. 358.
112. The first Eurockot launch was in March 2002 and put the two Grace satellites into orbit. A third launch was planned for early 2003 with microsats for Canada and the Czech Republic.
113. *Aviation Week & Space Technology*, 22 May 2000, p. 20.
114. *Spaceflight*, January 2001, p. 11.
115. http://www.sea-launch.com/index.html
116. *Spaceflight*, June 2000, p. 225.
117. *Spaceflight*, January 2001, p. 12.
118. *Spaceflight*, August 2000, p. 315.
119. *Spaceflight*, January 2001, p. 12.
120. http://www.spacenewsfeed.co.uk/2001/29July2001.html
121. http://www.spaceandtech.com/digest/flash2001/flash2001-063.shtml
122. *Spaceflight*, September 2001, p. 358.
123. *Spaceflight*, April 2003, p. 143.
124. *Spaceflight*, March 2002, p. 99.
125. *Spaceflight*, January 2002, p. 11.
126. http://www.fcc.gov/transaction/ico-globalstar.html
127. *Spaceflight*, May 2002, p. 184.
128. *Spaceflight*, May 2002, p. 184.

129. Alenia Spazio was the first company to sell a commercial Ka-Band satellite, Intelsat F1 in 1991.
130. http://www.ee.surrey.ac.uk/Personal/L.Wood/constellations/teledesic.html
131. http://web.archive.org/web/20021002112448/www.teledesic.com/newsroom/articles/10-01-2002.html
132. *Spaceflight*, December 2002, p. 490.

7

The Chinese experience

THE LONG MARCH

The development of the Long March 2C launch vehicle by China began in 1970, and exploited experience with military missiles. However, within seconds of lifting off from the Juiquan launch site in the Gobi desert of northwestern China for its début mission on 5 November 1974 it swayed precariously, and had to be destroyed.[1] The investigation of the telemetry and the wreckage concluded that a cable running between the gyroscope and the flight control system was broken by the vibration. The wiring was improved and subjected to more realistic testing, the quality control in manufacturing was improved, and the vehicle was successfully introduced a year later. It was able to place 2.3 tonnes into low orbit, and was used primarily with the Fanhui Shi Weixing (FSW) recoverable satellites.[2] On 6 October 1992 a Long March 2C carried Sweden's Freya satellite as a piggyback payload with an FSW satellite. In April 1993, after 12 straight successes, this launch vehicle was booked by Motorola to deploy one-third of its Iridium constellation.[3] These 22 satellites were to be launched in pairs using a 'smart dispenser' in a version named the Long March 2C-SD.[4] After testing the dispenser using dummy satellites on 1 September 1997, the deployments started on 8 December 1997, but ceased in 1999 when the company filed for bankruptcy. Having purchased a 5 per cent share, the China Great Wall Industry Corporation was rather disillusioned.

Long March 3
The Long March 3 was the Long March 2C with a cryogenic third stage, marking the first Chinese use of a hydrogen-burning engine. It could place 1.5 tonnes into geosynchronous transfer orbit.[5] On its maiden flight from the Xichang launch site in the mountains of south-central China on 29 January 1984, the cryogenic stage fired to enter parking orbit but failed to reignite for the geosynchronous transfer orbit manoeuvre. Even so, the Shiyan Tongbu Tongxin Weixing (STTW) experimental geostationary communications satellite was released to facilitate testing its basic

systems. On 8 April 1984 the Long March 3 was launched with another such satellite and established China's first geostationary relay. As China's economy developed, so too did its need for sophisticated telecommunications, much of which was most appropriately provided by satellites. Initially, China relied on Intelsat to provide 'trunk' telephone and television distribution. The first privately owned regional satellite carrier was the Asia Satellite Telecommunications Company, formed by the China International Trust and Investment Corporation, Cable and Wireless, and Hutchinson Whampoa.[6] This was soon joined by the Asia Pacific Telecommunications (APT) Satellite Corporation of Hong Kong and by the China Orient Telecommunications Satellite Company. China's operational geostationary satellites were known by both the Dong Fang Hong (DFH) series name and as ChinaSat. They were not as sophisticated as their western counterparts and their low power limited their utility for trunk telephony, so Intelsat continued to provide television broadcasting and data transmission. On 7 April 1990 a Long March 3 flew China's first commercial mission by inserting AsiaSat 1 (an HS-376) into geostationary orbit for the Asia Satellite Telecommunications Company.[7,8] When a Long March 3 set off on 28 December 1991 with one of the second series of DFH satellites, the third stage suffered a helium leak during the burn for geosynchronous transfer orbit and lost thrust, stranding its payload in a useless orbit.[9] Nevertheless, on 21 July 1994 a Long March 3 successfully deployed ApStar 1 (an HS-376) for APT. In late 1992, with the first generation of DFH satellites retired and one of five second-generation satellites lost, China bought Spacenet 1 (a GE Astro Space bus launched in 1984) from GTE Spacenet and operated it as ChinaSat 5 while the third generation of DFH satellites were being built.

Losing Optus B2

The Long March 2E was the Long March 2C two-stage launch vehicle augmented by four liquid strap-ons. Able to put 9.5 tonnes into low orbit or insert 3.5 tonnes into geosynchronous transfer orbit, it had a capability approaching that of the Ariane 4 or Proton.[10,11] On its maiden flight on 16 July 1990 it carried Pakistan's first satellite, Badr. On 14 August 1992 it deployed Optus B1, which was an HS-601 for the Australian Telecommunications Company (formerly Aussat, now Optus). As it was a two-stage vehicle that released its payload in low orbit, a kick-motor was required to insert the satellite into geosynchronous transfer orbit, and the HS-601s used the PAM-D2, which used a Thiokol Star 63 solid rocket motor.[12] When a Long March 2E released its payload on 21 December 1992, the Optus B2 satellite failed to respond. On reviewing footage of the ascent, an anomalous plume was seen to emerge from the shroud at T + 48 seconds. It appeared that the payload had suffered a fatal mishap at the moment the vehicle was subjected to the greatest aerodynamic stress. In the absence of proof that the shroud had failed, the Chinese decided that the interface between the upper stage and the payload (which used parts from both Hughes and the China Great Wall Industry Corporation) was insufficiently robust and had suffered a resonance that destroyed the vehicle.[13,14,15] The Long March 2E returned to service on 27 August 1994 with Optus B3.[16] Although it had blamed the payload adapter for the loss of Optus B2, on 20 January 1995 the Chinese acceded to

A Long March 2E lifts off with an Optus satellite.

a request from the EchoStar Communications Corporation to reinforce the shroud of the Long March 2E prior to launching that company's satellite.[17]

Introducing the Long March 3A
The Long March 3A introduced on 8 February 1994 had more sophisticated avionics and an upgraded third stage that increased the geosynchronous transfer orbit payload to 2.6 tonnes.[18] It launched the first of the third generation of DFH satellites on 30 November 1994 as ChinaSat 6A, but the satellite's apogee motor shut down early, leaving it in an orbit ranging between 6,400 and 36,000 kilometres. The satellite was able to limp into geostationary orbit by firing its thrusters, but doing so exhausted its propellant, with the result that it was unable to maintain its position for long and had to be abandoned.[19,20] It was to have relayed six television channels and up to 8,000 simultaneous telephone calls.

A disputed failure
On 26 January 1995 the Long March 2E carrying ApStar 2 (an HS-601) exploded at T + 50 seconds and, according to *Xinhua*, the New China News Agency, the wreckage fell onto a village 7 kilometres downrange killing six people and wounding two dozen others.[21,22,23,24,25] The video of the night launch showed that the explosion had initiated at the *top* of the vehicle. As the rocket flew through 26,000 feet it had gone supersonic and endured its maximum aerodynamic stress. The China Great Wall Industry Corporation concluded that the 150-kilometre-per-hour wind shear at this altitude that was produced by the winter jetstream over the site in the mountains of southern Sichuan Province had induced a resonance in the payload adapter.[26,27] The government-controlled newspaper *Ta Kung Pao* went further: "The satellite's explosion caused the rocket's explosion, which was entirely the responsibility of the US-made

satellite."[28] Hughes made its own investigation, and came to a different conclusion.[29,30] Since no HS-601 had exploded on any other type of vehicle, after losing Optus B2 Hughes had installed instrumentation to report on the stresses imposed by this type of launch vehicle. Two lines of evidence led Hughes to the conclusion that the shroud had failed. The telemetry from the 'break wire' sensors – which were in place to verify that the shroud had released – indicated that the shroud was disrupted a split second prior to the payload suffering the crushing pressure of being exposed to the airflow. In retrospect, it seemed likely that Optus B2 had been lost in this way. The launch criteria had failed to adequately account for the wind shear. The fact that the Chinese conducted their investigation in private did nothing to placate the insurance industry.[31] China was attracting customers by undercutting the fees of its competitors by up to 20 per cent, but the premiums were always high, and were increased significantly after a failure. Because the competition to supply transponders to the rapidly growing Asian market was fierce, APT ordered an HS-376 as a gap-filler, and asked Space Systems/Loral to build a 1300 series for ApStar 2R.[32,33,34] The Long March 2E resumed service on 28 November 1995 with AsiaSat 2. Originally set for December 1994, this had been twice postponed, first by the likely explosion of Telstar 402 on 8 September 1994 (both were Lockheed Martin Astro Space 7000 series) and then by the failure of the Long March 2E in January 1995, and was therefore a year late.[35,36] It became the first commercial satellite to use the FG-46 solid rocket motor made by the Chinese for the geosynchronous transfer orbit insertion burn.[37] (In view of the Chinese insistence that the losses of the HS-601s with American motors had been self-inflicted, AsiaSat was sitting pretty, especially because the shroud had been reinforced too.) On 28 December 1995 a Long March 2E successfully deployed EchoStar 1, which was another 7000 series communications satellite.

Long March 3B failure

The Long March 3B had a larger fairing and four liquid propellant strap-ons (as used by the Long March 2E) to enable it to insert 5 tonnes into geosynchronous transfer orbit. The Chinese intended it to supersede the Long March 2E as its primary commercial launcher.[38,39] Unfortunately, within two seconds of lifting off on its inaugural mission on 14 February 1996, it veered off course and fell to Earth, killing several people.[40,41,42,43] As the China Great Wall Industry Corporation reported: "There has been no damage to the launch facilities. However, the living facilities and the nearby residential houses suffered damage to varying

A Long March 3B on display (left) and in flight.

A Long March 3B topples over within seconds of lifting off on 14 February 1996.

degrees. There were a few casualties."[44] In the west, this mishap was dubbed the St Valentine's Day massacre.[45] Because the first stage shared technology with the Long March 2E, this failure also affected that vehicle, and withdrawing both vehicles excluded China from the commercial launch business for the foreseeable future. This time, the investigation was conducted more openly. The report, issued in September, identified the cause as a fault in the inertial reference system, which would have to be re-designed.[46,47,48,49,50] The failure was eventually traced to the deterioration of gold–aluminium wiring connections within the power amplifier for a motor in the inertial measurement unit.[51] The frustration for the insurance industry was that the US ban on the export of 'dual use' technologies to China – which might improve ballistic missiles – prevented the Chinese from purchasing better guidance systems. The China Great Wall Industry Corporation lost four bookings after this débâcle.[52] Unfortunately for Intelsat, whose Intelsat 708 had just been lost, the contract obliged it to complete the $60 million payment.[53] It withdrew Intelsat 709 from the repeatedly delayed Ariane V, and launched it on an Ariane 4 on 15 June 1996 to plug the gap in its service.

Losing ChinaSat 7

After losing ChinaSat 6A, the China Telecommunications Broadcasting Corporation – a division of the Ministry of Post and Telecommunications – ordered an HS-376 directly from Hughes (the first US satellite to be built specifically for China) to relay voice, fax, data, television and other C-Band services for the domestic market while an improved third-generation DFH satellite was developed incorporating Ku-Band systems supplied by Daimler–Benz Aerospace.[54,55,56] This satellite, ChinaSat 7, was launched by a Long March 3 on 18 August 1996. After making its 430-second initial burn, the cryogenic third stage was to have coasted for 368 seconds and reignited for a 328-second burn, but the sequencer restarted it 40 seconds early, and 280 seconds later (according to the investigation) "an abnormal phenomenon appeared in the engine control gas line, which caused [a] drop in the third stage engine thrust and the shut off about 48 seconds earlier than programmed", at which point the sequencer released the payload into an orbit with an apogee that fell significantly short of geosynchronous altitude.[57] Although the satellite was capable of manoeuvring to geostationary orbit, this would have exhausted its propellant and left it unable to maintain its station, so it was declared a total loss. The People's

Insurance of China paid the claim.[58,59] As ChinaSat 7 was similar to the Westar and Palapa satellites that had been retrieved by a Shuttle in 1984, and had propellant to circularise its orbit at 400 kilometres for such a collection, it was suggested that it be rescued. This would have presented a significant diplomatic opportunity (especially if a Chinese payload specialist was flown to oversee the retrieval) but NASA dismissed the suggestion.[60] In 1997, China purchased Spacenet 2 secondhand and renamed it ChinaSat 5R.

Crisis

The new confidence in the Long March 3 from the orbiting on 3 July 1996 of ApStar 1A (the HS-376 that was to plug the gap in APT's service until the replacement for ApStar 2 became available) evaporated with the loss of ChinaSat 7. This was the third time in nine flights that the cryogenic engine had malfunctioned, on each occasion in attempting to enter geosynchronous transfer orbit.[61,62] Since January 1995, three of six Long March missions had failed. The fact that these were *different* failures made the series uninsurable.[63,64] Hence, by this point China had a precarious presence in the commercial launcher market.[65] In investigating the spate of Long March 3 failures, the China Great Wall Industry Corporation identified 45 issues in need of attention, and invited international oversight of its improved quality assurance programme. The insurance market considered that at least three straight successes would be required to restore confidence.

SQUEEZED OUT

Long March 3 resumes service

The cryogenic third stage was reintroduced on 12 May 1997 when a Long March 3A successfully put ChinaSat 6B, the second of the third generation of DFH satellites, into geostationary orbit to replace its predecessor.[66] The Fen Yung 2A satellite was to have been China's first geostationary meteorological satellite, but it was lost on 2 April 1994 in an explosion while being loaded with propellant prior to being mated with its launch vehicle. One technician was killed and several dozen others were injured. It took three years to redesign the propulsion system to ensure that such an accident could not happen again.[67] On 10 June 1997, a Long March 3 deployed its replacement, but this ceased to transmit imagery on 8 April 1998 due to a fault in its S-Band antenna.[68,69] In deploying the Orion 1 communications satellite on 29 November 1994 for Orion Network Systems, an Atlas IIAS from Canaveral inserted its payload into "an unusual" transfer orbit with a 120,000-kilometre apogee, which was one-third of the distance to the orbit of the Moon, whereupon the Eurostar 2000 bus built by Matra Marconi Space fired its liquid-propellant engine four times over a two-week interval to manoeuvre into geostationary orbit. Because the orbital period of the transfer orbit exceeded 24 hours, this was called 'super-synchronous'. This made cancelling the inclination while manoeuvring into geostationary orbit more efficient.[70,71,72] When the Long March 3B finally made its reappearance on 20 August 1997, the third stage placed the Agila 2 communications satellite (which had been built by Aerospatiale for Philippine Agila Satellite Incorporated) into a super-synchronous transfer orbit with a 44,500-kilometre apogee. Then, on 16 October 1997 a Long March 3B deployed ApStar 2R. As China did not have the capability to build high-powered satellites for direct-to-home television and mobile communications, on 18 August 1995 the China Orient Telecommunications Satellite Company ordered an A2100 series satellite from Lockheed Martin, and this was launched as ChinaStar 1 by a Long March 3B on 30 May 1998.[73,74]

China's hopes are dashed
It seemed as if China had recovered the confidence of the insurance market, but on 18 June 1998 the House of Representatives of the US Congress formed a committee under the chairmanship of Christopher Cox to investigate whether, in the provision of commercial launch services, China had gained access to technologies that could be used to improve its ballistic missiles.[75,76,77,78,79] The report, written in January 1999, but not released until May, upheld this assertion.[80,81,82] The export restrictions imposed effectively inhibited the Chinese from launching satellites that were either made in America or constructed from American components.[83,84] Of course, this outcome was welcomed by the other launch providers.

In the 1980s Deutsche Aerospace helped China's Astronautics Ministry to design the DFH-3 series. In November 1993, after the Ministry of Space Technology had spun off the China Aerospace Corporation, Deutsche Aerospace signed up to cooperate on further projects.[85,86] In July 1994 they established Eurospace to develop satellites for the Sino Satellite Corporation, a Chinese company that provided telecommunications services to the People's Bank of China. The SinoSat satellites were to be built by Aerospatiale using a Spacebus 3000 design.[87,88,89,90,91,92] The launch of SinoSat 1 by a Long March 3B on 18 July 1998 effectively marked the end of the Chinese participation in the commercial launcher market.

NOTES

1. *China's Space Program*, B. Harvey, Springer–Praxis, 2004, p. 82.
2. http://space.cgwic.com/launch/vehicles.htm
3. *Aviation Week & Space Technology*, 26 August, 1996, p. 25.
4. http://space.cgwic.com/launch/vehicles.htm
5. http://www.astronautix.com/lvfam/lonmarch.htm
6. *Flight International*, 31 May–6 June 1995, p. 55.
7. *Aviation Week & Space Technology*, 26 February 1996, p. 68.
8. In fact, this satellite had been deployed by a Space Shuttle in February 1984 as Westar 6, but its PAM stage had fizzled and it had been retrieved by a Shuttle in November of that year, refurbished, and sold to AsiaSat, which booked it for relaunch by China.
9. *Aviation Week & Space Technology*, 13 January 1992.
10. http://www.astronautix.com/lvfam/lonmarch.htm
11. http://space.cgwic.com/launch/vehicles.htm
12. *Flight International*, 12–18 October 1994, p. 38.

13. *Aviation Week & Space Technology*, 18 January 1993, p. 28.
14. *Aviation Week & Space Technology*, 23 August 1993.
15. *Aviation Week & Space Technology*, 26 February 1996, p. 68.
16. *Aviation Week & Space Technology*, 26 February 1996, p. 68.
17. *Aviation Week & Space Technology* 3, July 1995, p. 22.
18. *Aviation Week & Space Technology* 21 October 1996, p. 22.
19. *Flight International*, 11–17 January 1995, p. 18.
20. *Flight International*, 18–24 January 1995, p. 22.
21. *Aviation Week & Space Technology*, 6 February 1995, p. 62.
22. *Aviation Week & Space Technology*, 19 February 1996, p. 25.
23. *Flight International*, 8–14 February 1995, p. 22.
24. *Flight International*, 15–21 February 1995.
25. *Aviation Week & Space Technology*, 26 February 1996, p. 62.
26. *Aviation Week & Space Technology*, 13 November 1995, p. 75.
27. *Flight International*, 9–15 August 1995, p. 20.
28. *Flight International*, 15–21 February 1995, p. 26.
29. *Aviation Week & Space Technology*, 13 February 1995, p. 21.
30. *Aviation Week & Space Technology*, 31 July 1995, p. 24.
31. *Aviation Week & Space Technology*, 26 February 1996, p. 68.
32. *Aviation Week & Space Technology*, 10 July 1995, p. 25.
33. *Flight International*, 15–21 March 1995, p. 24.
34. *Flight International*, 28 June–4 July 1995, p. 26.
35. *Flight International*, 11–17 January 1995, p. 18.
36. *Aviation Week & Space Technology*, 6 February 1995, p. 62.
37. *Aviation Week & Space Technology*, 26 September 1994, p. 88.
38. *Flight International*, 12–18 October 1994, p. 38.
39. *Aviation Week & Space Technology* 21 October 1996, p. 22.
40. *Aviation Week & Space Technology*, 19 February 1996, p. 25.
41. *Aviation Week & Space Technology*, 26 February 1996, p. 68.
42. *Aviation Week & Space Technology*, 11 March 1996, p. 21.
43. *Aviation Week & Space Technology*, 2 September 1996, p. 228.
44. *Aviation Week & Space Technology*, 26 February 1996, p. 62.
45. *Aviation Week & Space Technology*, 26 August 1996, p. 24.
46. *Flight International*, 16–22 October 1996, p. 25.
47. *Aviation Week & Space Technology*, 24 March 1997, p. 24.
48. *Aviation Week & Space Technology*, 11 November 1996, p. 25.
49. *Flight International*, 28 August–3 September 1996, p. 19.
50. *Aviation Week & Space Technology*, 16 September 1996, p. 21.
51. *Aviation Week & Space Technology*, 31 May 1999, p. 30.
52. *Flight International*, 28 August–3 September 1996, p. 19.
53. *Aviation Week & Space Technology*, 9 March 1998, p. 25.
54. *Aviation Week & Space Technology*, 19 August 1996, p. 32.
55. http://www.skyrocket.de/space/doc_sdat/zx-7.htm
56. *Aviation Week & Space Technology*, 22 July 1996, pullout p. S8.
57. *Aviation Week & Space Technology*, 26 August 1996, p. 24.
58. *Flight International*, 16–22 October 1996, p. 25.
59. Hughes bought ChinaSat 7 back, and renamed it HGS 2.
60. *Aviation Week & Space Technology*, 23 September 1996, p. 21.
61. *Aviation Week & Space Technology*, 26 August 1996, p. 24.

62. *Flight International*, 28 August–3 September 1996, p. 19.
63. *Aviation Week & Space Technology*, 26 February 1996, p. 68.
64. *Aviation Week & Space Technology*, 19 May 1997, p. 26.
65. *Aviation Week & Space Technology*, 26 August 1996, p. 24.
66. *Aviation Week & Space Technology*, 19 May 1997, p. 26.
67. *China's Space Program*, B. Harvey, Springer–Praxis, 2004, p. 142.
68. *Aviation Week & Space Technology*, 7 July 1997, p. 15.
69. http://www.fas.org/spp/guide/china/earth/fy-2.htm
70. *Aviation Week & Space Technology*, 5, December 1994, p. 26.
71. *Flight International*, 21 December 1994–3 January 1995, p. 30.
72. *Aviation Week & Space Technology*, 13 February 1995, p. 61.
73. *Flight International*, 30 August–5 September 1995, p. 24.
74. http://www.tbs-satellite.com/tse/online/sat_chinastar_1.html
75. *Aviation Week & Space Technology*, 22 June 1998.
76. *Aviation Week & Space Technology*, 29, June 1998.
77. *Aviation Week & Space Technology*, 6 July 1998.
78. *Aviation Week & Space Technology*, 13 July 1998.
79. *Aviation Week & Space Technology*, 8 November 1999, p. 34.
80. *Aviation Week & Space Technology*, 10 May 1999, p. 28.
81. *Aviation Week & Space Technology*, 31 May 1999, p. 26.
82. *Aviation Week & Space Technology*, 31 May 1999, p. 82.
83. *Aviation Week & Space Technology*, 22 February 1999, p. 24.
84. *Aviation Week & Space Technology*, 24 May 1999, p. 29.
85. *Aviation Week & Space Technology*, 3 October 1994, p. 63.
86. *Aviation Week & Space Technology*, 18 July 1994, p. 84.
87. *Space News*, 11–17 July 1994, p. 1.
88. *Spaceflight*, September 1994, p. 296.
89. *Aviation Week & Space Technology*, 3 October 1994, p. 63.
90. *Aviation Week & Space Technology*, 5 December 1994, p. 22.
91. *Flight International*, 18–24 January 1995, p. 22.
92. *Aviation Week & Space Technology*, 28 August 1995, p. 66.

8

The current crop

ARIANE V

When the Ariane 4 was introduced in 1988 it was able to launch the 'leading edge' communication satellites two at a time, but the trend was towards ever heavier satellites and so Arianespace promptly set out to develop the Ariane V to supplement and later to succeed it. The HM-60 Vulcain engine developed by the Société Nationale d'Etude et de Construction de Moteurs d'Aviation (SNECMA) together with the Société Européenne des Propulsion (SEP) for the hydrogen powered core had 112 tonnes of thrust.[1] It was to be augmented by two solid boosters for a lift-off thrust of 1,140 tonnes. These SEP-built motors were 10 times larger than any previously made in Europe, being comparable to those of the Titan IVA.[2] The storable-propellant Aestus second-stage engine supplied by Daimler–Benz Aerospace had 2.7 tonnes of thrust.[3,4] The Ariane V was to put 18 tonnes into low orbit, or 6 tonnes into geosynchronous transfer orbit.[5,6] The plan was to share the launch fee of about $130 million between two payloads. Competitiveness on cost was essential, because when the development began Arianespace had expected the competition to be the 'old' US vehicles, whereas since then the Americans had announced plans to build a new range of launchers with a view to cutting the cost to $80 million per flight, and there was new competition from Russia's Proton and China's Long March, both of which seemed likely to be able to offer bargain prices.[7] However, it appeared that the booming market would be able to accommodate them all. When it was conceived, the Ariane V was also to have carried astronauts on the Hermes spaceplace, and therefore was required to have a high level of reliability – the goal was for a 98.5 per cent overall success rate, with the figure for the first stage being 99.95 per cent.[8,9] However, Hermes was cancelled in 1993 for financial reasons. In April 1994, with Ariane V engine trials about to begin, Arianespace hoped to fly the début mission in October 1995, and make the first commercial flight in early 1996.[10] The company was so confident in the vehicle that it issued a guarantee: "If a satellite is lost during the launch phase – whether the failure is caused by the

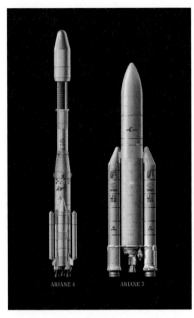

ARIANE 4   ARIANE 5

A comparison of the configurations
of the Ariane 4 and Ariane V.

launcher, or the satellite – the customer will be granted a free launch for a replacement satellite similar to the lost spacecraft."[11],[12]

Disastrous début

In June 1995, after faulty software delayed the final static test firings of the Vulcain engine, the first Ariane V was postponed from November 1995 to January 1996, and in September it was slipped to late April 1996.[13],[14] The second one was to launch in September with Intelsat 709 as a demonstration funded by Arianespace, as a precursor to the first commercial mission with PanAmSat 6 in early 1997. The mounting delay was due to oil and propellant leaks suffered in testing the first stage, some of which necessitated significant modifications. The payload for the first vehicle was a stack of four Cluster satellites, each weighing 1.2 tonnes, which had been developed by the European Space Agency to monitor the Earth's magnetosphere.[15] Fortunately, they did not need to be launched on a specific date.[16] After the loss of Intelsat 708 on a Long March 3B on 14 February 1996, Intelsat 709 was withdrawn and dispatched on an Ariane 44P on 15 June to plug the gap, and the European Space Agency's Atmospheric Re-entry Demonstrator assigned in its place.[17]

The début mission on 4 June 1996 appeared to have an excellent start, with the vehicle lifting off when the solids lit some 6 seconds after the main engine had ignited, but 30 seconds later it suddenly toppled over, and the auto-destruction system destroyed it at a height of 12,000 feet. In fact, the vehicle had already started to disintegrate as a result of the aerodynamic stress imposed by the increasing angle of attack, which was in response to the three engine nozzles being commanded to slew 'hard over' as a result of the failure of the inertial reference system. As the European Space Agency's press release wryly observed, the flight "did not result in validation of Europe's new launcher".[18] This is an interesting study in software engineering, due in part to an unwarranted reliance on heritage from its exceptionally reliable predecessor. Early in the development of the Ariane V, it was decided to reuse as many components of the Ariane 4 as possible in order to cut costs. In particular, the inertial reference system that had proved itself was carried over. The part of the software at issue was used prior to launch to initialise the inertial reference system and also, on the Ariane 4, to facilitate a rapid realignment of that system in the event of a last-minute hold in the countdown. While this realignment function served no rôle on the Ariane V, it was retained for commonality reasons, and allowed (as on the Ariane 4) to operate for approximately

40 seconds after lift-off.[19] Simulations had indicated that the software would function properly in the new vehicle, but no tests were made using real hardware to confirm this. It was discovered that flaws in the specification and design of the software had resulted in the total loss of guidance and attitude data.[20,21,22,23] In particularly, no thought had been given to the values that certain variables might take. Also, as the Ariane V rose more rapidly than its predecessor, its horizontal acceleration was five times faster, and a variable conversion between a 64-bit floating point value and a 16-bit signed integer resulted in an overflow condition as a result of producing a number higher than could be represented this way, and the software crashed. The software was written in Ada, and only some of the software variables were protected against overflow – a decision that had been agreed to by all the project partners. The others were expected to be either physically constrained, or small enough with a large margin of safety. This logic proved to be faulty for the Horizonal Bias, the variable that caused the exception in the inertial reference units. As guidance system flaws are difficult to identify (because hardware cannot be accelerated to high speed in a laboratory!) this represented another classic mistake. The investigation noted that the 'culture' was based on protecting against random hardware failures, and both of the inertial reference units had the same flawed software. Consequently, in the space of 20 milliseconds at T + 30 seconds, the backup and then the primary unit both raised an overflow exception and the guidance system, which was left to 'fly blind', ordered the Vulcain engine of the core and the nozzles of the strap-ons to gimbal over to their limit.[24,25] The investigation issued 40 recommendations, including revising the management to assign Arianespace overall responsibility for the embedded software.[26] The modifications included redesigning the shroud for smoother aerodynamics, structural reinforcement to eliminate the buffeting that had been observed near the base of the vehicle, and more effective thermal protection for the solid strap-on separation points.

Trying again
The launch of the second qualification flight of the Ariane V had been scheduled for June 1997, in time for the Paris Air Show, but by early 1997 it was clear that making the modifications would take longer.[27,28] The flight on 30 October was a partial success, in that although the upper stage performed the geosynchronous transfer orbit burn, the orbit did not have the planned parameters. The payload comprised two satellites supplied by the Centre National d'Etudes Spatiales in France, one of which had the physical characteristics of a typical large communications satellite and was fitted with sensors to record the stresses of launch.[29]

Ariane 4 finale
In view of the delay in introducing the Ariane V, an order was placed to manufacture a further 10 Ariane 4 launchers, although the largest payloads would either have to wait or be offloaded to other providers.[30,31] As it happened, however, the Ariane 4 was idle for four months in mid-1998 as a result of the late delivery of the satellites.[32] Nevertheless, Arianespace managed to catch up by flying 10 missions in the final four months of the year. In May 1999 the company announced that it would phase

out the Ariane 4 in 2002 or 2003, because with the majority of commercial satellites now exceeding 2–2.5 tonnes it was no longer able to launch them in pairs, and doing so one at a time was not cost-effective. In contrast, the Ariane V was capable of being upgraded to carry two 5-tonne satellites.[33],[34] Meanwhile, Arianespace once again had some 'no shows' as the delivery of satellites was delayed by manufacturing problems.[35],[36] For the second successive year, therefore, the Kourou launch site fell idle. However, this dearth of payloads meant that Telstar 7 was able to be accepted at short notice and launched on 25 September 1999. The final Ariane 4 on 15 February 2003, which deployed Intelsat 907, concluded a 15-year career during which 114 launches of the various forms of the vehicle had successfully placed 182 satellites into orbit.[37]

Ariane V success
After the third Ariane V on 21 October 1998 carried another demonstration satellite, the first operational mission on 10 December 1999 deployed the XMM astrophysical satellite for the European Space Agency. The next one, on 21 March 2000, was the first to carry a pair of satellites, in this case Insat 3B for India and AsiaStar for the WorldSpace Corporation. In succession, the next vehicles deployed Astra 2B for SES and GE 7 for GE Americom on 14 September, PanAmSat 1R on 17 November, Astra 2D and GE 8 on 20 December, and EuroBird (also known as Eutelsat W1R) on 8 March 2001.

Stranding Artemis
When an LE-7A engine suffered a hydrogen leak during a test firing in July 2000 necessitating a redesign that would postpone the demonstration flight of Japan's H-2A to July 2001,[38],[39] the European Space Agency reassigned its Artemis communications satellite to the Ariane V for launch in the summer of 2001. This did not pose any problems as Artemis had actually been designed for the Ariane V.[40] Unfortunately, when it was launched on 12 July 2001 the Aestus upper stage malfunctioned and stranded the satellite in the wrong orbit. This was a near-replay of the test in 1997 that failed to achieve the desired transfer orbit. A combusion instability at engine ignition had reduced the thrust, and resulted in the early

The launch of the 3rd Ariane V on 21 October 1998.

depletion of the propellant. The strong pressure variation that was responsible for this instability was attributed to a dynamic coupling between the propellant feed and the internal parts of the combustion chamber. The most probable cause of the high-frequency instabilities was the presence of water vapour in the propellant lines. The investigation recommended that the hydraulic conditions be dynamically modelled mathematically and the ignition phase made steadier and smoother.[41] A new ignition system was developed, and the procedures revised to improve the regulation of humidity within the engine during its preparation.[42,43] On the positive side, the exhaustive tests confirmed the robustness of the engine in *nominal* conditions. By the end of 2001, it was hoped to resume Ariane V operations in early 2002, and on 1 March 2002 one placed Envisat into Sun-synchronous orbit, marking the first time that this vehicle had aimed for polar orbit.[44]

Upgrading the Ariane V
The first stage of the upgraded Ariane V was to have the more powerful Vulcain 2 engine and improved strap-ons.[45] With the existing Aestus stage, this would increase the capacity to geosynchronous transfer orbit by 1 tonne to 7 tonnes. With the new ESC-A second stage employing the HM-7B cryogenic engine of the Ariane 4, this would be increased to 10 tonnes. In the fullness of time the larger ESC-B would boost this to 12 tonnes.[46,47,48,49,50,51]

Eutelsat had a history of booking début launches, balancing a bargain fee against the reduced insurance coverage.[52] This had worked well with Eutelsat 2F3 on the Atlas II in 1991, Eutelsat W3 on the Atlas IIAS in 1999, and Eutelsat W4 on the Atlas IIIA in 2000. In 2001 it assigned a Hot Bird television satellite to the first upgraded Ariane V.[53]

Unfortunately, on its début flight on 11 December 2002 the Ariane V–ESC-A had to be destroyed by the range safety officer after the Vulcain 2 engine developed a fault and caused the vehicle to veer off course, taking with it Hot Bird 7 and Stentor, an experimental communications satellite for France Telecom.[54] The initial anomaly occurred in the engine's cooling circuit at $T + 96$ seconds. At $T + 178$ seconds the engine performance started to fluctuate and a flight control perturbation developed. The shroud was jettisoned on time at $T + 187$ seconds, but the vehicle's attitude was incorrect. When the vehicle continued to display erratic behaviour it was destroyed at $T + 456$ seconds at an altitude of 69 kilometres, with the débris falling 800 kilometres offshore. The investigation determined that the chain of events was initiated by the failure of the coolant tubes of the engine nozzle, which led to the destruction of the engine.[55,56] In the redesign, a nickel-based alloy jacket was welded outside the upper part of the nozzle, with axial stiffeners to resist the bending moments caused by dynamic overshoot at ignition. In addition, the rate of flow of liquid hydrogen through the coolant tubes was increased, and an yttrium zirconate coating was applied inside the upper nozzle by a plasma deposition technique to act as a thermal barrier.

Back to the original
An indirect casualty of the Vulcain 2 problem was the Rosetta mission, intended for launch in January 2003. The postponement caused it to miss the 10-day window

In addition to two communications satellites, the Ariane V launched on 27 September 2003 carried the SMART 1 spacecraft.

Ariane V deployed Anik F2 which, at 6 tonnes, was its heaviest single geosynchronous transfer orbit payload to date.

The future
Arianespace had hoped to phase out the standard Ariane V in 2003 and employ the Ariane V–ESC-A until the ESC-B was introduced in 2005, but owing to the grounding of the Vulcain 2 engine and the depressed state of the market it decided to "consolidate", and in May 2003 it ordered 30 new vehicles, most of them in the upgraded form.[61],[62] The Ariane V–ESC-A did not fly again until 12 February 2005, carrying a demonstration payload. [63],[64],[65],[66] Its

needed to undertake the assigned cometary rendezvous. It was grounded while a new flight plan was developed. The storage and other costs imposed by this delay amounted to some 70 million Euros, which was almost as much as the cost of entire SMART 1 mission![57] The launch of Galaxy 12 for PanAmSat and Insat 3A for India had been set for 28 February, but was delayed by the investigation into the loss of the Ariane V–ESC-A. On being rescheduled to 31 March, it was postponed to 10 April to allow additional checks.[58],[59] An Australian and a Japanese satellite rode up together on 11 June, and then on 27 September E-Bird for Eutelsat and Insat 3E were joined by the European Space Agency's SMART 1 spacecraft, which later lit its ion engine to spiral slowly out to the Moon.[60] The Rosetta spacecraft was finally dispatched on 2 March 2004. On 18 July an

An artist's impression of the Rosetta spacecraft in the proximity of a cometary nucleus.

first operational use would be to launch the inaugural Ariane V Transfer Vehicle (ATV) named *Jules Verne* with supplies for the International Space Station.[67]

DELTA III

In May 1995 McDonnell Douglas said that it intended to upgrade its Delta II to the Delta III in order to compete with the Atlas in the high end of the 'intermediate' market. Although this was not being kick-started by a government contract, Hughes had booked 10 launches.[68],[69],[70],[71] The new vehicle was to have a new cryogenic second stage, larger strap-ons, and a payload shroud that was 4 metres in diameter – 1 metre wider than the Delta II. The first stage would use the same RS-27A engine as the Delta II, and the oxidiser tank would be 2.44 metres in diameter, as previously, but the kerosene tank above would match the shroud, as indeed would the second stage. In July, the contract for the second stage was given to Pratt & Whitney, in preference to engines proposed by Aerojet and Rocketdyne.[72] It would have a single RL-10 engine, the nozzle of which would have a large carbon–carbon extendable exit cone. In effect, therefore, the Delta III would be a Centaur-upgraded version of the Delta II. The contract for the strap-ons

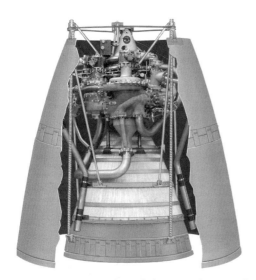

The RL-10-B-2 engine of the second stage of the Delta III introduced a 'crank down' extension to its nozzle bell.

went to Alliant Techsystems (formerly Hercules). Being 1.1 metres wider and 1.2 metres longer than the strap-ons of the Delta II, the new GEM-46 motors would give 25 per cent more power. Although the core of the first stage would be the same diameter as before, it would still be able to accommodate nine strap-ons. Three of the six ground-lit motors were to have flex-seal nozzles to enable them to be vectored for greater control authority. The Delta III was to have twice the capacity to geosynchronous transfer orbit of the model 7925 Delta II.[73] The simplicity of the design, and the use of systems developed for the Delta II, were expected to reduce operating costs and increase reliability. At that time, the Delta II had a reliability rating of 99 per cent, and there seemed to be no reason why this should not carry over to its successor.

Disastrous début

The inaugural launch of the Delta III from Canaveral on 26 August 1998 failed. The vehicle broke up at T + 72 seconds, destroying Galaxy 10, which was an HS-601 satellite for PanAmSat.[74] When a slow oscillating roll developed in the first minute of the flight, the control system tried to correct it by vectoring the RS-27A and the three strap-ons that had active nozzles, but the system overcompensated and the remedial action *contributed* to the instability. Ironically, once the hydraulic fluid for the thrust-vectoring system was expended the oscillation smoothed out, but then the disabled gimbals pitched the vehicle over, causing the self-destruct system to intervene. The investigation found that the roll oscillation was due to unpredicted solid rocket motor dynamics. "It took us four days to confirm the guidance problem," said Joseph Palsulich, the programme's senior manager for business development. Boeing attributed the loss to "improper analytical assumptions in the dynamic models, and poor communication between two design engineering groups".[75] The control system software was revised to compensate for this effect.[76,77]

Two down

When the second Delta III was launched on 4 May 1999 the new RL-10-B-2 engine on the second stage made its first burn successfully, but upon reigniting for the 162-second burn to enter geosynchronous transfer orbit it shut off after only 3.4 seconds, leaving the vehicle tumbling in an orbit with an apogee of 1,378 kilometres.[78] The telemetry indicated that the engine had suffered two shocks, the first occurring 4.2 seconds after the first firing, and the more violent second one 3.5 seconds into the second burn, in this case accompanied by a rapid increase in temperature.[79] An analysis determined that the engine had suffered "a 67-square-inch diamond-shaped breach of its combustion chamber".[80] The investigation focused on a fabrication process that had recently been introduced by Pratt & Whitney in which the soldered joints of the four segments of the hourglass-shaped chamber were reinforced by brazed seams of silver wire.[81,82,83] Boeing blamed "poor manufacturing process control" and "improper quality oversight".[84] Pratt & Whitney accepted the criticism. In fact, the breach was found to have developed at a seam that had leaked during a static firing, and the flawed brazing had left air pockets that allowed the joint to split. The fact that the repaired seam had survived a dozen later tests and its first in-flight firing, only to fail upon *restarting*, prompted Pratt & Whitney to argue that the engine, with its large new deployable nozzle, must have been subjected to unexpected torsional stress, but Boeing said that there was no evidence of such stress.[85,86] In future, the seams would be plated in place instead of being brazed, and then each unit would be put into an oven and baked.[87] An independent mission assurance review led by Sheila E. Widnall, a former Secretary of the Air Force, concluded in November that the task of developing the Delta III had been underestimated by Boeing, which was ironic considering that it was to have been a low-risk upgrade of the Delta II. Nevertheless, the company ruled out schedule and financial pressure as factors in either of the two malfunctions.[88,89]

As regards the Orion 3 HS-601HP communications satellite that had been stranded, Loral accepted $247 million in insurance, and Hughes approached NASA on behalf of the underwriter with a proposal that the satellite be retrieved by a Shuttle in 2001 and fitted with a solid motor for the geosynchronous transfer orbit burn.[90,91] However, the review of the rescue of Intelsat 603 in 1992 had concluded that the effort had not been cost-effective.[92,93]

Centaur grounded

As a precaution after the failure on 4 May 1999, Pratt & Whitney decided to make an ultrasound inspection of the 25 flight-ready RL 10 engines, which temporarily grounded the Atlas.[94] The Centaur resumed flying on 23 September with the launch of EchoStar 5 on an Atlas IIAS.[95] Meanwhile, because the maiden launch of the Atlas III, which also used a Centaur, had been postponed to 2000, Telstar 7, its intended payload, was shifted to Arianespace and sent up on an Ariane 44LP on 25 September.[96,97] As the Terra satellite for the Earth Observing System was to operate in polar orbit, it had to be launched from Vandenberg. The Atlas IIAS that was to carry it would be the first such vehicle to depart from the west coast. Owing to programme delays and concern over the RL-10, the vehicle stood on SLC-3E for 27 months! It was finally launched on 18 December 1999 with the heaviest payload yet carried by an Atlas–Centaur.[98]

An artist's impression of the Orion 3 satellite.

Partial success

The third Delta III was to have carried an ICO satellite, but when that company filed for bankruptcy in August 1999 this was withdrawn.[99,100,101] Boeing offered to launch the Artemis satellite for the European Space Agency, but this was declined.[102,103] With no prospect of a commercial payload, Boeing had no option but to mount a demonstration flight. This was made on 23 August 2000, and although the final orbit fell short of the intended apogee the mission was rated a success.[104]

Delta III doubts

When the development of the Delta III started, the plan was to introduce it in 1998 as a companion to the Delta II, but during the struggle to iron out the bugs the development of the Delta IV had started and the Delta III was now seen simply as a means of testing the new cryogenic stage, after which it would be phased out without ever serving as a commercial launcher.[105]

The launch of the Delta III on 23 August 2000.

A RE-ENGINED ATLAS

Upon the collapse of the Soviet Union in 1991, the US launch vehicle makers sought commercial relationships with their eastern counterparts. In October 1992 Energomash agreed to divide in half its four-nozzle RD-170 engine, which Pratt & Whitney would market as the RD-180.[106] In June 1993 Aerojet approached NK-Engines, the commercial arm of the Samara State Scientific and Production Enterprise with an invitation to make available the kerosene and oxygen engine that it had developed under the guidance of Nikolai Kuznetsov (hence the designation 'NK') in the 1960s.[107] The first stage of the ill-fated N-1 moonrocket had been powered by 30 of these units. Unfortunately, all four launches suffered first-stage failures.[108,109] When the N-1 was finally cancelled, 100 engines were placed in storage.[110,111,112,113] On 23 May 1995 Aerojet signed a protocol to enable it to manufacture the engine under licence as the AJ-26–NK-33A, with the 'A' showing that they were to be modified, chiefly by the replacement of the valve actuators and electrical wiring, and by the installation of gimballing for thrust vectoring.[114] By using a different design approach, Kuznetsov had obtained a high performance without having to push the technical limits as far as was done in America.[115] "It is what we would call a design-to-cost approach," observed Marc T. Constantine, Aerojet's director of strategic and space propulsion, adding, "even though they were well funded." Design-to-cost was to be the American strategy in the emerging and highly competitive international launch vehicle market. Although it had a thrust of only 340,000 to 380,000 pounds, the NK-33 had a large thrust-to-weight ratio and two of them could match the three-engined Rocketdyne power plant of the Atlas. In July Aerojet imported its first engines and sent them to its Sacramento test facility to verify their performance.[116,117] The Americans were surprised that 100 members of the original technical staff were still working for the company. Having survived long enough to witness a renewal of interest in his engine, Kuznetsov passed away on 30 July 1995. As Energomash reported to the Russian Space Agency and NK-Engines reported to the Russian Defence Ministry, the space agency opposed the NK-33 deal, arguing that it would threaten the RD-180, but the Russian government refused to intervene and left the ventures to compete.[118]

Launch services was a fiercely competitive niche market, and the development of a new rocket by one company influenced its competitors. Just as positive feedback drove the continuous development in the communications satellite business, the race to install global telecommunications services was driving the development of new, more effective launchers.[119] As the market shifted to payloads exceeding the capacity of the Atlas IIAS, Lockheed Martin decided to phase out the Atlas I, for which it was receiving ever fewer orders, and to develop a more powerful vehicle to supersede the Atlas IIAS.[120] As a sign of the company's confidence, whereas McDonnell Douglas had announced its Delta III only after Hughes booked 10 launches, Lockheed Martin initiated development prior to securing orders. Just as the Delta III was to enable McDonnell Douglas to compete with the Atlas in the 'intermediate' market, the new Atlas was to enable Lockheed Martin to compete more effectively with the heavier Chinese, Japanese and European launchers. It was clear that a significant improvement of the

Atlas would require a departure from its lineage. The core of the Atlas IIAR ('R' for 're-engined') was to have a new engine able to exceed the performance of the Atlas IIAS with its strap-ons, and the Centaur would employ either one or two engines to suit the requirements of the payload.[121,122,123] Later, the core would be fitted with strap-ons.[124] The plan was to introduce the Atlas IIAR in 1998, and phase out the Atlas II, Atlas IIA and Atlas IIAS by 2000. However, on reflection the company decided to build an additional eight Atlas IIAS vehicles to assure operations through 2002 just in case the development of its successor was delayed.[125] There were three candidates for the new engine: the RD-180, the NK-33, and an upgrade of the Rocketdyne MA-5A that was then in use.[126,127,128] As they all burned kerosene and oxygen, propellant was not a factor in the decision. However, Lockheed Martin had specified that the new engine should be throttleable down to 50 per cent to enable the new vehicle to serve as a 'one size fits all' launcher for a range of payload masses, and to give its payload a 'softer' ride. (Only two of the four strap-ons of the Atlas IIAS were ignited for lift-off, the others were air-started in order to limit the acceleration loads imposed on the payload.) The NK-33 and the RD-180 could be throttled, but the MA-5A could not.[129,130,131] In fact, Rocketdyne did not have the resources to devote to a major upgrade of the Atlas power plant, because it was building the throttleable RS-68 cryogenic engine for the Delta IV,[132] and in a letter dated 28 September it withdrew from the competition.[133] However, Lockheed Martin did not reveal this until the closing date for proposals on 16 November 1995,[134] the issue of concern presumably being that if the Russians discovered that they had a clear field, they might push for a better deal. The decision between the NK-33 and the RD-180 was to be made early in the new year. Meanwhile, there was another factor to be considered.

THE EVOLVED EXPENDABLE LAUNCH VEHICLE

In 1994 the Air Force conceived the Evolved Expendable Launch Vehicle (EELV) programme in order to replace the Delta, Atlas and Titan with a single family of launch vehicles that would satisfy its requirements through to 2020. One of the goals was to cut launch costs by 25–50 per cent compared to the Delta and Atlas. In November 1994 a cap of $2 billion was put on the programme.[135] The Request for Proposals was issued in May 1995.[136,137] As the White House desired a 'dual use' launcher that would be capable of competing commercially in addition to serving the government's needs, the proposers would *not* be obliged to work to full military specifications. In September the Air Force gave Lockheed Martin, McDonnell Douglas, Alliant Techsystems and Boeing $30 million each to submit concepts by September 1996.[138] In view of the longstanding rule that only 'All-American' rockets could launch government satellites, the Air Force warned that it would consider a proposal involving Russian engines only if these were manufactured in the USA.

Lockheed Martin selects the RD-180
In January 1996 Lockheed Martin selected the RD-180 to power the first stage of the Atlas IIAR.[139] The contract was contingent on the Russian government issuing an

Country	UNITED STATES			RUSSIA				FRANCE	CHINA					JAPAN
Reliability	Delta	Atlas	Titan 4	Tsykion	Soyuz	Proton	Zenit	Ariane 4	LM-4	LM-2C	LM-2E	LM-3	LM-3A	H-2
Total	94.6%	90%	94.4%	94.9%	93.1%	88.8%	78.9%	94.6%	100%	100%	75%	78%	100%	100%
	(212/224)	(226/251)	(170/180)	(224/236)	(1008/1083)	(190/214)	(15/19)	(35/37)	(2/2)	(14/14)	(3/4)	(7/9)	(1/1)	(2/2)
Last 20 Flights	100%	85%	90%	100%	100%	95%	78.9%	95%	100%	100%	75%	78%	100%	100%
	(20/20)	(17/20)	(18/20)	(20/20)	(20/20)	(19/20)	(15/19)	(19/20)	(2/2)	(14/14)	(3/4)	(7/9)	(1/1)	(2/2)
Last 5 Years	100%	84.2%	85.7%	97.8%	100%	95.6%	62.5%	93.8%	100%	100%	75%	75%	100%	100%
	(40/40)	(16/19)	(12/14)	(44/45)	(133/133)	(43/45)	(5/8)	(30/32)	(1/1)	(3/3)	(3/4)	(3/4)	(1/1)	(2/2)
Last 10 Years	98%	86.5%	86.7%	97.5%	100%	93.2%	78.9%	94.6%	100%	100%	75%	86%	100%	100%
	(48/49)	(32/37)	(26/30)	(117/120)	(339/339)	(96/103)	(15/19)	(35/37)	(2/2)	(9/9)	(3/4)	(6/7)	(1/1)	(2/2)

AVIATION WEEK & SPACE TECHNOLOGY/February 20, 1995

The world's launch vehicles circa 1995, as compiled by *Aviation Week & Space Technology* using data from the Department of Defense.

export licence. With 362,880 kilograms of thrust, the RD-180 was not only more powerful than the previous power plant, it also could maintain a constant specific impulse as it throttled down to 37 per cent. This flexible performance enabled it to be proposed for the EELV, which required a vehicle capable of placing payloads with masses ranging from 2,500 to 45,000 pounds into low orbit as well as being able to deliver 13,500 pounds directly into geostationary orbit.[140] Lockheed Martin praised the RD-180 as "literally the world's greatest rocket engine". Although it had a single combustion chamber this had a pair of nozzles, giving the impression that it was a twin engine. It operated at high pressure, but because it operated oxygen-rich it ran cooler than US high-performance engines – a factor that enhanced its reliability. The first test firing in the USA was on 15 November 1996.[141] In June 1997 Lockheed Martin placed a $1 billion order for 101 engines.[142] Energomash and Pratt & Whitney set up a joint venture to make it under licence in West Palm Beach, Florida, independently of the production plant at Khimsky near Moscow.[143,144]

The unlucky NK-33

The NK-33 engine was later chosen by Kistler Aerospace for its proposed two-stage reusable K-1 launch vehicle for placing lightweight satellites into low-to-medium orbit, but despite Space Systems/Loral booking 10 launches the venture stalled on the collapse of the market for satellite-based mobile communications.[145,146,147]

The rationale for a 'Heavy' EELV

In August 1996, the Commercial Space Transportation Advisory Committee of the Department of Transportation's Office of Commercial Space Transportation issued two reports.[148] First, it predicted a sharp increase in the size and number of satellites, and warned that unless the USA developed large launchers it would cede up to 50 per cent of the 'heavy comsat' market to Arianespace's Ariane V, China's Long March 3B, Russia's Proton and the Ukraine's Zenit 2. While it was true that Lockheed Martin was marketing the Proton and Boeing owned most of Sea Launch, these were considered to be 'foreign rockets'. The US workhorse for the heaviest of communications satellites was the Atlas II – that is, the Atlas–Centaur. The committee also urged the Air Force, in choosing the EELV, to select a launch vehicle that would be able to serve the heavy end of the commercial market, as opposed to simply choosing one that suited its own requirements. In addition, the National Reconnaissance Office came under pressure from Congress to desist from using a few very large satellites and instead to develop constellations of smaller ones.[149,150] Consequently, the rationale for the Heavy version of the EELV was shifting from supporting primarily government payloads to mainly commercial payloads.

EELV concepts

The four contenders submitted their detailed concepts for the EELV in September 1996. McDonnell Douglas proposed to use the RS-68 hydrogen-burning engine that Rocketdyne was developing for the Delta IV. This was the first large liquid-propellant engine development by the USA since the Shuttle. It was derived from the SSME but had fewer parts, which made it simpler and cheaper.[151,152,153] The second stage would use the Pratt & Whitney cryogenic engine that was already in development for the Delta III. Lockheed Martin proposed to use the RD-180 engine that it was introducing with the Atlas IIAR. Boeing proposed to develop a recoverable and reusable form of the Block-II SSME, and Alliant Techsystems based its design on the Solid Rocket Motor Upgrade that it had produced for the Titan IVB. Apart from the manufacturing cost, a critical factor for the Air Force was the time that it would take to prepare and launch a vehicle – that is, the operating cost – because time is most definitely money in the preparation of a launch vehicle.[154]

Boeing flexes its muscles

After developing the first stage of the Saturn V in the 1960s, Boeing withdrew from the launcher business, and supplied only upper stages – most notably the IUS for the Air Force. In 1994, soon after it was awarded the $6.3-billion prime contractorship to supply the hardware for the International Space Station, the company took the strategic decision to dominate space transportation as it does commercial aviation in the USA. Its first step was to create the Sea Launch consortium in April 1995, with a view to initiating operations in 1998. In early December 1996, while awaiting news of its EELV proposal, it made itself a major Space Shuttle contractor by completing the acquisition of Rockwell International's space operations, thereby gaining not only Rocketdyne, which built the SSMEs, but also a half share in the United Space Alliance, which was a joint venture with Lockheed Martin that serviced the Shuttle. One week later, on 15 December, Boeing announced its intention to buy McDonnell Douglas for $13.3 billion. On 4 August 1997, when this was finalised, Boeing not only augmented its International Space Station contract by the $2.1 billion worth of work that had been allocated to McDonnell Douglas – thereby making itself the principal player in human spaceflight – but its acquisition of the Delta II also gave it a solid presence in the medium-lift market.[155,156]

EELV finalists

On 20 December 1996, the Air Force issued McDonnell Douglas and Lockheed Martin each $60 million for pre-development work on their proposals. The schedule called for a decision in June 1998 as to whether the Delta IV or the Atlas V would receive the 8-year production contract worth $1.5 billion for sufficient vehicles for 20 years of government launches.[157] The fact that Boeing's proposal was rejected was not a setback since the company was in the process of buying McDonnell Douglas. The real loser was Alliant Techsystems.

Two are better than one

As a result of the robust commercial launcher market, in November 1997 the Air Force decided to subsidise the development of the Delta IV *and* the Atlas V and share the production order. In the expectation that the companies would recoup some of their costs from commercial operations, and hence be able to sell their vehicles to the Air Force at a lower price, it reduced the value of the contract to about $1 billion.[158] This twin-track approach also offered the advantage of having a backup in case one of the vehicles was to be delayed in development or temporarily grounded in service.[159]

Even if a 'winner' had been chosen in 1998, the 'loser' would very likely have been able to raise the funding to complete the development and operate independently of the EELV contract.[160] Indeed, in 1997 this might have been seen as the *safer* option, because the 'winner' would have been concerned by the prospect of the Air Force contract being axed due to budgetary constraints!

On 14 October 1999 the mobile service structure and umbilical tower at Pad 41 were demolished.[161] Since its introduction in 1965, this pad had dispatched 17 Titan III and 10 Titan IV missions. By early 2000 the assembly of new 300-foot-tall mobile service structures was underway on this pad for the Atlas V, and on Pad 37 (which was formerly a Saturn IB facility) for the Delta IV.

ATLAS III

In April 1998 Lockheed Martin renamed the Atlas IIAR the Atlas III.[162] It had been hoped to make the first flight before the end of 1999, but the manufacturing fault in the RL-10 engine (revealed by a Delta III on 4 May 1999) meant that the Centaur stage of the Atlas was also grounded for four months.[163,164] Meanwhile, the Atlas IIIA that had been undergoing test on Pad 36B in May for launch in June was removed.[165,166] The plan was to re-erect it in January 2000 and launch it a few months later with Eutelsat W4, which was a Spacebus 3000 built by Alcatel.[167,168] It took five attempts, but it was finally launched on 24 May.[169] The throttleable RD-180 proved its versatility: it lifted off at 74 per cent thrust in order to minimise any damage to the pad, and as it cleared the umbilical tower it increased its power to 92 per cent. Its 'stately' lift-off was reminiscent of a Saturn V. At $T + 33$ seconds it throttled down to 64 per cent while the vehicle went supersonic, and 30 seconds later, once it was safely through Max-Q, it increased to 87 per cent. Once it had achieved an acceleration of 5.5 g, the engine slowly tailed off to maintain this load. As it was more powerful, the first stage shut down at $T + 3$ minutes, which was some 2 minutes earlier than the Rocketdyne-powered variant.[170] The nozzle bell of the single-chamber RL-10 on the Centaur was cranked down without incident.[171],[172] This marked the 50th consecutive success for the Atlas. There was a rich irony in the fact that a vehicle which had been conceived to strike at the Soviet Union was now powered by a Russian engine, and the 'enemy' had become Boeing!

With a single-chamber Centaur, the Atlas IIIA could insert a payload of 4 tonnes into geosynchronous transfer orbit. The two-chamber Centaur would increase this by half a tonne. The first Atlas IIIB had been expected to be launched by the end of 2001, but it was delayed, and did not depart from Canaveral with EchoStar 7 until 21 February 2002.[173],[174] The next launch on 12 April 2003 was also an Atlas IIIB, and it deployed AsiaSat 4. On 18 December an Atlas IIIB deployed the 11th HS-601-based UFO satellite for the Navy.[175] Next was an Atlas IIIA on 13 March 2004 with a satellite for the Mobile Broadcasting Corporation of Japan. In contrast to the Delta III, the Atlas III was proving itself to be a reliable launcher.

There was about 80 per cent commonality of systems between the Atlas III and the 'common booster core' that was being developed for the Atlas V.[176] (There was no Atlas IV, perhaps because the company saw the new vehicle as a successor to the Titan IV.[177]) Nevertheless, the Atlas III was a transitionary vehicle in that it had the original 'balloon' tankage – the new structurally stable tankage would be introduced with the Atlas V.[178] In mid 2000, the Air Force decided that the Heavy EELV rôle would be served by the Delta IV.[179],[180] Nevertheless, Lockheed Martin continued with the development of the Atlas V-Heavy in order have a vehicle to match the upgraded Ariane V. ILS planned to phase out the Atlas III as soon as the Atlas V entered commercial service, which was expected to be in 2003–2004.[181]

The first Atlas IIIA lifts off on 24 May 2000 with Eutelsat W4.

ATLAS V

There were high hopes for the Atlas V as a commercial launcher. "Reliability is the key buying criterion," observed ILS president, Mark Albrecht.[182] Khrunichev expressed concern that ILS would support its Atlas V at the expense of the Proton – the shrinking commercial market was undermining the production line.[183] While it would be cheaper to import the RD-180 engines from Russia, these would have to be made in the USA to achieve the 'All-American' status required for EELV contracts for government payloads. Pratt & Whitney had intended to start production at its plant in Florida as soon the Atlas V entered service, but when it became clear that the vehicle would initially be operated as a commercial launcher, local engine production was postponed to 2007.[184]

The first Atlas V arrived at Canaveral in June 2001 amid speculation as to whether it would be able to be launched ahead of the Delta IV.[185,186,187] In early 2002 it was rolled out, but its launch was slipped from the planned 9 May to 29 July. The launch was again postponed to late August to allow further tests of the umbilical retraction system, and Countdown Demonstration Tests were conducted in March, May and July.[188,189,190,191] On 21 August 2002 it was successfully launched from Pad 41 carrying Hot Bird 6 for Eutelsat.[192,193] This was the *sixth of six* Atlas variants to have a successful maiden launch. On 13 May 2003 the second Atlas V carried Hellas-Sat for the eponymous Greek Cypriot-led consortium.[194,195]

Phasing out the Atlas II
The final Atlas II was launched on 16 March 1998, the last Atlas IIA on 4 December 2002, and the last Atlas IIAS on 31 August 2004.[196,197,198] Since 1991, 63 Atlas II, IIA and IIAS vehicles had carried satellites for commercial customers, the Department of Defense and NASA.[199] "This is an awesome accomplishment, being the only US expendable launch vehicle series to have had 100 percent success throughout its entire lifespan," observed ILS president Mark Albrecht.

DELTA IV

Boeing's plan to introduce the Delta IV in April 2001 was dashed by the protracted development of the RS-68 engine, which was late partly as a result of the design-to-cost approach.[200,201] At 109 per cent of its nominal performance, the SSME gave a thrust of 418,660 pounds with a combustion chamber pressure of 3,000 psia. The RS-68 for the Delta IV had a thrust of 650,000 pounds at 100 per cent – making it the most powerful hydrogen-burning engine yet developed – at the relatively modest pressure of 1,410 psia, and when it was throttled down to 59 per cent this dropped to a benign 836 psia.[202,203] In a static test of the RS-68 at the Stennis Space Center, a blade in the fuel pump failed due to high cycle fatigue, but the engine shut down safely.[204] In early 2001 Boeing integrated its Delta II, III and IV programmes into a single organisation "to bring production into a single entity to ensure efficient management".[205] In late 2001 the launch was scheduled for 30 April 2002,[206] and if a

The inaugural Atlas V lifts off on 21 August 2002 with Hot Bird 6.

On 17 July 2003 an Atlas V augmented by two strap-on rockets lifts off with Rainbow 1.

The inaugural Delta IV lifts off on 20 November 2002 with Eutelsat W5.

A Delta IV lifts off on 29 August 2003 with DSCS-III-B6.

commercial satellite could not be signed up then it was to fly with a dummy payload.[207,208,209,210] On 30 April 2002 the first vehicle was set up on Pad 37B.[211] The launch was first slipped to July, and then to late October in order to provide more time to prepare the vehicle.[212,213] Meanwhile, Boeing decided that the Delta III would not fly again – in fact, four had been cannibalised for parts in support of the Delta II – and the commercial payloads earmarked for the Delta III were to be transferred to the Delta IV.[214] A software glitch in the final minutes of a Countdown Demonstration Test on 30 August prompted a slip to November.[215] The first Flight Readiness Firing was successfully conducted on 14 October.[216] When launched on 20 November 2002 it successfully deployed Eutelsat W5. In this initial configuration, the core was augmented by a pair of GEM-60 strap-ons.[217] Turning to the EELV contracts, on 10 March 2003 the second Delta IV dispatched DSCS-III-A3 to geostationary orbit, and on 29 August 2003 the third launch added DSCS-III-B6 to complete that constellation.[218,219,220] Both of these missions used the 'clean' core, without strap-ons. This configuration could insert just over 8 tonnes into low orbit or 4.2 tonnes into geosynchronous transfer orbit, and the use of GEM-60 strap-ons raised these figures to 11.5 tonnes and 6.5 tonnes respectively.

By this point, the Air Force had delayed the introduction of the Atlas V as an EELV, using the Delta IV instead.[221] In effect, because the depressed market was insufficient to support both, one was attempting to operate commercially while the other was providing government-funded launches. However, the Atlas V did pick up one EELV contract. On 3 September 1999 the National Reconnaissance Office had selected Boeing to supply the next generation of imaging satellites – a decision that shocked Lockheed Martin, which had led the development of classified imaging satellites since 1958.[222] In accordance with the new mantra, the latest satellites were to be smaller and cheaper. In 2004 the Air Force awarded ILS a contract for a launch in 2006.[223] This was to penalise Boeing for a case of "unethical conduct" in bidding for the initial EELV contract, when the company had exploited information improperly gained from Lockheed Martin.[224,225]

Meanwhile, the inaugural flight of the Delta IV-Heavy had slipped into 2003, and it had yet to be decided whether it would carry a dummy payload.[226,227] This had three 47-metre-long common booster cores arranged in parallel, and a second stage powered by an uprated RL-10-B-2 engine. With a capability of placing 23 tonnes into low orbit or 13 tonnes into geosynchronous transfer orbit, it was to supersede the Titan IVB. In addition to Pad 37 at Canaveral, this configuration was to use SLC-6 at Vandenberg – one of the *least used* pads in history.[228,229] By the end of 2003 the launch had been postponed to May 2004.[230] No sooner had this been slipped to September than the Cape suffered three hurricanes in the space of a few weeks and all operations were temporarily halted. After a series of technical issues, the vehicle was dispatched on 21 December, but the fact that all three main stages shut down 8 seconds early due to a common fault involving propellant quantity sensors, obliged the second stage to extend its parking-orbit burn, with the result that it ran dry attempting to place its payload (which was, after all, a 6.5-tonne demonstration satellite) into geostationary orbit.[231]

Preparing the inaugural Delta IV-Heavy on the Pad 37B at Canaveral.

The Delta IV-Heavy lifts off on 21 December 2004.

PHASING OUT THE TITAN IVB

The Titan IVB was to be retired once the Delta IV-Heavy became available, but the development of this vehicle was running late, and in early 2001 the Air Force decided to postpone the phase-out of the Titan IVB beyond 2002.[232] With Pad 41 being rebuilt for the Atlas V, the Titan IVB was using Pad 40 at Canaveral and SLC-4E at Vandenberg, and despite the improvements to simplify the preparation of this vehicle the turnaround between launches on a given pad was still no shorter than six months. Lockheed Martin shipped its last Titan IVB core in 2002.[233] The vehicle that lifted off on 14 February 2004 had the final IUS and the penultimate DSP satellite.[234] This was the 37th launch, of which 26 had departed from Canaveral and 11 from Vandenberg. The *final* Titan IVB was scheduled for Vandenberg in 2005. The final DSP was to ride a Delta IV in 2005.

The verdict on the Titan IV was that it was "one of the Air Force's most successful" launch vehicles. However, it had not yielded the expected operational efficiency. Instead of each launch costing about $100 million, it had averaged five times that amount, which verged on the cost of launching a Shuttle.[235] In part, this failure to drive down costs was a consequence of so few vehicles being ordered, which in turn reflected the decline in the demand for heavy military satellites after the end of the Cold War, and the fact that those that were sent up proved to be long-lived. Perhaps the most damning incident was that – despite the hope of achieving a rapid turnaround – when one vehicle sat on the pad for over 1,000 days a frustrated Air Force commander threatened to mount a plaque on it that added up the $3.5-million-per-day bill to the US taxpayer. It was hoped that the new heavyweights would prove more cost-effective than their predecessors.

NOTES

1. *Flight International*, 17–23 October 1990, p. 39.
2. *Aviation Week & Space Technology*, 4 April 1994, p. 48.
3. *Aviation Week & Space Technology*, 4 April 1994, p. 55.
4. *Aviation Week & Space Technology*, 20 February 1995, p. 48.
5. *Flight International*, 14–20 June 1995, p. 61.
6. *Aviation Week & Space Technology*, 12 June 1995, p. 115.
7. *Aviation Week & Space Technology*, 21 March 1994, p. 25.
8. *Flight International*, 23–29 April 1997, p. 25.
9. *Aviation Week & Space Technology*, 6 May 1996, p. 60.
10. *Aviation Week & Space Technology*, 4 April 1994, p. 45.
11. *Flight International*, 16–22 November 1994, p. 22.
12. *Flight International*, 17–23 May 1995, p. 18.
13. *Flight International*, 21–27 June 1995, p. 17.
14. *Flight International*, 4–10 October 1995, p. 6.
15. *Aviation Week & Space Technology*, 25 March 1996, p. 48.
16. *Aviation Week & Space Technology*, 6 May 1996, p. 60.
17. *Aviation Week & Space Technology*, 25 March 1995, p. 51.

18. *Aviation Week & Space Technology*, 5 August 1996, p. 74.
19. *Report by the Inquiry Board into the failure of Ariane V flight 501*, Arianespace, Paris, 19 July 1996.
20. *Aviation Week & Space Technology*, 24 June 1996, p. 77.
21. *Aviation Week & Space Technology*, 29 July 1996, p. 33.
22. *Flight International*, 23–29 April 1997, p. 26.
23. *Spaceflight*, May 2003, p. 196.
24. *Aviation Week & Space Technology*, 9 September 1996, p. 79.
25. *Aviation Week & Space Technology*, 16 September 1996, p. 55.
26. *Aviation Week & Space Technology*, 23 June 1997, p. 26.
27. *Aviation Week & Space Technology*, 27 January 1997, p. 62.
28. *Aviation Week & Space Technology*, 23 June 1997, p. 26.
29. *Spaceflight*, January 1998, p. 13.
30. *Flight International*, 9–15 April 1997.
31. *Flight International*, 23–29 April 1997, p. 25.
32. *Aviation Week & Space Technology*, 31 August 1998, p. 29.
33. *Aviation Week & Space Technology*, 13 December 1999, p. 61.
34. *Spaceflight*, July 2003, p. 280.
35. *Aviation Week & Space Technology*, 21 June 1999, p. 46.
36. *Spaceflight*, August 1999, p. 312.
37. *Spaceflight*, April 2003, p. 137.
38. *Aviation Week & Space Technology*, 14 June 1999, p. 212.
39. *Spaceflight*, July 2001, p. 5.
40. *Spaceflight*, March 2001, p. 94.
41. *Spaceflight*, October 2001, p. 400.
42. *Spaceflight*, March 2002, p. 92.
43. *Spaceflight*, May 2002, p. 182.
44. *Spaceflight*, May 2002, p. 182.
45. *Spaceflight*, July 2002, p. 268.
46. *Aviation Week & Space Technology*, 28 June 1999, p. 37.
47. *Spaceflight*, August 2001, p. 314.
48. *Spaceflight*, March 2002, p. 92.
49. *Spaceflight*, May 2002, p. 182.
50. http://www.astronautix.com/lvs/arie5eca.htm
51. *Spaceflight*, March 2002, p. 99.
52. *Spaceflight*, February 2002, p. 50.
53. *Spaceflight*, October 2001, p. 406.
54. *Spaceflight*, February 2003, p. 54.
55. *Spaceflight*, June 2004, p228.
56. *Spaceflight*, July 2004, p. 273.
57. http://www.planetary.org/news/2004/rosetta_launch-delay1.html
58. *Spaceflight*, May 2003, p. 187.
59. *Spaceflight*, June 2003, p. 226.
60. *Spaceflight*, September 2003, p. 357.
61. *Spaceflight*, April 2003, p. 137.
62. *Spaceflight*, August 2003, p. 314.
63. http://news.bbc.co.uk/1/hi/sci/tech/3747808.stm
64. http://news.bbc.co.uk/1/hi/sci/tech/4054329.stm
65. *Spaceflight*, June 2003, p. 226.

66. *Spaceflight*, July 2003, p. 268.
67. *Spaceflight*, July 2004, p. 273.
68. *Aviation Week & Space Technology*, 15 May 1995, p. 28.
69. *Space News*, 15–21 May 1995.
70. *Aviation Week & Space Technology*, 5 June 1995, p. 68.
71. *Aviation Week & Space Technology*, 5 August 1996, p. 55.
72. *Flight International*, 12–18 July 1995, p. 18.
73. *Aviation Week & Space Technology*, 11 March 1996, p. 64.
74. *Spaceflight*, November 1998, p. 440.
75. *Spaceflight*, February 2000, p. 47.
76. *Spaceflight*, January 1999, p. 9.
77. http://www.boeing.com/defense-space/space/delta/delta3/d3inv.htm
78. *Aviation Week & Space Technology*, 10 May 1999, p. 30.
79. *Spaceflight*, September 1997, p. 357.
80. *Spaceflight*, January 2000, p. 5.
81. *Spaceflight*, October 1999, p. 406.
82. *Delta 269 (Delta III) Investigation Report*, Boeing, 16 August 2000.
83. http://www.boeing.com/defense-space/space/delta/delta3/d3_report.pdf
84. *Aviation Week & Space Technology*, 1 November 1999, p. 31.
85. *Aviation Week & Space Technology*, 27 September 1999, p. 38.
86. *Aviation Week & Space Technology*, 1 November 1999, p. 31.
87. *Aviation Week & Space Technology*, 9 August 1999, p. 79.
88. *Aviation Week & Space Technology*, 22 November 1999, p. 17.
89. *Spaceflight*, February 2000, p. 47.
90. Note that Orion 2 was scheduled for October 1999 on an Ariane 4; that is, they were going up out of sequence.
91. *Spaceflight*, November 1999, p. 444.
92. *Aviation Week & Space Technology*, 16 November 1992, p. 22.
93. *Aviation Week & Space Technology*, 23 August 1999, p. 36.
94. *Aviation Week & Space Technology*, 5 July 1999, p. 22.
95. *Spaceflight*, December 1999, p. 491.
96. *Aviation Week & Space Technology*, 19 July 1999, p. 25.
97. *Spaceflight*, December 1999, p. 491.
98. *Aviation Week & Space Technology*, 1 January 2000, p. 38.
99. *Spaceflight*, June 2000, p. 225.
100. *Spaceflight*, July 2000, p. 268.
101. *Spaceflight*, September 2000, p. 355.
102. *Spaceflight*, November 2000, p. 445.
103. *Spaceflight*, January 2001, p. 5.
104. *Spaceflight*, December 2000, p. 491.
105. *Spaceflight*, September 1999, p. 357.
106. *Aviation Week & Space Technology*, 14 August 1995, p. 52.
107. *Aviation Week & Space Technology*, 27 June 1993, p. 22.
108. *Flight International*, 8–14 November 1995, p. 32.
109. *Challenge to Apollo*, A.A. Siddiqi, NASA, 2000.
110. *Aviation Week & Space Technology*, 30 March 1992, p. 21.
111. *Aviation Week & Space Technology*, 25 October 1993, p. 29.
112. *Aviation Week & Space Technology*, 24 October 1994, p. 25.
113. *Aviation Week & Space Technology*, 24 April 1995, p. 40.

114. *Flight International*, 4–10 October 1995, p. 32.
115. *Aviation Week & Space Technology*, 14 August 1995, p. 51.
116. *Aviation Week & Space Technology*, 24 April 1995, p. 46.
117. *Aviation Week & Space Technology*, 14 August 1995, p. 52.
118. The advantage of the RD-180 for the Atlas, would be that it would need only one engine, rather than two NK-33s.
119. *Aviation Week & Space Technology*, 7 April 1997, p. 41.
120. *Aviation Week & Space Technology*, 6 November 1995, p. 63.
121. *Flight International*, 15–21 November 1995, p. 20.
122. Since its development in the early 1960s, the RL-10 had been continually upgraded, and included the RL-10A-3-3A (Titan–Centaur), the RL-10A-4 (Atlas II), the RL-10A-4-1 (Atlas IIAS, III and V) and the RL-10B-2 (Delta III and V).
123. http://www.spaceandtech.com/spacedata/engines/rl10_sum.shtml
124. *Aviation Week & Space Technology*, 5 August 1996, p. 18.
125. The production of eight Atlas IIAS as a contingency was in addition to the 62 that had been built over the years.
126. *Aviation Week & Space Technology*, 12 September 1994, p. 55.
127. *Aviation Week & Space Technology*, 6 November 1995, p. 65.
128. *Aviation Week & Space Technology*, 5 December 1994, p. 50.
129. http://www.spaceandtech.com/spacedata/engines/nk33_specs.shtml
130. http://www.spaceandtech.com/spacedata/engines/rd180_specs.shtml
131. http://www.boeing.com/defense-space/space/propul/atlas.html
132. *Flight International*, 4–10 October 1995, p. 32.
133. *Aviation Week & Space Technology*, 20 November 1995, p. 35.
134. *Spaceflight*, October 2001, p. 406.
135. *Aviation Week & Space Technology*, 14 November 1994, p. 63.
136. *Aviation Week & Space Technology*, 15 May 1995, p. 30.
137. *Spaceflight*, November 2002, p. 450.
138. *Flight International*, 6–12 September 1995, p. 19.
139. *Flight International*, 24–30 January 1996, p. 22.
140. *Aviation Week & Space Technology*, 23 June 1997, p. 23.
141. *Aviation Week & Space Technology*, 25 November 1996, p. 17.
142. *Aviation Week & Space Technology*, 23 June 1997, p. 23.
143. *Aviation Week & Space Technology*, 10 February 1997, p. 101.
144. *Aviation Week & Space Technology*, 7 April 1997, p. 40.
145. *Aviation Week & Space Technology*, 7 April 1997, p. 41.
146. http://www.astronautix.com/lvs/kislerk1.htm
147. http://www.kistleraerospace.com/
148. *Aviation Week & Space Technology*, 5 August 1996, p. 18.
149. *Aviation Week & Space Technology*, 26 February 1996, p. 62.
150. *Aviation Week & Space Technology*, 8 July, 1996, p. 24.
151. *Flight International*, 15–21 January 1997, p. 22.
152. http://www.spaceandtech.com/spacedata/engines/rs68_sum.shtml
153. http://www.boeing.com/defense-space/space/propul/RS68.html
154. *Aviation Week & Space Technology*, 6 January 1997, p27.
155. *Aviation Week & Space Technology*, 23/30 December 1996, p. 13.
156. In the four-digit scheme used for the Delta variants, the digits specify in turn the series, the number of strap-ons, the second stage motor ('0' indicating an AJ-10-118F, '1' a TR-201, '2' a cryogenic stage that was not pursued, and '3' an AJ-10-118K) and the third

stage ('0' indicating no third stage, '2' an FW-4D, '3' a Star 37D, '4' a Star 37E, '5' a Star 48B or '6' a Star 37FM). The Delta III became the 8000 series, with a '2' for the cryogenic second stage.

157. *Flight International*, 15–21 January 1997, p. 22.
158. *Spaceflight*, November 2002, p. 450.
159 *Aviation Week & Space Technology*, 9 March 1998, p. 24.
160. *Aviation Week & Space Technology*, 13 December 1999, p. 54.
161. *Aviation Week & Space Technology*, 18 October 1999, p. 22.
162. http://www.spaceline.org/rocketsum/Atlas-IIIa.html
163. *Aviation Week & Space Technology*, 9 August 1999, p. 79.
164. *Aviation Week & Space Technology*, 27 September 1999, p. 38.
165. *Aviation Week & Space Technology*, 19 July 1999, p. 25.
166. *Aviation Week & Space Technology*, 25 October 1999, p. 23.
167. *Aviation Week & Space Technology*, 27 September 1999, p. 38.
168. *Aviation Week & Space Technology*, 22 November 1999, p. 52.
169. http://www.spaceandtech.com/digest/flash-articles/flash2000-026.shtml
170. *Aviation Week & Space Technology*, 29 May 2000, p. 28.
171. *Spaceflight*, July 2000, p. 312.
172. *Aviation Week & Space Technology*, 22 May 2000, p. 35.
173. *Spaceflight*, May 2001, p. 182.
174. ILS had previously launched satellites for the EchoStar Communications Corporation on Atlas II and Proton rockets.
175. *Spaceflight*, March 2004, p. 96.
176. *Spaceflight*, July 2000, p. 312.
177. http://www.space.com/businesstechnology/technology/atlas_delta_020320-1.html
178. *Spaceflight*, February 2002, p. 50.
179. *Spaceflight*, June 2000, p. 225.
180. *Spaceflight*, January 2001, p. 5.
181. *Spaceflight*, October 2001, p. 406.
182. *Spaceflight*, October 2001, p. 403.
183. *Spaceflight*, August 2002, p. 316.
184. *Spaceflight*, July 2002, p. 274.
185. *Spaceflight* August 2001, p. 316.
186. *Spaceflight*, November 2002, p. 450.
187. *Spaceflight*, February 2002, p. 50.
188. *Spaceflight*, June 2002, p. 224.
189. *Spaceflight*, August 2002, p. 316.
190. *Spaceflight*, September 2002, p. 357.
191. *Spaceflight*, November 2002, p. 450.
192. http://www.ilslaunch.com/atlas/atlasv/
193. Hot Bird 6 was an Alcatel Spacebus 3000B3 with Ka-Band and Ku-Band transponders for TV and radio broadcasting. It joined its partners to broadcast to Europe, North Africa and the Middle East.
194. *Spaceflight*, August 2003, p. 313.
195. Hellas-Sat was an Astrium-built Eurostar 2000-Plus bus.
196. *Spaceflight*, March 2003, p. 99.
197. http://www.astronautix.com
198. http://www.ilslaunch.com/atlas/atlasii/
199. http://www.spaceflightnow.com/atlas/ac167/

200. *Aviation Week & Space Technology*, 13 December 1999, p. 54.
201. *Aviation Week & Space Technology*, 5 June 2000, p49.
202. *Aviation Week & Space Technology*, 5 June 2000, p. 49.
203. *Aviation Week & Space Technology*, 19 July 1999, p. 28.
204. *Spaceflight*, April 2001, p. 138.
205. *Spaceflight*, May 2001, p. 182.
206. *Spaceflight*, January 2002, p. 16.
207. If the first Delta IV-Medium was successful, the second launch would carry DSCS-III-A3 for the Air Force in 2002 and the third, later in 2002, would be commercial. Next in line was to be the first test of the Delta IV-Heavy, hopefully with a commercial customer, followed in 2003 by the second, this time from Vandenberg, that would place the DMSP 17 metsat into polar orbit for the Air Force.
208. *Spaceflight*, July 2001, p. 272.
209. *Spaceflight*, October 2001, p. 403.
210. *Spaceflight*, October 2001, p. 406.
211. *Spaceflight*, November 2002, p. 446.
212. *Spaceflight*, July 2002, p. 274.
213. *Spaceflight*, September 2002, p. 357.
214. *Spaceflight*, June 2002, p. 229.
215. *Spaceflight*, November 2002, p. 446.
216. http://www.spaceflightnow.com/delta/delta4/021112rs68/
217. http://www.astronautix.com/lvs/delium42.htm
218. *Spaceflight*, November 2003, p. 445.
219. http://www.forrelease.com/D20030829/nyf062.P1.08292003203909.02852.html
220. http://www.spaceflightnow.com/delta/d300/
221. *Spaceflight*, October 2003, p. 402.
222. *Aviation Week & Space Technology*, 13 September 1999, p. 25.
223. *Spaceflight*, March 2004, p. 97.
224. *Spaceflight*, June 2004, p. 228.
225. *Spaceflight*, December 2004, p.453.
226. *Spaceflight*, July 2002, p. 269.
227. *Spaceflight*, January 2001, p. 5.
228. *Flight International*, 11–17 January 1995, p. 43.
229. SLC-6 was built in the 1960s for the Titan IIIM for the Air Force's Manned Orbiting Laboratory, but this was cancelled in 1969. The facility was rebuilt for the Space Shuttle, but launches from Vandenberg were cancelled in 1986 after the loss of Challenger from Canaveral.
230. *Spaceflight*, November 2003, p. 445.
231. http://www.spaceflightnow.com/delta/d310/
232. *Spaceflight*, May 2001, p. 184.
233. *Spaceflight*, July 2002, p. 269.
234. *Spaceflight*, April 2004, p. 139.
235. http://www.floridatoday.com/news/space/stories/2004a/021404titan.htm

Part Two
Satellites and space probes

9

Failure and redundancy

TYPES OF FAILURE

The term 'anomaly' is widely used in spacecraft operations to denote the unexpected behaviour of a system. An anomaly sometimes indicates an extant or imminent failure, but not always. In this book, we generally restrict our attention to cases where something actually goes seriously wrong. Not all failures are fatal to the spacecraft,[1] and, in fact, the most impressive feats of space systems engineering tend to be those that overcome failures. Impressive though such recoveries may be, they nevertheless represent undesirable conditions, and are usually costly.

For purposes of discussion we have categorised failures, but it must be recognised that these distinctions are somewhat arbitrary, and depend on context. For example, one study defined radiation-induced failures in spacecraft prior to 1972 as environmental failures, and as design failures thereafter – in fact, the hazard due to the van Allen radiation belts was evident all along.[2] A component might fail because it is too hot, and its thermal condition exceeds its tolerance. In some cases, such as an external sensor on a planetary probe, it might be too hot because its environment is hotter than expected. In others, the temperature of the component is to be maintained within tolerable limits by the thermal control system, in which case it fails as a result of the underperformance (i.e. failure) of the thermal control system, which in turn may be the result of a design failure. Furthermore, failures propagate, and thus may straddle a number of categories. For example, the Mars Observer spacecraft probably suffered structural failure as a result of a rapid spin, but in this scenario (others scenarios are possible, because the cause of a failure is rarely determined with certainty) this spin was caused by rapid venting of propellant and pressurant through a hole in a pipe, making it a propulsion system failure. The hole was caused by fuel and oxidiser inadvertently accumulating and mixing, in turn due to the plumbing being colder than originally designed. Thus the root cause of the failure may be considered to be a thermal design failure. While in some sense the root cause is the most important classification because it marks the start of the chain, it is often useful to bear in mind the downstream failures, since these may represent areas where a more robust design could arrest the propagation of failure and prevent loss of the mission. An environment failure occurs when the conditions in which the system is to perform are inadequately predicted. Radiation and spacecraft charging hazards are typical examples for satellites around the Earth, as are the atmospheres of planets for deep space missions. The operating environment is usually defined in the spacecraft's system specification as a model – often provided by the scientific community from available data – in the form of a set of curves, numerical tables, or software, usually with a nominal expectation and the minimum and maximum profiles which the environment is not expected to exceed. Occasionally, however, the environment might exceed these expectations, and even though a system satisfies its specification, it fails due to the specification being inadequate. Although a spacecraft is principally designed for the space environment, it must survive launch. If the acceleration, vibration and acoustic environment of a launcher exceeds the specification against which a spacecraft has been designed, then, although the root cause may pertain to the launch vehicle, it nevertheless represents an environment failure from the point of view of the payload. Since there is little protection against environment failure, one strategy is to acknowledge the uncertainties in the validity of the model, and incorporate large margins in the design; another strategy is to apply systems engineering tactics towards the models and seek independent evaluations (i.e. redundancy). A design failure occurs when a system fails despite the fact that the environment remains within expectations and each subsystem functions correctly, the failure often being the result of unexpected *interactions* between subsystems. These most insidious of failures are only preventable by careful design, simulation and testing.

In 1992 a survey was published of 2,500 spacecraft failures between 1962 and 1988.[3] The pattern of failures by subsystem to some extent reflects the relative complexity and testability of the elements. That is, structures are usually straightforward to verify, and mechanical engineering is, by and large, a well-established discipline. On the other hand, the myriad range of states of command computers and attitude control systems are difficult to define and test methodically, with the result that these tend to be responsible for the majority of the failures.

Table – Types of failure

Cause	%
Assigned	
Design	24.8
Environment	21.4
Operations	4.7
Random	
Parts	16.3
Quality	7.7
Other	6.3
Unknown	18.9

Table – Distribution of failures

	%
Spacecraft Bus (total 73.3)	
Telemetry, command and control	24.6
Guidance and navigation	13.6
Electric power	13.2
Data system	9.1
Thermal control	5.6
Propulsion	3.7
Structures	3.5
Payload (total 26.7)	
Visual/IR sensors	13.1
Communications payload	5.2
Special payloads	4.9
Navigation payloads	3.5

REDUNDANCY

Some failures may be expected, and indeed quantitative expectation of component failure is the basis for the rational application of 'redundancy'. Just as a human can live with only one kidney, evolution has refined the design to furnish each individual with a pair of kidneys. This decision has an evolutionary cost (growing and nourishing a kidney draws metabolic activity at the expense of other growth) but it conveys an advantage by offering a superior peak processing capability (but one that kidney donors evidently can manage without) and, more particularly, superior reliability. Spacecraft design engineers often have to trade-off the consumption of resources against the increase in reliability of incorporating two units for a given subsystem. The calculus of redundancy relies on the assumption of *uncorrelated failure* – that systems fail at a random time, if not from a random mechanism, and that there is no strong expectation of both systems failing at the same time from the same mechanism. It is acknowledged, however, that this assumption does not always hold! This is particularly so with environment failures. Some radiation failures are random, but others relate to total dose effects, and redundancy will obviously offer little protection if both units are exposed to the same hazardous dose; and similarly, if both units are too hot or too cold, both may fail. The configuration of redundant units requires care.

Often systems may be operated in parallel without mutual interference (for example, telemetry systems can transmit the same data at different frequencies) but in other cases this is not so (for example, if two or more control systems are run in parallel they may well provide signals that conflict). The challenge with dual control systems is to identify a failure: if units have differing signals, which one is incorrect? One approach is to provide *three* units and use a voting logic in which the signal from one unit is rejected if it differs from the other two. Redundancy therefore imposes complexity. A notable failure of this kind of arrangement was that of Phobos 2. This spacecraft was launched by the Soviet Union in 1988 with computers that proved to be faulty. One processor expired *en route* and another began to malfunction intermittently soon after the spacecraft entered orbit around Mars.[4] Unfortunately, the voting logic was not itself reprogrammable, and with two failed units voting 'no' the sole functioning processor was unable to make its 'yes' vote heard! On 27 March 1989, as the spacecraft was manoeuvring in the vicinity of the eponymous Martian moon, it failed to commence a planned downlink. Some fragmental signals were subsequently received which indicated that the supposedly 3-axis stabilised spacecraft was spinning in a state of uncontrolled precession, and beyond recovery. In one sense, the proximate cause was component failure but the inflexibility of the voting logic contributed to the loss of the spacecraft by permitting correlated failures to accumulate. Of course, the third processor may well have failed long before the spacecraft was able to complete its mission.

An artist's impression of the Phobos 2 spacecraft manoeuvring in close proximity to Mars's primary moon.

Our perception of spacecraft failures – like that of other disasters such as air crashes or earthquakes – tends not to be completely rational. Although our attention may be caught by some high-profile failures at any given moment, the fact is that broadly speaking the number of failures is decreasing, and their significance is generally becoming less serious.[5] Clearly some lessons are being learned, but others are not, and we hope this book will help to improve the corporate memory of the space community.

NOTES

1. For example, of 50 anomalies recorded in Goddard Space Flight Center spacecraft in 1989, only one was catastrophic and one was 'major': 30 per cent were classed as 'minor' and the rest as 'negligible', but even these impose a mission cost and must be documented because they may indicate incipient failure. *Orbital Anomalies in Goddard Spacecraft for CY 1989*, W.G. Elsen, GSFC Office of Flight Assurance.
2. 'Reliability during space mission concept exploration', H. Hecht, in *Space Mission Analysis and Design*, W.J. Larson and J.R. Wertz (eds), Microcosm, 1992.
3. 'Reliability during space mission concept exploration', H. Hecht, in *Space Mission Analysis and Design*, W.J. Larson and J.R. Wertz (eds), Microcosm, 1992.
4. *The Making of a Soviet Scientist*, R. Sagdeev, Wiley, 1994, p. 317.
5. Appendix B to 'The cosmos on a shoestring: small spacecraft for space and Earth science', L. Sarsfield, RAND Document No: MR-864-OSTP, ISBN: 0833025287, 1998.

10

Propulsion system failures

THE DIFFICULTY OF TESTING

A liquid-propellant rocket motor will force a fuel and an oxidiser into a combustion chamber to react energetically and produce hot gas which yields thrust by expanding and accelerating in a nozzle. High performance requires that the combustion occur at high pressure, which in turn requires that the propellants are either pumped in or forced in by yet higher pressure in the propellant tanks. In most cases, launch vehicles pump their propellants. Spacecraft use a pressure-fed system that uses a (usually inert) pressurant gas that is either maintained at a constant pressure in the propellant tank by a regulator, or introduced while allowing its pressure to decrease as the propellant is expelled and the pressurant volume increases; this is known as the 'blowdown' mode. As the propellant flows through a series of valves and filters, it may be exposed to various temperatures (possibly causing changes in its properties, such as viscosity), and dynamic effects may occur, such as slosh in the propellant tank or water hammer in the pipes.

The violent nature of rocket propulsion systems makes them very difficult to test as part of a complete space system in a high-fidelity simulation. In addition, the production of a large volume of gas is difficult to accommodate in a space simulation chamber, and the propellants and/or their residues are often corrosive and otherwise hazardous. Although rocket motors – and propulsion systems as a whole – can be tested in (usually) open-air facilities, doing this does not reproduce the gas dynamic or thermal environment of space. System-level tests in which other elements such as translational or rotational dynamics, mechanisms and sensors come into play are not usually feasible. Individual components such as valves can be tested, and plumbing systems can be exercised using inert fluids. However, many propulsion system elements involve pyrotechnic devices which, by their nature as single-shot devices, are impossible to test directly. A project must rely on batch testing to develop a statistical measure of reliability and then hope for the best.

Although textbooks on spacecraft systems usually describe various categories of propulsion system, their treatment is often cursory. Detailed diagrams of a few examples are sometimes given, but without comment on what is actually happening, why all these components are required and, thus, what can go wrong.

TRIALS AND TRIBULATIONS

Surveyors

Launched on 20 September 1966 by an Atlas–Centaur, Surveyor 2 was lost when one of its three thrusters failed to ignite for the 10-second mid-course correction 16 hours later, and the asymmetric thrust set the vehicle tumbling. It crashed onto the Moon on 23 September.[1] The transmission from Surveyor 4 ceased on 17 July 1967, two seconds before its retro-rocket was to burn out and be jettisoned at an altitude of 11 kilometres above the lunar surface. It is possible that the motor had some kind of a structural fault that caused the solid propellant to explode. For more than 25 years, until Mars Observer was lost in similar circumstances in 1993, Surveyor 4 remained the last US probe to have catastrophically failed in flight.

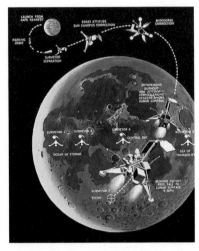

A depiction of the phases of Surveyor mission to land on the Moon.

Mariner 9, for a few dollars more

After three flyby missions which, between them, imaged 10 per cent of the surface of Mars, on 14 November 1971 Mariner 9 became the first spacecraft to settle into orbit around the planet, and by the time it exhausted its attitude control gas on 27 October 1972 it had mapped the entire planet. This was a tremendous success, but as Carl Sagan recalled a few years later:

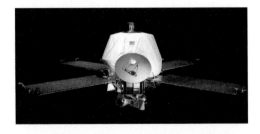

A model of the Mariner 9 spacecraft.

About a year before the Mariner 9 mission was launched, the possibility was raised that the spacecraft would run out of control gas. A solution was proposed: That the propulsion tanks be connected to the attitude control gas system – a kind of spacecraft anastomosis. Excess propulsion gas could then be used for attitude control in case the attitude control nitrogen was exhausted. This possibility was rejected – largely because of its expense. It would have cost

$30,000. But no one expected Mariner 9 to last long enough to use up its attitude control gas. Its nominal lifetime was ninety days – and it lasted almost a full year. The engineers had been overly conservative in assessing their superb product. In retrospect, it sounds very much like false economy. With an adequate supply of attitude control gas, the spacecraft might have lasted another full year in orbit around Mars. About $150 million of science might have been bought for $30,000 of pipe. Had we known that the spacecraft would die from a lack of nitrogen, I am almost certain that the planetary scientists involved would have raised the $30,000 themselves.[2]

However, the main engine burned nitrogen tetroxide and monomethyl hydrazine and the attitude control thrusters squirted cold nitrogen gas, so the design change would have been rather more involved than a simple piece of pipe with a valve. This capability was built into the propulsion system of the follow-on Viking spacecraft.

Viking pressurisation anomaly
In 1975 NASA launched two orbiter/lander spacecraft to Mars. As Viking 1 made its approach the following summer, it suffered a major propulsion system anomaly. It had a regulator valve to pressurise the tanks of nitrogen tetroxide and monomethyl hydrazine at 179 kilograms per square centimetre. A 64-centimetre-diameter tank stored the helium pressurant at 2,530 kilograms per square centimetre. If this gas was dumped unchecked into the propellant tanks it would rupture them. It was isolated by a set of pyrotechnic valves to guard against regulator leakage. The helium could flow to the regulator via lines A, B or C through pyrovalves (P1, P2, P4), (P3, P4) and (P5) respectively. Valves P2 and P4 started out open, and firing would close them, but P1, P3 and P5 were shut at launch and would be opened by being fired. P1 was fired before the first mid-course correction, and that path (line A) was then closed by firing P2. On 7 June 1976, P3 was opened to pressurise the propellant tanks in pre-paration for the Approach Course Manoeuvre. However, telemetered data showed that the tanks continued to rise in pressure after the regulator threshold had been reached. The leak-age rate was some 0.16 kilogram per hour.[3]

The obvious solution was to fire P4 to cut off line B and thereby prevent the pressure in the propellant tanks from building up during the 12 days remaining to the Mars Orbit Insertion (MOI)

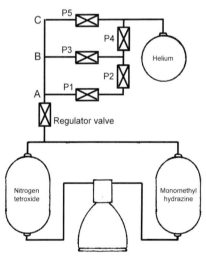

A schematic of the pressurisation section and valve ladder of the main propulsion system of a Viking orbiter.

burn. However, this would leave only line C, and turn P5 into a 'single point failure' in that if it failed to open when commanded, the entire mission would be lost because, with the engine unusable, the spacecraft would sail past Mars. The alternative solution, which was adopted, was to increase the volume of helium in the tanks in order to slow the rate at which the pressure rose. (That is, a given mass flow rate leaking through the regulator corresponded to an ullage – 'headspace' – volume multiplied by the pressure increase rate; therefore an increase in the ullage volume would slow the pressure rise.) By the expedient of splitting the Approach Course Manoeuvre scheduled for 9 June into two separate burns of 50 and 60 metres per second on 10 and 15 June respectively – while also, in the process, increasing the amount of propellant used – the pressure in the tanks was able to be held below rupture levels for the MOI burn, and thus preclude the spacecraft being reliant on line C for the rest of the mission. However, this change delayed the MOI burn by six hours, which displaced the longitude of periapsis away from the planned landing site, so it was decided to enter an initial orbit with a period of 42.6 hours designed to place the spacecraft over the landing site 49.2 hours later, at which time an additional burn put the spacecraft into the desired orbit with a period of 24.6 hours (i.e. one local day, or sol). After successfully delivering its lander on 20 July, Viking 1 returned data until 7 August 1980, when it exhausted its attitude control propellant and was ordered to shut down.[4]

A model of a Viking spacecraft, with the engine at the top, the lander within the large bulbous bioshield, the high-gain antenna and (on the end of a solar panel) the antenna for receiving the signal from the lander.

Viking 2, a month behind, was reprogrammed to delay the pressurisation of its tanks until 12 hours before the MOI burn, and while there was some evidence that its regulator had also leaked, the data was inconclusive. The report on this anomaly concluded that the regulator leakage could have been caused by a particle as small as 1 micron in diameter preventing the correct seating of the valve.[5]

Mars Observer lost

The loss of the Mars Observer spacecraft is an example of a good idea gone bad. In the mid-1980s NASA decided to reduce the cost of its deep space missions by standardising on two types of bus. The Mariner Mark II vehicles were for the outer Solar System, and the Observers were for the inner Solar System. The Comet Rendezvous and Asteroid Flyby and the Cassini mission to Saturn were to be the first Mariner Mark IIs, but when Congress cancelled the former in 1992 the latter became a one-off. The Observer programme was to use the bus and electronics derived from the Satcom 4000 series communications satellites and DMSP 5D2 and Advanced-TIROS-N Earth satellites built by the Astro Space division of General Electric (originally RCA Astro Space, and soon to become Lockheed Martin Astro

Space). The first mission, Mars Observer, was launched on 25 September 1992 by the final Commercial Titan. It was to conduct a global survey of the Martian surface and atmosphere over the course of a local year of 687 terrestrial days to produce a composite global portrait to match the unprecedented perspective of the Earth gained from satellites such as Landsat. The 2.5-tonne, 3-axis stabilised spacecraft was to control its orientation using four reaction wheels, deriving attitude information from a horizon sensor (defining the nadir), a star mapper (for inertial attitude), gyroscopes and accelerometers (for angular rates and linear accelerations), and a suite of Sun sensors. The X-Band communications system used a 1.5-metre-diameter articulated high-gain antenna on a 6-metre-long boom. In Sun-synchronous orbit around Mars,

An artist's depiction of the Mars Observer spacecraft in its operational configuration in orbit around Mars.

a six-segment solar array with a total area of 24 square metres was to be unfolded to generate between 1.1 and 1.5 kilowatts. It had hydrazine and bipropellant propulsion systems for interplanetary trajectory manoeuvres, initial Mars orbit insertion and circularisation, and subsequent orbit maintenance.[6]

Unfortunately, contact with Mars Observer was lost on 22 August 1993, three days before it was due to enter orbit around its objective.[7] Its transmitter – a commonly used travelling-wave-tube amplifier (TWTA) – had been turned *off* in preparation. A TWTA is an electronic device in the sense that it manipulates electrons in an electric field, like an old-fashioned thermionic vacuum tube, and as such it incorporates a tiny filament that is heated by current to incandescence so as to produce the free electrons. However, a hot filament is fragile; if a lamp is stressed while weakened by heat during operation, it can fail. Because the TWTA had not been qualified against the shock from firing the pyrotechnic valves that were to open the propulsion lines in preparation for the orbital insertion burn, it had been decided to switch it off as a precaution. Because contact was not re-established, the investigation was unable to conclusively identify a single cause of the failure, but opinion has tended to weigh on the first of the following six possibilities:[8],[9]

- An inadvertent mixing of the hypergolic oxidiser and fuel in the pressurisation system lines causing the uncontrolled venting of pressurant and propellant, resulting in loss of attitude control and communication capability.

- The failure of the series-redundant regulator to close during the pressurisation event, leading to over-pressurisation and rupturing of the propellant tanks.
- The expulsion of a NASA Standard Initiator (NSI) – a squib – from one of the pyrotechnic valves fired during the pressurisation sequence, leading to impact damage and rupture of the fuel tank.
- A primary power system failure due to a short to ground within the power system electronics.
- Loss of computational function on the spacecraft due to electronic part latch-up caused by chassis currents induced by the pyro events during the pressurisation sequence.
- An inability to reactivate the transmitters due to electronic part latch-up in the telecommunication subsystem.

The central point of the leading theory is that valves are not perfect; they leak. In particular, the 'soft' seat of an electrically operated check valve (that is, a valve to provide one-way flow) will allow vapour to diffuse through the material itself, and 'hard' seats tend to not seal properly – especially when a particle of contaminant is present, even if this is only a few microns in size. Such leakage is of little concern on a satellite that uses its main propulsion system only for a few days while it manoeuvres to its operating station. But Mars Observer was to be in space for 11 months prior to using its engine. Furthermore, it would double its distance from the Sun during the interplanetary cruise, and therefore would be operating in a very different environment. Not only would the plumbing system chill down, but its temperature would not be uniform. There were also some 'dead ends' where vapour could accumulate and condense, and resist being driven into the oxidiser tank when the helium pyro valve P7 was opened. One test established that 40 milligrams of nitrogen tetroxide vapour could leak through the check valve seal in only 200 hours. A

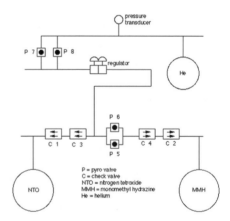

A schematic showing the arrangement of the valves in the main propulsion system of Mars Observer.

substantial amount of oxidiser may well have accumulated in the year or so that elapsed between filling the tank and preparing the engine for use. When the fuel-side pyro valve P5 was opened five minutes or so after P7, some of this accumulated oxidiser may have met the monomethyl hydrazine. The convenience of these chemicals as rocket propellants, is that they ignite on contact, simplifying the engine. An inadvertent fire in the plumbing could have heated and burst the pipes, spewing out the fuel and helium pressurant. The thrust of this venting would have set the spacecraft spinning, and the cloud of fuel would have corroded and/or shorted out electrical equipment.

However, this pipe-explosion scenario could not be replicated by ground tests simulating an oxidiser leak. Of 13 tests, most generated pressures in the fuel line of only 200 psi, whereas the pressurisation lines were rated at 10,000 psi. Only one test attained a pressure as high as 4,000 psi, and that required 10 grams of oxidiser, which was perhaps a factor of 10 more than was likely to have accumulated.

The second most likely mode, the failure to close off the pressurisation regulator, was further investigated by an analysis of the heritage of regulators for another programme a few years later.[10] This established that nitrogen tetroxide vapour can corrode the braze (AMS 4774) used in the flow restrictors in the pressurisation system – a reaction that would not necessarily have been evident to short-term compatibility tests. This alternative series of events is as follows:

- Both stages of the regulator open during the cruise when the tank pressures drop well below regulator set points.
- Oxidiser vapour permeates the check valves and diffuses into the flow restrictors, corroding the braze material.
- Pyro shock upon firing P5 dislodges the corrosion products.
- The corrosion products are forced into the regulator, holding it in a near-open state.
- After about 2 minutes the overpressurised oxidiser tank bursts.

The fact that the line-rupture model was never reproduced, and regulator failure was prominent in the Viking problems, lends some support to this second theory.

A third possibility is interesting to mention. The pyrovalve is opened by the firing of a small explosive charge – the NSI. If the threads holding the NSI in the valve assembly failed, the initiator would be expelled from the valve at 200 metres per second, sufficient to puncture the fuel tank. This rather improbable-sounding possibility did actually occur in ground testing for the European Space Agency's Cluster mission.[11]

Both of the favoured scenarios have the same root cause, namely the application of a design that worked satisfactorily on Earth-orbiting missions but failed when applied to a long cruise in deep space. In view of the loss of Mars Observer, this series of missions was cancelled.

Picking up the pieces

Most of Mars Observer's sophisticated suite of instruments were reflown on a flotilla of orbiters, including Mars Global Surveyor and Mars Odyssey. The unfortunate loss of Mars Climate Orbiter dealt a particular blow to some investigators, notably those involved in the pressure-modulated infrared radiometer. To have devoted many years to building an instrument, have it launched and then lost a few days before arriving at its objective is tragic, but for a team to have this happen twice must be heartbreaking.

A close call for Galileo

Fears of similar problems arose as the Galileo spacecraft completed its long cruise to Jupiter. After the spacecraft had released its probe on 13 July 1995 on a trajectory that would make it dive into the planet's atmosphere on 7 December, the main vehicle made a manoeuvre on 27 July to ensure that it did not share the fate of its probe. In preparing the engine for this deflection burn, the flight controllers were puzzled by a telemetry indication of a difference in pressure between the fuel and oxidiser tanks. The fact that the pressure in the oxidiser tank was the same as in the helium pressure regulator suggested that the check valve (across which there ought to have been a 350-millibar differential) had stuck in an open position.[12] One possibility was that a teflon bushing in the valve had swollen due to long-term contact with oxidiser vapour. Although the *downstream* side of the valve was expected to be exposed, the upstream side (where the bushing was) was not expected to be exposed, and hence had not been so extensively tested. Details of the valve design were difficult to retrieve, since that work had been done two decades earlier and the key individuals had either retired or died. In fact, this is a common problem in the exploration of the outer Solar System, in which missions take many years to get off the ground, and many more years to reach their objectives.

As the risk was of temperature changes in the propulsion system forcing nitrogen tetroxide into the pressurisation system, the flight controllers initiated a programme of thermal management to avoid such thermal pumping by using small electrical heaters to minimise temperature differences across the propulsion system. Because the temperature differences were only 3–4°C (compared to 25–30°C in the case of Mars Observer) the engineers were confident that they would be able to limit the leakage of oxidiser, especially as the next firing was only four months away, and, indeed, the Jupiter Orbit Insertion burn on 7 December was made without incident.

Nozomi's struggle

Ironically, one 'lesson' learned from Mars Observer may well have led to the first in a catalogue of anomalies to befall Japan's first planetary mission, Nozomi (which means Hope), known as Planet B prior to launch on 4 July 1998. The third stage of the M-5-III launch vehicle attained a parking orbit ranging between 146 and 417 kilometres, and at apogee the fourth stage made this circumlunar. The total mass of the probe at launch (including 282 kilograms of propellant) was 540 kilograms. It had a bipropellant propulsion system to which an additional valve had been added to prevent migration of the oxidiser. After using lunar flybys on 24 September and 18 December to increase the apogee of its orbit, Nozomi was to make a 7-minute burn during its perigee passage on 20 December to inject itself onto an escape trajectory for Mars, which it was to reach on 11 October 1999, enter a highly eccentric orbit, and conduct a long-term study of the planet's atmosphere.[13,14] However, during the escape burn the anti-back-flow valve did not open fully, causing the engine not to deliver the required thrust.[15] Two further burns the next day improved the trajectory, but in the process left the spacecraft with insufficient propellant to enter orbit around Mars. Working over the holiday season, the mission analysis team devised a recovery strategy. By flying past Mars in 1999, and remaining in heliocentric orbit,

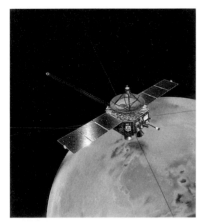

An artist's impression of the Nozomi
spacecraft in orbit around Mars.

the spacecraft would be able to exploit Earth
flybys in December 2002 and June 2003 to
encounter Mars again in December 2003 at a
slower relative velocity than originally, which
it would be able to overcome with the
propellant available. The penalty, however,
would be spending an additional four years in
space. Although Nozomi rode out a solar flare
in April 2002, the bombardment of charged
particles from a series of intense flares as it
approached its objective in late 2003 damaged
a power control circuit. This current switch
(itself a protective device – possibly a current
limiter to protect the circuit from a short)
controlled power to both a telemetry mod-
ulator and the heaters for the fuel tanks. What
had seemed to be an *efficient* design feature of
using one switch to handle two functions had in fact exposed the system to a single-
point failure. It might have been possible to recover from compromising one or the
other of these systems, but losing both was crippling. Despite power-cycling the system
in excess of 1,000 times in an effort to clear the power switch failure, this remained
inoperable, and, with the heaters off, the propellants and piping of the main engine
froze solid.[16] As the spacecraft neared Mars on 9 December 2003, the engineers were
obliged to abandon their attempt to enter orbit and, since the probe had not been
sterilised, instead adjusted the trajectory by firing the attitude control thrusters (the
propellant for which had not frozen) in order to minimise the chance of it hitting the
planet. On 14 December the largely
derelict probe sailed past Mars at a
range of 1,000 kilometres, into heliocentric oblivion.

So NEAR
Dispatched on 17 February 1996
by a Delta II, the Near Earth
Asteroid Rendezvous (NEAR)
spacecraft developed by the Applied Physics Laboratory of Johns
Hopkins University in Baltimore,
was the first in NASA's Discovery
series of 'faster-better-cheaper'
missions. To facilitate extensive
manoeuvres, 318 kilograms of its
805 kilograms were propellant.
Spectacular results were achieved
during the flyby of asteroid 253

An artist's impression of the NEAR spacecraft.

Mathilde on 27 June 1997 at a heliocentric distance of 2 AU. The bipropellant engine was fired for the first time on 3 July 1997 in a deep space manoeuvre to establish an Earth flyby on 23 January 1998 that would place the craft on course to rendezvous with asteroid 433 Eros. On 20 December 1998, as it approached its objective, the spacecraft was to fire its thrusters to settle the propellants in their tanks and then fire its main engine in order to match velocities. However, as the investigation reported:

> Almost immediately after the main engine ignited, the burn aborted, demoting the spacecraft into safe mode. Less than a minute later the spacecraft began an anomalous series of attitude motions, and communications were lost for the next 27 hours. Onboard autonomy eventually recovered and stabilised the spacecraft in its lowest safe mode (Sun-safe mode). However, in the process NEAR had performed 15 autonomous momentum dumps, fired its thrusters thousands of times, and consumed 29 kilograms of fuel (equivalent to about 96 metres per second in lost delta-V capability). The reduced solar array output during periods of uncontrolled attitude ultimately led to a low-voltage shutdown in which the solid-state recorder was powered off and its data lost. After reacquisition, NEAR was commanded to a contingency plan and took images of Eros as the spacecraft flew past the asteroid on 23 December. The NEAR team quickly designed a make-up maneuver that was successfully executed on 3 January 1999. The make-up burn placed NEAR on a trajectory to rendezvous with Eros on 14 February 2000, thirteen months later than originally planned. The remaining fuel is sufficient to carry out the original NEAR mission, but with little or no margin.[17]

The engine was supposed to burn for 15 minutes to produce a delta-V of 650 metres per second, but it was shut down within a fraction of a second.[18] The accelerometers had sensed a lateral acceleration that greatly exceeded the 0.10 metre per second per second threshold that had been defined to detect uncontrolled lateral thrust during any burn. One possibility was that the oxidiser in two tanks that were by this point approximately half-full, may have sloshed in such a manner as to exceed this threshold. Telemetry from the deep space manoeuvre in 1997 documented a similar large transient at ignition, but the significance of this does not appear to have been appreciated. The situation worsened. The computer started to

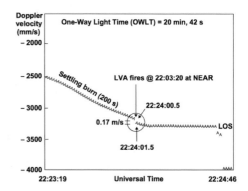

A plot of the radial velocity component as measured by the Doppler shift on the signal from the NEAR spacecraft showing the settling burn, the moment that the bipropellant 'large-velocity adjust' burn to rendezvous with asteroid 433 Eros was aborted, and the loss of signal 37 seconds later. (Failure report.)

execute a script to slew to a safe Earth–Sun-pointing orientation, doing so using the thrusters (as was normal during a burn) instead of the reaction wheels, and when it disengaged the thrusters and re-enabled the precision-pointing system the reaction wheels were unable to halt the rapid rate of rotation imparted by the thrusters and the spacecraft overshot its intended Sun-pointing attitude. Having failed to acquire the Sun within the allotted 300 seconds, the computer switched to a new mode. Noting that it had a high angular rate, it powered up the thrusters again in order to dump angular momentum. However, a software error caused the speed of the reaction wheels when at their maximum speed to be reported as zero. Additional software errors meant that there was a mis-synchronisation between thruster enablings, GNC mode switches, and switches between the two Attitude Interface Units, prompting the thrusters to fire repeatedly to reduce the (perceived) angular momentum of the system. Of course, these interactions might have been exacerbated by the several seconds it took to reboot the Attitude Interface Unit processors. Despite over 100 simulations, the investigation was unable to determine the sequence of events in the interval from about 47 minutes after the initial failure through to the regaining of control over the spacecraft 26 hours later, at which time it was safely Sun-pointed. However, it was clear that a large amount of propellant had been consumed during several momentum dumps. Although the investigation discovered 17 errors in the 80,000 lines of code between the main computer and the Attitude Interface Units, it was unable to state that the momentum dumps were solely due to software, since thruster leaks, structural oscillations or propellant slosh may have been contributory factors.

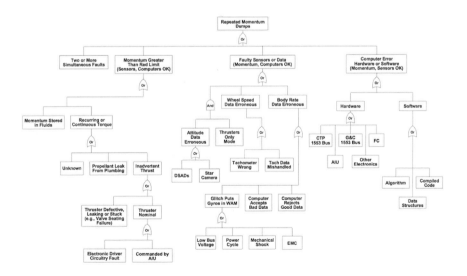

Fault tree of possible causes of the repeated momentum dumps by the NEAR spacecraft after it aborted its burn. Certain events result from AND-ing or OR-ing other events. (Failure report.)

The NEAR failure review board was scathing regarding the challenges that unrecovered data presented to their diagnosis:

> The NEAR team is to be commended on the speed with which they determined the proximate cause of the abort itself. The rapid diagnosis of the precipitating events, coupled with quick work from Mission Operations and the rest of the NEAR team, enabled the successful re-burn two weeks later that rescued the mission. The NEAR team won an American Institute of Aeronautics and Astronautics Space Operations and Support Award for the successful recovery effort. On the other hand, the rôle Mission Operations played in causing the RND1 failure cannot be ignored. Furthermore, Mission Operations was less than successful in protecting and acquiring spacecraft data needed to support a diagnosis: certain data that might have been valuable to the diagnosis were lost because Mission Operations overwrote spacecraft command memories, failed to downlink processor memory images, or made other operational errors. There was no plan or procedure in place for recovering from an anomalous event, despite several previous entries into safe modes during cruise.

Despite this mishap, the spacecraft, now renamed NEAR–Shoemaker in memory of the recently deceased Gene Shoemaker, successfully rendezvoused and on 14 February 2000 fired its engine to enter orbit around Eros. To end its science programme, it was manoeuvred into a grazing impact at a speed of about 1.75 metres per second which enabled it to soft land on 12 February 2001, where it remained in communication until 28 February.[19,20]

The loss of CONTOUR

The Comet Nucleus Tour (CON-TOUR) mission was developed jointly by Cornell University in Ithaca and the Applied Physics Laboratory of Johns Hopkins University in Baltimore for NASA's Discovery programme. After launch by a Delta II on 3 July 2002, it remained in a highly elliptical orbit awaiting the burn on 15 August that would send it to a flyby of Encke's comet on 12 November 2003.[21] This 1,920-metre-per-second burn was to be made by a Star 30 solid rocket motor built by Alliant Techsystems (formerly Thiokol) that was embedded in the centre of the spinning spacecraft. (This configuration was similar to the Giotto spacecraft that the European Space Agency had sent

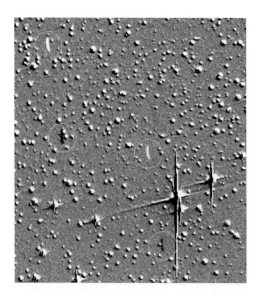

Spacewatch imagery captured trails of what could only be fragments of the CONTOUR spacecraft.

The CONTOUR spacecraft being mated with its solid rocket motor, showing that it was deeply embedded in the structure.

on a flyby of Halley's comet in 1986, although in that case there were clamshell doors that closed over the nozzle after firing in order to guard against dust impacts during the encounter.) At 04:49 EDT, the start of the burn, CONTOUR was 225 kilometers above the Indian Ocean and out of sight of the antennas of NASA's Deep Space Network, but contact was expected to be established at 05:35. However, there was no signal.[22],[23] Ninety-six hours after its last receipt of communications from the ground, the spacecraft would have gone into a mode in which it would seek communication, but nothing was received. Attempts to contact the spacecraft continued until December, without success. In fact, optical imaging on 16 August by a 1.8-metre telescope on Kitt Peak used by the University of Arizona's Spacewatch programme to search for Earth-crossing asteroids found two objects just beyond the Moon's orbit that were moving on trajectories similar to, but slightly slower than, that predicted for the post-burn CONTOUR trajectory.[24] At NASA's request, the search was widened and a third object was detected. The implication was that the burn had failed towards the end and the spacecraft had broken up, ejecting fragments at a speed of at least 6 metres per second. It was subsequently revealed that a US Air Force 'asset' (presumably an infrared telescope, possibly airborne) had noted an anomalous infrared emission during the burn. Again, the lack of telemetry at a crucial time was criticised by the failure board. This 'loss without trace' phenomenon – which had already claimed Mars Climate Orbiter and Mars Polar Lander – led NASA to require communications during injection and insertion burns, atmospheric entry, descent and landing, and other critical events, even if doing so increased the risks. For example, the Mars Exploration Rovers were required to land during the early afternoon in order to have a clear line of sight for telemetry to reveal how they were lost if they failed, even though the winds would be at their strongest at that time; fortunately they reached Mars successfully.

The investigation determined that the most likely proximate cause of the loss of the CONTOUR spacecraft was overheating resulting from the exhaust plume from the solid rocket motor. A deeper cause was poor systems engineering and review, and the reliance on analysis by similarity (a recurrent issue with propulsion system failures, for example Mars Observer). The failure board took note of the fact that the motor was more deeply embedded in the spacecraft than was typical. Independent analyses commissioned by the failure board indicated that the heating at the base of the spacecraft would be higher than the 65 kilowatts per square metre (i.e. 50 solar constants) that the designers had assumed to be the worst case. Furthermore, the heating may have been enhanced by positive feedback. For example, extreme heating of the surfaces, or even direct impingement of particles or gas in the plume (recall that a plume will expand in all directions after leaving a nozzle in the vacuum of space) might have altered the optical absorptivity and emissivity of the surfaces and caused them to absorb more heat. The details of what happened next are not clear, but structural weakening may have prompted dynamic instabilities in the spinning spacecraft, causing it to break up. (It was not simply spinning for stability, but spinning rapidly in order to even out any asymmetry in the thrust from the solid motor.) Even if it had not failed, it may have been rendered useless because revised thermal analyses suggested that some components – such as the forward low-gain antenna – would have reached melting point.[25]

Hot thrusters on GOES 8

On 13 April 1994, an Atlas–Centaur launched GOES 8, the first of a new series of Geostationary Operational Environmental Satellites for service with the National Oceanic and Atmospheric Administration. Its development had been delayed by a variety of problems.[26,27] The bus was base on a Space Systems/Loral communications satellite, and was to use a 490-newton bipropellant engine to raise its orbit. During the first thruster firing, engineers de-

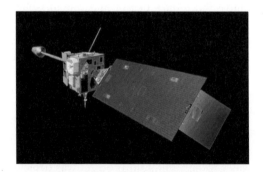

An artist's impression of GOES 8.

tected an unacceptably high temperature on a thruster mounting flange. However, careful analysis showed that in fact the temperature limits (set for other satellites using the same thrusters in a different configuration) were too conservative and could be relaxed. This analysis was confirmed during a 15-minute characterisation burn. The third firing, however, had to be aborted when the attitude control thrusters in three axes inadvertently fired, perhaps as a result of an upset in the attitude control computer.[28] Nevertheless, subsequent burns nursed the satellite into geostationary orbit at the GOES-EAST slot at 75° west longitude.[29,30]

Galileo's overheating thrusters

In November 1988, during confidence testing by Messerschmitt–Bölkow–Blohm of the 10-newton thrusters it had supplied for the Galileo spacecraft, rapid increases in the temperature of the combustion chamber were sometimes noted immediately after the ignition of the engine. These 'hot starts' necessitated manual intervention to halt the test before the overheating could damage the engine. The problem was determined to be a 30-kilohertz acoustic instability of the combustion chamber and injector, almost certainly triggered by severe ignition transients. The transients were alleviated by relocating the engine trim orifices downstream of the engine valve, and changing the flight operating mode of the thrusters. It was also determined that the original orifice configuration had tended to saturate the propellant with bubbles, which were also a factor in triggering the instability. The mission had to be redesigned to avoid continuously firing the thrusters, and to restrict pulse-mode duty cycles.[31] Interestingly, this fault was discovered two years after Galileo was *supposed* to have been launched – the grounding of the Shuttle after the loss of Challenger meant that the mission missed its 1986 launch window, and this delay may well have saved it from discovering this subtle, but potentially catastrophic thruster problem in space.

Deep Space 1's teething problems

Deep Space 1 was the first mission of the New Millennium series set up by NASA as technology demonstrators, in this case featuring a variety of new technologies including concentrator-style solar arrays and techniques for autonomous spacecraft operations and the automatic targeting of instruments. However, the feature that attracted the headlines in the popular press was the ion propulsion system.[32] While this was not the first time that an ion engine had been employed, it was the first time for this design, and the first use of such an engine on an interplanetary mission. The 486-kilogram spacecraft was built by the Jet Propulsion Laboratory in Pasadena in collaboration with Spectrum Astro, and was based, in part, on a design for the MSTI satellite that had been built for the Ballistic Missile Defense Organisation, with most of its electronic systems being installed on the outside rather than the inside of the structure. On 24 October 1998

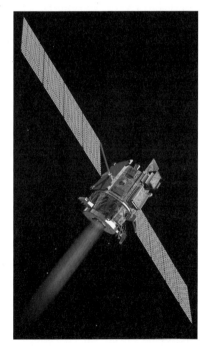

An artist's impression of the Deep Space 1 spacecraft.

the second stage of a 'Med-Lite' Delta II entered parking orbit, coasted for a short time, raised the apogee and then released the upper stage, and this fired its Star 37FM solid motor to accelerate the spacecraft away from the Earth. There was a delay in establishing contact with the spacecraft after separation due to the radiation in the van Allen belts upsetting the Stellar Reference Unit. In comparison to the 50 kilonewtons of the escape stage, the subsequent manoeuvring would be much more sedate. By electrically accelerating ions of xenon gas through a pair of molybdenum grids set 0.66-millimetre apart, across which a potential difference of about 1,000 volts (the actual value depending on the throttle) was held, the ion engine could produce a thrust in the range 18 to 92 millinewtons. However, by firing continuously this engine was able to sustain an acceleration of 13 metres per second per day. A normal part of operating such an engine is that the ion beam will sputter atoms from the grids, and this 'erosion' will gradually wear them out. In addition, dust particles can occasionally become caught in a grid, shorting it out. When this occurs, the response was to interrupt the engine for 1 second or so in order to allow the particle to escape. The first firing, on 10 November, was to have lasted 17 hours but was interrupted after only 4.5 minutes, and after the automatic 'recycle' failed to clear the obstruction, the engine shut itself down. The spacecraft was commanded to alternately point the thruster at the Sun and then into shadow, to enable the difference in thermal distortion between the inner and outer grids to work the particles free, and on the next attempt to operate it two weeks later the engine worked correctly. The engine settled down to a nominal 1 or 2 recycles per week of operation, and went on to set records for the longest powered flight by any spacecraft and the largest change in velocity by any space propulsion system.[33,34,35,36] On most interplanetary missions, the spacecraft is placed into a state of semi-hibernation for the cruise phase, but because Deep Space 1's primary function was to demonstrate new technologies it was kept very busy. One problem that caused the spacecraft to safe itself early on was an erroneous signal from a Sun sensor resulting from the fact that a narrow zone at the edge of the sensor's field of view suffered a geometric distortion which made the Sun appear to be in a slightly different position to where it really was. This fault was able to be overcome by a software fix, and cost only one day of operations.

Xenon Ion Propulsion Systems failures
An ion drive combines a high specific impulse with a low propellant mass. Because station-keeping propellant is a major life-limiting (and therefore revenue-limiting) factor for a geostationary satellite, and because communications satellites usually have plenty of power available, ion thrusters were an attractive option for the north–south station-keeping rôle. The HS-601HP introduced the Xenon Ion Propulsion Systems (XIPS – pronounced 'zips').[37,38] The first satellite to carry it was PanAmSat 5, which was launched on a Proton-K by ILS on 27 August 1997.[39] However, failures with these systems led to several satellites (Galaxy 4R, PanAmSat 5, Galaxy 8i and PanAmSat 6B) being taken out of service prematurely. These satellites also had chemical propulsion systems, but with propellant only for three or four years of operation.[40] The version of system on the follow-on HS-702 series performed more reliably.

SMART 1's slow route to the Moon

When an Ariane V released two communications satellites into geosynchronous transfer orbit on 27 September 2003, it also dropped off the first Small Mission for Advanced Research and Technology (SMART) spacecraft for the European Space Agency. This ignited its xenon ion propulsion system to start slowly spiralling out to the neutral point between the Earth and the Moon, at which time it would become captured by the Moon. The downside of using a low-acceleration engine as a means of escaping from the Earth was that it meant spending a *long time* transiting the van Allen belts. Having been dropped off in a 12-hour orbit that had its perigee below the belts and its apogee above the belts, SMART 1 was repeatedly exposed to this radiation. Although this had been taken into account by its designers, a series of exceptional solar flares prompted the spacecraft's star tracker to malfunction, which made it difficult to maintain the proper orientation,[41] and single-event upsets in an opto-coupler in the Power Processing Unit caused

An artist's impression of PanAmSat 5.

An artist's impression of the SMART 1 spacecraft.

eight uncommanded shutdowns of its engine.[42] Nevertheless, the spacecraft was able to continue, and successfully slipped into lunar orbit in November 2004.[43,44]

Hipparcos recovery

The High-Precision Parallax Collecting Satellite (Hipparcos) was developed by the European Space Agency to make precision measurements of the positions of stars on the sky.[45] It was launched by an Ariane 4 on 8

An artist's impression of Hipparcos.

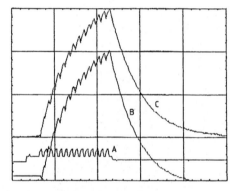

A plot of the electrical currents while attempting to fire the solid rocket motor to circularise Hipparcos's orbit.

August 1989 (together with the TVSat 2 satellite) and released into geosynchronous transfer orbit. A MAGE 2 solid rocket was to have fired at apogee to enter a geostationary position from which the satellite would be able to be operated from a single ground station, but when the firing command was issued on 10 August Doppler tracking showed no change in the satellite's velocity, indicating that the burn had failed to occur. After the firing commands had been verified, several further attempts were made to start the motor, without success. Reformatting the telemetry to report the currents in the pyrotechnic relay unit 200 times per second confirmed that the relays were functioning. Attention then turned to the two pyrotechnic igniters, which were to fire in succession for redundancy. The ignition train began with a through-bulkhead initiator, and the ignition event ought to have been propagated to the motor propellant by a 'flexible explosive transfer assembly' similar to a detonation cord. The investigation fired 40 initiators and compared their electrical characteristics with that in the telemetry, which confirmed that the initiators had indeed fired. This meant that, for some unknown reason, the pyrotechnic ignition event had failed to propagate from the initiators to the motor propellant.[46]

To rescue the mission, the satellite's attitude control system was reprogrammed to enable it to determine its orientation in the unplanned 10-hour orbit ranging between 526 and 36,000 kilometres, additional ground stations were recruited to augment the original site at Odenwald in Germany to provide continuous communications, and new software was written to process the data that was returned. However, the fact that the satellite was repeatedly passing through the van Allen belts, rather than orbiting far above them, meant that the solar arrays deteriorated much more rapidly than otherwise would have been the case.[47] Nevertheless, Hipparcos was able to undertake observations from 26 November 1989 to 15 August 1993, and the astrometric catalogue published in 1997 far exceeded the original expectation.[48,49,50,51]

TDRS 9 failure and recovery

On 8 March 2002 an Atlas IIA from Canaveral launched a communications satellite for NASA's Tracking and Data Relay System (TDRS). The early satellites in this series had been built by TRW, and been delivered to geostationary orbit by the two-stage IUS. The contract for the advanced model (of which this was the second in the series of three) had been awarded to Hughes Space and Communications (now part of Boeing). The satellite, TDRS 9,

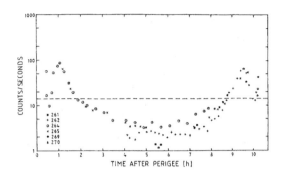

A plot of the spurious radiation counts suffered by Hipparcos as it repeatedly passed through the van Allen belts.

was released by the Centaur into a transfer orbit ranging between 247 and 29,135 kilometres inclined at 27.1 degrees to the equator. As a derivative of the HS-601 (of which 48 had launched) it had a bipropellant system that burned monomethyl hydrazine in nitrogen tetroxide, with two tanks for each propellant. Such a satellite achieves its assigned geostationary slot by making a series of burns over a 10-day interval. On 11 March, a burn raised the orbit to 433 by 29,146 kilometres.[52,53] On 13 March a perigee burn increased the apogee to geosynchronous altitude. The next day, fuel from one of the tanks ceased to flow, apparently because its helium pressurisation valve had failed to open. Reports indicated that photographs taken during assembly suggested that the valve had been wired incorrectly.[54]

Nominally, 84 per cent of the propellant was required to achieve geostationary orbit, with an additional 7 per cent being required for station-keeping during a planned 15-year life; the remainder being margin. The potential denial of 50 per cent would therefore be catastrophic. Fortunately, the engineers developed a workaround. Once the fuel in tank 1

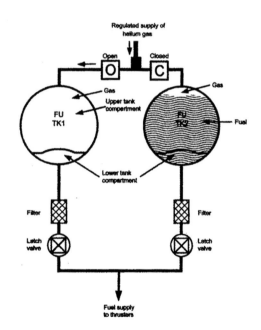

The propulsion system of TDRS 9 enabled helium pressurant to be fed from a nearly empty tank (left) into the other tank (right) after the failure of the pressurisation valve for that tank to open.

had been consumed, helium would be allowed to flow into the unpressurised tank 2 by opening the downstream latch valves to both tanks. In other words, helium would pass through the empty tank, follow the fuel path through the latch valve to the point where the fuel lines from both tanks joined, then flow 'backwards' along the fuel line from tank 2 and *into* that tank. At this time, the tank 1 latch valve would be closed, and tank 2 would operate in a 'blowdown' mode until its pressure dropped. The cycle would then be repeated, with the blowdown burns lasting progressively longer as the pressurisation headspace above the fuel in tank 2 increased. This ingenious approach relied on the fact that the tanks included 'propellant management devices'. In some designs a propellant tank uses a pressurised bladder to force the fuel into the drain, and this keeps the fuel and pressurant separate. A propellant management device is a shaped structure inside the tank that exploits capillary action to cause the fuel to gather near the drain in the absence of gravity. This feature enabled the helium to bubble through the fuel to accumulate 'above' it to exert pressure. This workaround also relied on the valves being able to be operated many more times than on the nominal mission, and on a sufficient supply of pressurant.

The recovery started on 19 March, with a burn that raised the perigee to 3,510 kilometres. After analysing the performance of the burn, the engineers concluded that the fuel was almost free of bubbles, and that its purity would not be an issue.[55] Raising the orbit took rather longer than the two months initially thought. A major issue was that the exhaustion of tank 1 displaced the vehicle's centre of mass, which resulted in a burn that not only propelled the satellite forward but also made it yaw. The spin rate had to be adjusted accordingly. Fortunately, as the satellite was to serve as an in-orbit spare, the delay in achieving its operating station did not degrade TDRS operations for the Shuttle and International Space Station.[56]

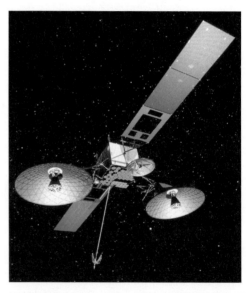

An artist's impression of an HS-601-based TDRS satellite.

Landsat 6 and Telstar 402

Landsat 6 was to be the first of the 'next generation' of Earth-resources satellites. On 5 October 1993 a Titan IIG from Vandenberg released it into a suborbital trajectory with an apogee at an altitude of 740 kilometres, at which point a Star 37 solid rocket motor was to fire to circularise in Sun-synchronous orbit, but this manoeuvre failed. The object initially identified as Landsat 6 by satellite trackers was found to be the

ERS 1 satellite operated by Europe.[57,58] The investigation was hampered by the absence of telemetry, as this system was not to have been activated until after the spacecraft was in orbit so as to guard against the transmission interfering with the operation of the second stage of the launch vehicle.[59] The initial suspicion was that the motor had failed to fire, but the investigation attributed the loss to a ruptured hydrazine manifold in the satellite that rendered its reaction control system useless by preventing propellant from reaching the thrusters, which in turn caused the spacecraft to tumble during the burn and therefore fail to build up sufficient velocity to attain orbit. The manifold is thought to have been ruptured by the detonation of some hydrazine fuel as a result of pyrovalve blow-by.[60,61] The Earth Observation Satellite Company, EOSAT, continued to operate as well as it could using Landsat 4 (launched in 1982, and at one time intended to be replenished by a Shuttle), and Landsat 5 (launched in 1984), both of which were well beyond their three-year life expectancy.[62,63]

An Ariane 42L placed AT&T's $200-million Telstar 402 into geosynchronous transfer orbit on 9 September 1994. Analysis of the telemetry showed that the satellite was lost 10 minutes later, when two redundant pyrovalves in the propulsion system were fired sequentially to open a fuel line. The investigation conducted ground tests involving over 50 test firings and concluded that some hydrazine must have accumulated on both sides of the second pyrovalve, and the shock of the firing and/or release of some pyro combustion gases ignited the hydrazine, resulting in an explosion and/or venting of corrosive fuel.[64,65] This satellite was the second 7000 series bus built by General Electric's Astro Space division (which Martin Marietta had bought in April 1993). The fact that this group had also built Landsat 6 and Mars Observer, both of which appeared to have had propulsion failures, suggested that there were endemic design and/or manufacturing problems.[66]

Engineering Test Satellite 6
The Engineering Test Satellite ETS 6 that was dispatched by a Japanese H-2 launch vehicle on 28 August 1994 was similar to the European Space Agency's Olympus in that it was to demonstrate technologies for communications satellites.[67] However, when a fuel valve opened only partially, the apogee engine gave just 10 per cent of its designed thrust due to inadequate combustion pressure. Two further burns were attempted to rectify the situation, but the engine continued to underperform, and further firings became impossible when the oxidiser ran out. As a result, the satellite was left in a transfer orbit with a perigee of some 7,500 kilometres, which subjected it to the radiation of the van Allen belts. Although some of the technology experiments were able to be conducted (e.g. the attitude control system and the nickel–hydrogen battery) the radiation damage to the solar arrays caused their power output to drop from 5.8 kilowatts to 5.3 kilowatts in just 10 days, with the power projected to fall to inoperable levels within a year.[68]

Lunar A
The development of the Japanese Lunar A mission, originally scheduled for launch in 1999 on an M-5-III launch vehicle, was delayed by various problems associated

with the design of the penetrators that it was to fire down into the surface of the Moon. Then, soon after it was finally rescheduled for the summer of 2004, the propulsion system had to be dismantled so that the thruster valves could be returned to the US manufacturer, Moog Inc., which had issued a recall notice in order to check for possible faults.[69,70] As of the time of writing, the launch has yet to be rescheduled.

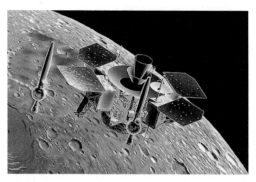

An artist's impression of the Lunar A spacecraft and one of its surface penetrator probes.

Westar 6 and Palapa B2

In February 1984 STS-41B deployed two HS-376 satellites, Westar 6 for Western Union and Palapa B2 for Indonesia. Each was to be inserted into geosynchronous transfer orbit by a Payload Assist Module using a Star 48B solid rocket motor. However, in both cases the motor fizzled out after 10 seconds, leaving the satellites in orbits with an apogee of 1,000 kilometres. The investigation examined motors from the same batch and found that the curing of the graphite epoxy composite wall of the nozzle had bubbles of gas trapped within it that would have resulted in a burn-through, and since combustion is strongly pressure-dependent this hole would have reduced the pressure in the motor sufficiently to end combustion.[71] The satellites were declared lost, and their title passed to the insurance consortium. By slashing the $225 million pool of money available for satellite insurance by a total of $180 million, this dual failure posed a grave risk to insurers.[72] However, after jettisoning their useless motors, the satellites were able to use their thrusters to lower themselves to an altitude from which they would be able to be retrieved. This was done by astronauts on STS-51A in November 1984. Westar 6 was refurbished and relaunched as AsiaSat 1 on 7 April 1990, on the first commercial flight of China's Long March 3 launch vehicle, and Palapa B2 was relaunched a week later by a Delta II as Palapa B2R.[73,74]

MAKING DO

Although in-space propulsion failures have caused a number of spacecraft losses, in some cases when a launcher strands a satellite in the wrong orbit the performance of its own propulsion system facilitates at least a partial recovery of the mission. Often, a geostationary communications satellite has propellant for many years of station-keeping, and substantial manoeuvres are able to be made at the expense of the overall service life. In some cases, the satellite has enough capability to recover the originally planned orbit; one such example being TDRS 1. In other cases, the propulsion system can allow only a modest amelioration of the situation, and the mission must make do with a less desirable orbit; the case of Hipparcos being an excellent example. In any event, since the initial orbit may deteriorate through atmospheric drag, or expose the satellite to radiation, the flight dynamics specialists often have to work against the clock to develop a creative trajectory and manoeuvre sequence.

The case of GEOS 1

An early example of a 'make do' type of failure was the European Space Agency's GEOS 1 launched on 20 April 1977.[75] This spin-stabilised satellite (which later formed the basis for the design of Giotto) was to investigate the magnetospheric environment in geostationary orbit. It was to have been put into a 10.5-hour transfer orbit that would be circularised at apogee using a solid rocket motor, but the second stage of the Delta 2914 underperformed badly and left the satellite in an orbit with a period of only 3.7 hours. Furthermore, the launcher was supposed to spin in the satellite to 90 revolutions per minute, but only achieved a distinctly wobbly 1.5 revolutions per minute. Within three days, the flight controllers had devised and implemented a new plan wherein the solid rocket was fired to achieve an elliptical 12-hour orbit to enable the satellite to intermittently sample the geostationary altitudes of interest while it was visible to its ground station.[76]

Artemis

An underperforming Ariane V on 12 July 2001 left the European Space Agency's Artemis communications technology satellite in a transfer orbit with an apogee of 25,664 kilometres, considerably below geosynchronous altitude.[77] The satellite's primary pro-pulsion system was used to raise the perigee to 37,000 kilometres in order to lift the satellite safely above the van Allen belts. After reprogramming the attitude and orbital control system by replacing about 20 per cent of the software (an uplink of some 15,000 words), the ion propulsion system that had been installed for north–south station-keeping was used to achieve the desired orbit, but doing this took almost a year, in part because three of the four thrusters failed. Climbing at

An artist's impression of the Artemis satellite.

only 15 kilometres per day on its single remaining thruster, Artemis limped into its assigned geostationary slot on 31 January 2003.[78]

KoreaSat 1 and GStar 3

When one of three air-started solid motors of a Delta II launched on 5 August 1995 failed to jettison, the failure was attributed to possible exposure of the explosive transfer lines (i.e. a sort of detonation cord) to high temperatures. As the other motors separated correctly, and as they all used the same electrical detonator, the system must have failed in the transfer assembly or later.[79] In any case, the launcher, laden with the 'dead weight', left KoreaSat 1, a Lockheed Martin 3000 series also known as Mugunghwa, in a transfer orbit with an apogee that fell about 5,000 kilometres short of geosynchronous altitude. The orbit was roughly circularised at this altitude by firing the apogee kick motor. An addditional 325 metres per second of delta-V was required to attain the geostationary slot. To use the conventional hydrazine thrusters would consume so much propellant as to cut the planned 10.5-year orbital life by 6 years. Using the more efficient Primex electrothermal hydrazine thrusters (EHT) that used electrical heating to raise the specific impulse of hydrazine from 220 seconds to 300 seconds would have recovered 12 months worth of propellant, but the turns required to point these thrusters in the right direction would have required the satellite to adopt a low power orientation that would cause several components to cool below their qualification temperature limits, and this, together with the complex series of manoeuvres that the EHT constraints would impose, argued for using the conventional thrusters and accepting the additional reduction in orbital life. By the end of August 1995, the satellite was on-station.

An artist's impression of KoreaSat.

In an earlier delivery of the same type of satellite, operators had no choice of which type of thruster to use. GStar 3 was inserted into geosynchrononous transfer orbit by an Ariane 3 on 9 September 1988. The circularisation burn using a Star 30BP motor was to last 55 seconds, but contact was lost due to severe attitude motion towards the end of the manoeuvre.[80] When the motor shut down, the satellite was spinning at 33 revolutions per minute in the wrong direction – the initial suspicion was that the nozzle had burned through, but after Palapa B2 and Westar 6 had been stranded by nozzle burn-throughs in their Star 48 perigee motors Thiokol had replaced the carbon–carbon nozzle with a carbon phenolic nozzle to avoid a recurrence.[81] However, a later analysis found that the torques, which were an order of magnitude larger than the expected worst case, arose from an offset between the thrust vector and the centre of mass, which was apparently the result of an unequal distribution among four tanks holding the 158 kilograms of hydrazine for a decade of north–south and east–west station-keeping in geostationary orbit.[82] This attitude divergence meant that the apogee motor produced 30 per cent less net delta-V than intended, yielding a 16-hour orbit with a perigee that was nearly 20,000 kilometres below geosynchronous altitude. The unaugmented hydrazine thrusters were capable of providing only 88 per cent of the 500-metre-per-second delta-V to circularise the orbit. However, with electrical heating (of some 400 watts per thruster) the specific impulse of the EHTs would provide up to 570 metres per second, which was sufficient to manoeuvre GStar 3 into its assigned geostationary slot and leave hydrazine for at least east–west station-keeping. There was therefore no choice but to use the EHTs, even though doing so entailed significant complications. First, the satellite had to be rolled on its side, because orbit-raising thrust must be along the velocity vector, which in this case was in an east–west direction, whereas the EHTs had been installed only for north–south station-keeping. Second, since the EHTs had very low thrust, 166 separate burns over a 12-month period were required to circularise at geosynchronous altitude. The residual inclination of 1.85 degrees complicated its subsequent operations somewhat. During the orbit-raising the flight controllers used a neat trick to redistribute hydrazine between the tanks, one of which became exhausted before the others. Operation with a near-empty tank would introduce low-performance and possibly damaging two-phase flow into the thrusters. Since the four tanks were connected as two pairs, alternately warming two of the tanks to expand the pressurant enabled fuel to be thermally pumped between them.

The HGS 1 experiment

Perhaps the most remarkable propulsion recovery was that of AsiaSat 3, an HS-601HP communications satellite that was launched by a Proton rocket on 24 December 1997. The Block-D stage entered geosynchronous transfer orbit, but on trying to circularise at apogee it shut down 1 second into the 110-second burn, leaving its payload in an orbit ranging between 365 and 36,000 kilometres at an angle of 51 degrees to the equator. The insurance underwriter paid the owner of the satellite, and sold it back to Hughes, where it was renamed HGS 1 (Hughes Global Services). Starting in April 1998, the satellite, which had 1,680 kilograms of propellant, performed a series of six burns to raise its apogee to 413,000 kilometres, beyond the orbit of the Moon. On 13 May, the satellite, which was spin-stabilised at 10 revolutions per minute with its solar panels still stowed and its payload inactive, flew by the Moon at an altitude of 6,248 kilometres, which stretched its apogee to 488,000 kilometres, raised its perigee and, most significantly, reduced its inclination. It encountered the Moon a month later, on 6 June, this time at a range of 34,300 kilometres. In a series of manoeuvres over the next two weeks, it was able to circularise at geosynchronous altitude. However, the fact it was still inclined at 8 degrees meant that it would nod north and south of the equator, which precluded it from being used in its original rôle. Nevertheless, it was leased and operated as PanAmSat 22 to relay maritime communications, which do not require such precise pointing. The $4-million operation was entirely repaid by the end of 2001. Meanwhile, in March 1999 a Proton had deployed the replacement satellite, AsiaSat

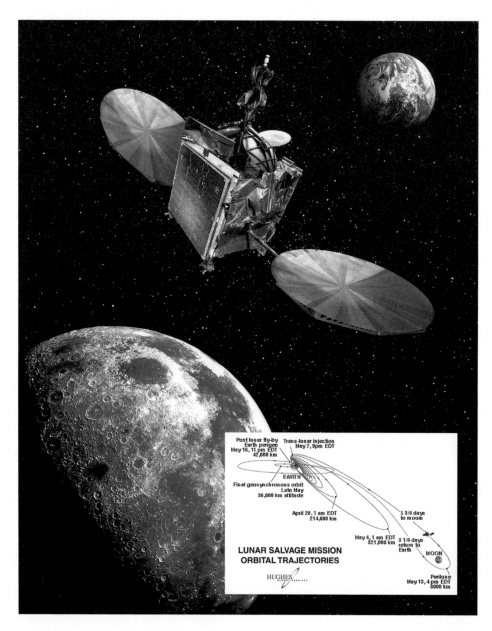

An artist's impression of HGS 1, an HS-601HP communications satellite, making one of its lunar flybys *en route* to geostationary orbit.

3S, for the original client.[83,84,85,86] Having consumed so much propellant early on, HGS 1 had a relatively brief life, and in 2002 was manoeuvred into 'graveyard orbit'.[87]

The innovative rescue of HGS 1 demonstrated the feasibility of using the Moon to facilitate manoeuvring from a high-inclination transfer orbit into geostationary orbit – however, although it would greatly reduce the amount of propellant that had to be loaded into a satellite, deliberate use of this method would require several systems to be redesigned.

NOTES

1. *Solar System Log*, A. Wilson, Jane's, 1987, p. 40.
2. *The Cosmic Connection*, C. Sagan, Coronet, 1975, p. 138.
3. *Spaceflight*, February 1976, p. 75.
4. *On Mars – Exploration of the Red Planet, 1958–1978*, E.C. Ezell and L.N. Ezell, SP-4212, NASA, 1984.
5. 'Propulsion lessons learned from the loss of Mars Observer', C.S. Guernsey, AIAA 2001-3630, 37th AIAA/ASME/SAE/ASEE, Joint Propulsion Conference, Salt Lake City, Utah, 8–11 July, 2001.
6. http://www.skyrocket.de/space/doc_sdat/mars_observer.htm
7. *Aviation Week & Space Technology*, 30 August 1993, p. 20.
8. http://klabs.org/richcontent/Reports/Failure_Reports/MarsObserverFailureSummary.htm
9. 'JPL Mars Observer in-flight anomaly investigation, with emphasis on attitude control aspects', R. Stephenson and D. Bernard, AAS 94-061, *Advances in the Astronautical Sciences*, vol. 86, p. 483.
10. 'Propulsion lessons learned from the loss of Mars Observer', C.S. Guernsey, AIAA 2001-3630, 37th AIAA/ASME/SAE/ASEE, Joint Propulsion Conference, Salt Lake City, Utah, 8–11 July, 2001.
11. 'JPL Mars Observer in-flight anomaly investigation, with emphasis on attitude control aspects', R. Stephenson and D. Bernard, AAS 94-061, *Advances in the Astronautical Sciences*, vol. 86, p. 483.
12. *Aviation Week & Space Technology*, 11 September 1995, p. 74.
13. http://www.astronautix.com/craft/planetb.htm
14. http://www.solarviews.com/eng/nozomi.htm
15. In fact, the decision to incorporate a valve in oxidiser pressurisation line was made prior to the loss of Mars Observer, but following that failure this decision seemed eminently sensible. H. Hayakawa, personal communication with Ralph Lorenz, 2004.
16. http://www.isas.jaxa.jp/e/snews/2003/1224_nozomi/1224_01.shtml
17. *The NEAR Rendezvous Burn Anomaly of December 1998. Final Report of the NEAR Rendezvous Burn Anomaly Board*, Applied Physics Laboratory, Johns Hopkins University, Baltimore, November 1999.
18. http://www.klabs.org/reports.htm#NEAR
19. http://nssdc.gsfc.nasa.gov/planetary/near.html
20. *Asteroid Rendezvous: NEAR-Shoemaker's Adventures at Eros*, J. Bell and J. Mitton (eds), Cambridge University Press, 2002.

21. http://www.space.com/missionlaunches/delta2_launch_020703.html
22. http://www.spaceflightnow.com/news/n0208/15contour/
23. http://www.space.com/missionlaunches/contour_020815.html
24. http://spacewatch.lpl.arizona.edu/contour.html
25. *Comet Nucleus Tour CONTOUR Mishap Investigation Board Report*, NASA, 31 May 2003.
26. *Flight International*, 10–16 July 1991, p. 21.
27. *Flight International*, 24–30 July 1991, p. 18.
28. *Aviation Week & Space Technology*, 25 April 1994.
29. http://www.earth.nasa.gov/history/goes/goes8.html
30. *Aviation Week & Space Technology*, 1 May 1995, p. 27.
31. NASA Public Lessons Learned Database # 0351: Galileo Messerschmitt–Bölkow–Blohm (MBB) Ten Newton Hot Start Phenomena.
32. http://www.boeing.com/defense-space/space/bss/factsheets/xips/nstar/ionengine.html
33. *Aviation Week & Space Technology*, 2 November 1998, p. 32.
34. *Aviation Week & Space Technology*, 30 November 1998, p. 20.
35. *Aviation Week & Space Technology*, 7 December 1998, p. 82.
36. 'The successful completion of the Deep Space 1 mission: important results without a flashy title', M.D. Rayman, *Space Technology*, vol. 23, 2003, p. 185.
37. http://www.boeing.com/defense-space/space/bss/factsheets/xips/xips.html
38. http://www.boeing.com/defense-space/space/bss/hsc_pressreleases/photogallery/xips/xips.html
39. http://www.boeing.com/defense-space/space/bss/hsc_pressreleases/97_08_28_pass5launch.html
40. *Space News*, 21 July 2003, p. 3.
41. http://www.space.com/scienceastronomy/astronomy/smart1_route_031028.html
42. Presentation by G. Racca at the European Geosciences Union in Nice, 26 April 2004.
43. http://sci.esa.int/science-e/www/object/index.cfm?fobjectid = 35925
44. http://news.bbc.co.uk/1/hi/sci/tech/4006293.stm
45. http://astro.estec.esa.nl/Hipparcos/
46. 'The Hipparcos mission - On the road to recovery', *ESA Bulletin*, no. 64, November 1990, p. 59.
47. http://astro.estec.esa.nl/Hipparcos/further_more.html
48. http://www.esoc.esa.de/external/mso/hipparco.html
49. *The Hipparcos and Tycho Catalogues*, SP-1200, ESA, June 1997.
50. http://www.to.astro.it/astrometry/Astrometry/HIPPARCOS.html
51. 'The Hipparcos mission – Four years after launch', *ESA Bulletin*, no. 75, August 1993, p. 7.
52. http://www.spaceandtech.com/digest/flash2002/flash2002-026.shtml
53. http://www.gsfc.nasa.gov/topstory/20020220tdrs_i.html
54. *Aviation Week & Space Technology*, 1 April 2002, p. 26.
55. *Aviation Week & Space Technology*, 6 May 2002, p. 31.
56. *Aviation Week & Space Technology*, 7 October 2002, p. 46.
57. *Flight International*, 27 October–2 November 1993, p. 34.
58. *Flight International*, 20–26 October 1993, p. 32.
59. *Jane's Space Directory*, vol. 10, 1994–1995, p. 400.
60. *Aviation Week & Space Technology*, 3 April 1995, p. 28.
61. NOAA Press Release 95-13, 10 March 1995.
62. *Aviation Week & Space Technology*, 12 May 1997, p. 47.

63. *Aviation Week & Space Technology*, 16 May 1994, p. 60.
64. *Aviation Week & Space Technology*, 19 September 1994, p. 70.
65. *Aviation Week & Space Technology*, 17 April 1995, p. 24.
66. The importance of pyrovalves in failures in 1990s was noted by B. Robertson and E. Stoneking in *Satellite GN&C Anomaly Trends*, AAS 03-071.
67. http://www.skyrocket.de/space/doc_sdat/ets-6.htm
68. *Aviation Week & Space Technology*, 3 October 1994, p. 66.
69. http://www.msnbc.msn.com/id/5676434/
70. http://home.kyodo.co.jp/all/display.jsp?an = 20040811250
71. *Aviation Week & Space Technology*, 21 May 1984, p. 18.
72. *Aviation Week & Space Technology*, 13 February 1984, p. 23.
73. http://www.skyrocket.de/space/doc_sdat/westar-4.htm
74. http://www.palapasat.com/technical.html
75. 'Lost in Space?', A. Smith *et al.*, *ESA Bulletin*, no. 117, February 2004, p. 55.
76. http://www.magnet.oma.be/sevem/ESA-GEOS1.html
77. 'Lost in Space?', A. Smith *et al.*, *ESA Bulletin*, no. 117, February 2004, p. 55.
78. 'Artemis finally gets to work', G. Oppenhauser and A. Bird, *ESA Bulletin*, no. 114, May 2003, p. 50.
79. 'Launch recovery mission for the KoreaSat Flight 1', H. Hwangho and S. Im, AAS 96-215, *Advances in the Astronautical Sciences*, vol. 93 pt. 2, p. 1697.
80. 'GStar 3 mission recovery progress following partial orbit injection event of September 11, 1988', R. Bennett, P. Misra and W. Upson, AIAA 89-3638.
81. *Aviation Week & Space Technology*, 19 September 1988, p. 18.
82. 'Attitude and orbit control for GStar 3 recovery mission', K. Raman and J. Swale, AIAA-90-2873.
83. http://www.skyrocket.de/space/doc_sdat/asiasat-3.htm
84. http://www.flug-revue.rotor.com/FRheft/FRH9809/FR9809h.htm
85. *Deep Space Chronicle: A chronology of deep space and planetary probes 1958–2000*, A.A. Siddiqi, NASA Monographs in Aerospace History no. 24.
86. http://www.boeing.com/defense-space/space/bss/hsc_pressreleases/98_06_17_hgs1ready.html
87. http://www.oosa.unvienna.org/COPUOS/stsc/2003/presentations/Johnson/tsld006.htm

11

Attitude control system failures

A CLASSIC CASE

The first American satellite, Explorer 1, was built by the Jet Propulsion Laboratory in Pasadena, and launched from Canaveral on 31 January 1958 by a modified Jupiter-C missile into an orbit ranging between 354 kilometres and 2,515 kilometres at an angle of 33 degrees to the equator. The scientific payload was integrated into the solid-propellant rocket that formed the fourth stage, with the overall package being 203 centimetres in length and 15 centimetres in diameter, with a mass of 14 kilograms.[1] Its main instrument was provided by James van Allen of the University of Iowa, and was designed to measure the radiation environment in space. Although the satellite is famous for having revealed that the Earth possesses a belt of trapped charged particles, it is not so well known that it suffered an attitude failure.

The satellite had four whip antennas mounted symmetrically about the mid-section, and was to be spin-stabilised on its long axis. Spinning objects tend to keep their spin axis aligned in a constant direction in space, and in the case of a rocket motor, even out thrust misalignments. However, once in orbit, periodic fade-outs of Explorer's radio signal showed that it was tumbling around a *perpendicular* axis instead of continuing its axial spin. The reason is simple – even in the absence of thruster firings or external torques from solar radiation pressure, aerodynamic forces or magnetic fields – a satellite will tend towards the minimum energy state in which it rotates about the axis with the maximum moment of inertia. In the case of a long thin body, this is a transverse axis. A spin around the long axis (i.e. the lowest moment of inertia) is not stable if energy loss can occur, as was occurring when the whip antennas flexed. Thus, in a straightforward demonstration of classical mechanics, Explorer 1 did not have a stable design.[2]

William Pickering (left, director of the Jet Propulsion Laboratory), James van Allen and Wernher von Braun hold up a full-scale model of Explorer 1. Insert: a cutaway showing that the payload was integrated with the final solid rocket stage.

SIGN ERRORS

The recurrence of sign errors in attitude determination and control systems is little short of staggering. This appears to be a particular problem in systems utilising magnetic coils ('magnetorquers') to control their attitude. In one case, the problem was due to the environment, but several failures appear simply to be sign errors introduced during construction.

The case of TOMS–EP

The Total Ozone Mapping Spectrometer in NASA's Earth Probe series (TOMS–EP) was a small satellite built by TRW for the Goddard Space Flight Center. It carried a single instrument to monitor the ozone layer. Similar instruments had been flown on NASA's Nimbus 7, on a Russian satellite in the Meteor 3 series, and on Japan's ADEOS 1. It was launched on 2 July 1996 by a Pegasus XL and manoeuvred into a Sun-synchronous orbit at 500 kilometres, but after the loss of ADEOS 1 in 1997 it was raised to 750 kilometres in order to continue that programme.[3,4] Although not officially listed as such, the project was subjected to the schedule and budgetary constraints of the 'faster-better-cheaper' programme. The contract was issued in 1991 with a view to launch in 1993, but this slipped due to problems with the launch vehicle. The integration and test programme lasted just nine months. The satellite was well equipped to maintain its nadir-pointing attitude, having Sun sensors, a magnetometer, gyros and two spinning Earth sensors ('scanwheels') for attitude determination, with thrusters, magnetorquers and reaction wheels to control its attitude. After concern about the polarity of the attitude sensors was raised at the Critical Design Review, a polarity-testing plan was developed, reviewed and approved by the contractor and government – polarity would be checked at component level, and a system-level polarity check would methodically test the sensors with known inputs and verify the responses of the actuators. However, rotating the spacecraft in the Earth's magnetic field to simulate tumbling was not possible, and the solar arrays did not possess the rigidity to withstand deployment in a 1-g environment, and these compromises led to failure.

The TOMS–EP project patch.

Shortly after launch, the Sun sensors were found to be providing anomalous outputs. After careful monitoring, it became evident that two of the sensors had been cross-wired. This was solved by switching the sensors in the software, but because the configuration table was stored in random access memory the update was lost when the computer was rebooted. The fault was a pin-assignment error in the connector between the satellite and one of the two solar arrays. It was not detected during testing because the sensors were not installed in their flight configuration, but were mounted on a test fixture near the spacecraft to allow convenient stimulus of the sensors and to preclude damaging the arrays. It was later found that the polarities of the magnetorquers were inverted. This was attributed to an error of interpretation – during testing, two axes had been determined to be incorrect, and so were reversed into what ultimately proved to be an incorrect configuration. This problem, too, could be resolved by a software update. One factor that might have contributed to both faults is that to cut costs the engineers who were familiar with the coordinate frames, the locations of the sensor, and the control laws to be applied, were removed from the programme when their participation in the design was completed, and only later rehired to support launch and early operations; if they had been present for the ground testing, they may have detected the faults.

While in service, TOMS–EP suffered radio interference from the ground and single-event upsets which interfered with its Earth sensor, and in 1998, on being forced into safe mode by a single-event upset, it developed a nutation. In restabilising itself by the use of its thrusters, it depleted its hydrazine and was left with only the magnetorquers, which had much lower control capability.[5]

The case of TIMED

The Thermosphere Ionosphere and Mesosphere Energetics and Dynamics (TIMED) satellite was launched from Vandenberg on 7 December 2001 by a Delta II and placed in a 675-kilometre orbit inclined at 74 degrees to study the atmosphere at altitudes between 60 and 180 kilometres.[6,7] Once in service, the satellite was to maintain itself in nadir-pointing attitude using reaction wheels, and sense its orientation using magnetometers, star trackers, Sun sensors and Inertial Reference Units (IRU). However, because the angular momentum immediately following the satellite's release was likely to be in excess of what the reaction wheels could absorb, the first order of business was to dump this momentum using magnetorquers. As TIMED could provide telemetry via NASA's Tracking and Data Relay System, the IRUs were able to be monitored in real-time. Unfortunately, this showed the momentum to be *increasing* rather than decreasing. As there were no gases on board that could be venting, the inescapable conclusion was that the magnetorquer was malfunctioning. There was no easy way to disable or to change the sign of the torquing, but because the attitude control system was distinct from the main command and data handling system, and was triggered off magnetometer data, switching off the magnetometer suppressed the unhelpful torquing commands. This was inelegant, but effective. A sign change was later made by substituting bogus (negative) scale factors into the magnetometer's software, in effect reversing the measured field direction. As

An artist's impression of the TIMED satellite.

soon as this was implemented and the torquers were re-enabled, the angular momentum began to decrease as intended. It is noteworthy that it took the operations team only 80 minutes to implement this temporary fix, and bring the satellite under control. Since the rate of tumbling never exceeded 2.5 degrees per second and the solar panels were able to provide sufficient power to prevent the battery discharging, the satellite was never in immediate danger, but "several operations personnel suffered minor premature ageing".[8] The root cause of the sign error is interesting. The polarity of the torquing rods had been checked during integration and test procedures – both under ground control and using the onboard software. However, this had been checked using a compass and the engineer had failed to appreciate that the end of a compass that points to the north magnetic pole is itself the south end! The sign error introduced as a result of these tests was implemented both in the flight software and in a dynamics simulation model on the ground.

Soon after recovering from this initial problem, the satellite began to gyrate about its desired Sun-pointing safe attitude. For some reason, it was trying to point the *wrong axis* at the Sun. Again, the satellite was in no danger because this attitude was providing plenty of power. Checks of the telemetry, hardware drawings and control software were all consistent with expectations. The problem was identified 19 hours into the mission, in photographs taken during tests at the launch site. Two Sun sensors were mounted 90 degrees from where they should have been. This misalignment occurred because these sensors were mounted on a panel that gave the main access to the interior of the satellite. For ease of access during integration and testing of the attitude control system, this panel and its two sensors were temporarily hung off to the side, with the result that their orientation did not represent flight conditions. This, however, was overcome by revising the software. Thereafter, the spacecraft maintained itself in a nadir-pointing orientation and performed its observational programme.

The case of TERRIERS

The Tomographic Experiment using Radiative Recombinative Ionospheric Extreme ultraviolet and Radio Sources (TERRIERS) microsatellite was developed by the Center for Space Physics at Boston University under NASA's Student Explorer Demonstration Initiative.[9,10] The satellite was put into Sun-synchronous orbit at 550 kilometres by a Pegasus XL on 18 May 1999. Unfortunately, it failed to face its solar array towards the Sun and promptly drained its battery. Within a week, analysis by a 'Tiger Team' had pruned

An artist's impression of the TERRIERS satellite. The solar array is hidden.

the likely fault tree from 12 possible causes down to a sign flip in one of the magnetorquer controllers (whether this was due to the compass confusion in the case of TIMED was not mentioned). The satellite had otherwise been in good health when it went to sleep, and simulations on an engineering model suggested that if it could be powered back up it would reboot correctly, at which time the problem could be resolved by a software update. The question therefore became whether the solar panel would ever be illuminated. However, the dynamics were difficult to model due to uncertainties in the satellite's inherent magnetic moment, centre of pressure, and eddy current–hysteresis damping characteristics. Nevertheless, as data from the radars of the NORAD tracking network indicated that the satellite's spin was decreasing at a linear rate, it was hoped that after several months instabilities would cause the inert craft to tumble sufficiently for the Sun to illuminate its array for intervals of up to 10 minutes, sufficient to enable it to start to transmit a carrier signal, and possibly restart its computer. Since satellites in polar orbits are most readily contacted from high-latitude sites, the team set up a station at Poker Flats in Alaska, but July and August passed with no result. In the hope that reflected sunlight over Antarctica in the southern summer might do the trick, a vigil was mounted at NASA's McMurdo Ground Station for nine days in December, but contact was never re-established. It is possible that the satellite experienced a failure beyond the simple sign error in the attitude control system.[11]

One has to wonder whether the builders of TIMED and TERRIERS were aware of the experiences of the TOMS–EP team.

The sad case of Lewis

The low-cost Lewis satellite of NASA's Small Spacecraft Technology Initiative was launched from Vandenberg by an Athena I on 23 August 1997 to make high-resolution multispectral observations of the Earth.[12] After being inserted into a 300-kilometre orbit, it was to climb to twice this altitude. However, when contact was established a few hours later, it was found that the spacecraft's computer had switched from its A-side processor to the B-side. Unfortunately, the investigation was hampered by the fact that the onboard solid-state recorder had failed. The

An artist's impression of the Lewis satellite.

processor switch could have resulted from a transient A-side failure, a gyro failure or a software error such as an unrealistically short time-out for an event while on the launch vehicle. After 45 hours, during which time the satellite correctly maintained its Earth-pointing mode, controllers activated the A-side processor and Lewis fell silent. When contact was re-established, it was discovered that the attitude control system had malfunctioned, the satellite had departed from the proper orientation and the batteries had partially discharged. Flight controllers initiated a Sun-pointing safe mode in which the spacecraft pointed its intermediate axis of inertia towards the Sun to light the array. However – in contrast to the TOMS–EP, on which the bus was based – when the axis of maximum moment of inertia was oriented at the Sun this attitude was not passively stable and required active control to maintain it. The cause of the anomalies was unknown, and after verifying that the A-side system was working, that the satellite was Sun-pointed, and that the batteries had charged, the operations crew retired to rest. When they returned, the satellite was in a flat spin, which was a stable spin about the axis of maximum moment of inertia. Later analysis indicated that excessive thruster firings caused by the satellite's autonomous attempts at control in the intermediate axis mode had prompted the A-side processor to disable the A-side thrusters and hand control to its B-side counterpart. Excessive thruster firings on the B-side had then caused the B-side processor to disable those thrusters, leaving the craft in an uncontrolled attitude in which it had drained most of its battery charge. The controllers hastily attempted a recovery on the next contact by commanding the B-side thrusters directly, but some commands were mis-addressed, the manoeuvre failed, and contact was never regained.

Fundamentally, the failure was due to poor design and inadequate testing of the Sun-pointing mode. Although this mode was simulated without difficulty, this was somewhat idealised, notably in assuming that thrusters fired perfectly and imparted an impulse only in the direction intended. Whereas the TOMS–EP (which had a *stable* Sun-pointed axis) used three gyros to completely determine attitude, Lewis (which did not) had only a two-axis gyro and so could not sense rotation around the Sun-pointing axis induced by slight thruster imbalance or misalignment. On Lewis, a time-out logic was used to disable the thruster electronics when the processor detected "excessive" thruster firings, which was defined to mean in excess of 225 milliseconds in any axis during any 61.4-second interval. This feature was to preclude an inadvertent vehicle spin-up and to preserve fuel, but in trying to suppress a motion induced by a thruster imbalance it behaved as if it was 'stuck in a loop', which doomed the satellite. Although the angular momentum vector of the satellite would have remained Sun-pointing, once the thrusters were disabled and control was lost, dissipation would have caused the satellite to tip over to align its stable axis, rather than its unstable axis, on that vector. With the solar array now edge-on, the battery would not have been able to recharge.

In addition to the flawed design of the attitude control system, which was in part due to over-reliance on its similarity with the TOMS–EP, the failure board cited the inadequate operations support as a contributory factor. As a cost-cutting measure there was only one small crew, which meant that for long periods the satellite was unmonitored, leaving it vulnerable to being overwhelmed by anomalous situations. There should have been more ground station availability during the critical early operations phase. The inadequacies in the design of the attitude control system may have derived from a cost-saving project reorganisation that relocated all the effort – except the attitude control system design and the ground operations – from Virginia to California. Even in the age of the internet, placing the attitude control system team so far away from the remainder of the project was hardly likely to result in a robust development. The board also noted fundamental structural problems with the project management, and how NASA and the contractor, TRW, understood their rôles and priorities in a 'faster-better-cheaper' project.[13] The project suffered a heavy turnover of staff, having four Project Managers in 14 months, and lost a Systems Engineer and a lead Integration and Test Engineer.

Not surprisingly, the development of the Clark satellite, which was to have served as the counterpart to Lewis, was cancelled soon thereafter.[14]

Field reversals

Satellites in geostationary orbit can use magnetorquers to control roll and yaw (as determined, for example, by an infrared Earth sensor) but because the strength of the Earth's magnetic field declines with the cube of distance from the planet, the control authority of a magnetorquer at geosynchronous altitude is so weak that the sum of the Earth's field, the solar field, and the effects of magnetospheric currents can *reverse* the polarity and cause the commanded magnetorquer actuation to exacerbate rather than counter an attitude excursion.[15] When this happens control is switched to thrusters. This occurred to the Anik B1 satellite twice during its 7-year life.

GYRO FAILURES

Many spacecraft use gyros to sense their attitude, or more specifically changes in, or rates of changes in orientation. This is particularly the case with astronomical or remote-sensing spacecraft, where precision pointing is required. Unfortunately, these components are particularly prone to failure.

In the most conventional sense, a gyro is a small flywheel that is set in motion by an electric motor to define an angular momentum vector, which will tend to remain fixed in space. Any motion of the gyro housing, and the spacecraft to which it is attached, can be sensed by comparison to this reference. For that reference to have an *absolute* meaning, a context based on the Sun, Earth or the stars is required. However, gyros are very useful for sensing *relative* attitude during manoeuvres, and for maintaining attitude knowledge during periods between external references, such as during an eclipse when Sun-sensing is impossible. They are particularly useful in measuring rates; for example, immediately after release from the launch vehicle, before the nominal operating attitude is achieved. The term 'gyro' is often applied in a functional sense to devices used to sense orientation changes – for example, 'solid-state gyros' using piezoelectric vibrating masses, or fibre-optic gyros and ring-laser gyros which use the change in optical path introduced by rotation to sense angular motion. Note that although they are conceptually similar, gyros differ in architecture (and usually in size, as well as in function) from other spinning masses such as reaction wheels and momentum wheels. Gyros are purely *sensing* devices, while the others are *actuators*. Reaction wheels and momentum wheels can generally be the same devices, although the latter are typically larger: reaction wheels nominally have a zero spin rate, being spun one way or another to impart a torque to the spacecraft, whereas momentum wheels are designed for a constant high spin in order to give the spacecraft a momentum bias, a large angular momentum as if the entire spacecraft were spinning. In anticipation of gyro failure, many spacecraft are fitted with spares, and many spacecraft have been able to continue operations using two, one, or even zero operating gyroscopes. To a large extent, for routine operations, gyro functions can be performed by software. When the motion between a pair of external fixes is sufficiently predictable, state estimators such as Kalman filters can make a robust guess at the projected attitude.

ASCA's troubles

The attitude control system of the 'Astro D' satellite launched by Japan on 20 February 1993 and then named the Advanced Satellite for Cosmology and Astrophysics (ASCA) incorporated a four-gyro (plus one backup gyro) inertial reference system, four reaction wheels, two star trackers and three magnetorquers.[16] When one gyro suffered a scale factor shift of 10 per cent on 7 September 1993 this resulted in a Sun-pointing error of 15 degrees and forced the satellite into safe mode.[17] Once this problem was rectified, the satellite resumed operations.[18]

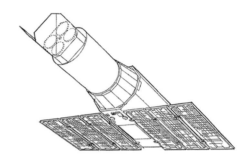

The ASCA satellite.

Although the solar heating of its structure caused attitude transients, this did not seriously affect its mission. However, on 16 July 2000 a strong solar flare heated the Earth's upper atmosphere, inflating it and increasing the air density at the altitude of the perigee of the satellite's orbit to a value 100 times greater than that for which its attitude control system had been designed to cope. When the aerodynamic torque on the asymmetric structure exceeded specified limits, the satellite tried to adopt a safe-hold attitude, but the magnetorquers were unable to compensate, and control was lost.[19,20,21] Although flares of this intensity can be expected to occur once or twice during an 11-year solar cycle, if the control thresholds had been set at more forgiving levels the satellite would probably have been recoverable.

Hipparcos wins through

The Hipparcos satellite, which required exceptional attitude knowledge to infer the position of stars, faced a considerably more complicated attitude determination task than originally intended. It was planned to operate from geostationary orbit, with a modest radiation count and predictable perturbation torques. Unfortunately, it was stranded in geosynchronous transfer orbit with not only a much heavier radiation dose but also a varying gravity gradient and other torques. However, a precision attitude model was devised, and the suite of onboard gyros was utilised to assist in attitude determination and control. Five gyros were carried for redundancy. Initially gyros 1, 2 and 4 were used. When 4 failed after one year, 5 was used. The reserve (number 3) suffered a partial failure after two years. By the time number 5 failed at the end of 1991, the package had met the specification defined for the original mission. The satellite limped along using various incomplete gyro combinations (1–2, 1–3, and 2–3) until its computers succumbed to the radiation in mid-1993. Most of these failures were attributed to the CMOS logic and opto-couplers in the motor drive circuitry, which experienced nearly 10 times their designed radiation dosage. Gyro number 4 had high and variable drag torque, which led to premature degradation. In fact, when it failed and spun down, the other gyros were sufficiently sensitive to note the tiny change in attitude as the failed unit's angular momentum was transferred to the rest of the satellite.[22]

Nursing the Röntgen satellite

The US–German–British Röntgen X-ray astronomy satellite (ROSAT) that was launched from Canaveral by a Delta II on 1 June 1990 lost a star tracker after only five months, but since it had redundancy it was able to continue its mission. However, over the next few months one of its four gyros failed completely, and the other three degraded seriously, with their drift and scale factors not only straying far beyond specification (in one case, the drift exceeded its specification by a factor of 10) but also varying with

An artist's impression of the ROSAT satellite.

time.[23] The attitude control system was reprogrammed (in the process using 94 per cent of the 12 kilobytes of random access memory on its computer). When memory dumps following repeated checksum errors showed that a single bit in

An artist's impression of the BeppoSAX satellite.

random access memory was flipping from '0' to '1' some 10 minutes after the revised code was uploaded, the software had to be rewritten to avoid this faulty memory location.

The case of BeppoSAX

The Italian X-ray astronomy satellite BeppoSAX was launched on 30 April 1996 by an Atlas–Centaur into an equatorial orbit at an altitude of 600 kilometres.[24] It had three gyros, plus three spares. When the last gyro failed in September 2001, it joined the club of 'gyroless' astronomy satellites.[25,26] It continued to make observations, albeit more laboriously, until battery failures degraded its electrical and thermal stability and it was shut down in April 2002.[27]

Magellan's jitters

On 4 May 1990 a Shuttle dispatched the Magellan spacecraft to Venus. Five months into its interplanetary cruise, it began to exhibit erratic motor current shifts in one of its gyros. This was initially thought to be due to vibration in its retainer ring, but was later found to be a manufacturing process error in which the bearing lubricant had been contaminated by a solvent; the unit was taken off-line.[28] Once Magellan was using its synthetic-aperture radar to map Venus through the planet's cover of cloud, it suffered a separate attitude anomaly in which the X-axis attitude error and body rates sporadically increased. This was diagnosed when the mapping activity was suspended while the planet passed through solar conjunction. At low angles, the solar panels could not quite converge and the algorithm commanded them to tilt back and forth by one step in search of the minimum angle, and this jittering excited the 7-hertz oscillation mode of the panels; the algorithm was fixed by a software patch.

Arabsat

In addition to solar array deployment difficulties, the Arabsat 1A communications satellite, an Aerospatiale Spacebus 1000 series that was launched on 8 February 1985 by an Ariane 3, suffered multiple gyro anomalies early on.[29]

Galileo

Even the long-suffering Galileo spacecraft had gyro problems. During its extended mission in orbit around Jupiter, the analogue-to-digital converter for one of its two gyroscopes is thought to have been damaged by the intense radiation. The converter used different components for different signs of X-axis signals, and one component produced digital outputs that were 20 per cent too high, making the gyro response asymmetric.[30] By January 1998 this had produced an off-Earth pointing error of 9 degrees, but it was overcome by a software patch.

The Hubble Space Telescope

A failure mode identified (after a year and a half of investigation) in the gyros of the Hubble Space Telescope was that the oil in which the wheel was suspended had been forced into the chamber by pressurised air, and it is believed that a small amount of oxygen that had dissolved in the fluid had reacted to create compounds which then corroded the very fine wires taking current to the motor.[31] Since the Hubble Space Telescope could be serviced by Shuttle astronauts, it was possible to eliminate the corrosion problem by replacing the failing units with ones in which the oil had been forced in by inert nitrogen gas.

STAR TRACKER FAILURES

Sensing the Sun is easy – except during an eclipse! In some of the cases documented in this book, this was achieved by measuring solar array currents, or even by measuring temperatures. However, sensing other stars requires rather more sensitive detectors, and sometimes these can cause problems by sensing things they should not, such as particles shed from the spacecraft.

Deep Space 1 overcomes adversity

As the first of NASA's New Millennium missions, Deep Space 1 was to break out of the chicken-and-egg situation in which new technologies were not used in spacecraft for the simple reason that they had not previously flown in space and were therefore deemed to be 'risky'. Some of the innovative technologies involved the autonomous operation of a spacecraft. The Jet Propulsion Laboratory hoped that after it had achieved its primary objectives, the spacecraft would be able to exploit the capabilities of its ion propulsion system to make a flyby of an asteroid. The asteroid chosen did not have a name, only the designator 1992KD, which reflected the date of its discovery. At only 2 kilometres in diameter, this was the smallest object to be assigned as a flyby target. The big day was to be 29 July 1999. A software problem caused the spacecraft to safe itself 16 hours prior to the encounter. Nevertheless, the team was able to reactivate the spacecraft, design and uplink the data for the final trajectory correction maneuver and restart its instruments, but the actual distance of closest approach was almost twice the planned 15-kilometre range.[32,33] One of the innovative technologies was the Miniature Integrated Camera–Spectrometer (MICAS) in which the optical system served four detectors. Another was the 'Autonav' software that acquired and processed the images to autonomously determine and correct the spacecraft's position and trajectory. With 27 minutes to go, the MICAS, as planned, ceased to use its CCD sensor (which was expected to be unusable when the bright target spanned too many pixels) and switched to the Active Pixel Sensor. However, as later analysis would determine, the facts that this new sensor had an unsuspected non-linearity under low-illumination and the target was dimmer than the worst-case predictions meant that the signal-to-noise was too low for the Autonav to point the instrument that required to be precisely aimed at the target. The other instruments required only to be pointed in the general direction of the target and were able to collect data throughout the encounter. About 15 minutes after the point of closest approach, the Autonav managed to resume fine pointing on the basis of a position estimate derived from CCD images that had been secured 70 minutes prior to the encounter. However, at a relative speed of 15 kilometres per second, by this time the range had opened to 14,000 kilometres and the asteroid had already diminished to a speck in the field of view. Several days earlier, the International Astronomical Union had announced that the asteroid was to be named 9969 Braille, after the inventor of readable material for the blind. The irony of this name was not lost on the flight team.[34]

The MICAS instrument was later to redeem itself following the failure of the Stellar Reference Unit (SRU), or star tracker, which was a commercial device. If the attitude calculated using the SRU and laser gyros differed by more than a given amount from that estimated from the less accurate Sun sensor, then software would declare a fault and place the spacecraft in a safe state. The SRU could detect stars brighter than magnitude 7.5 within a square field of view 8 degrees wide, and generate an attitude determination four times per second as a quaternion – a special type of vector favoured by attitude control engineers having four terms rather than three in order to avoid the computational singularity that can otherwise occur in certain orientations. It had suffered intermittent problems throughout the flight, and failed on 11 November 1999, about two months after the official end of the primary mission. The fault protection system attempted to power-cycle the device, and when this failed to rectify the fault the spacecraft adopted a Sun-pointed mode. However, in this mode, without the SRU, the spacecraft could not even point its main antenna Earthward. Simply devising a way to point the antenna took 2 months. Only then could the state of the SRU be diagnosed. As there was no redundancy (to save cost), and as the SRU was considered to be a mission-critical unit, the obvious course of action was to terminate the mission on the basis that the technology-based objectives had been achieved. However, the team were reluctant to give up on the mission extension, which was to attempt a comet flyby. It was decided to try to use the MICAS as a substitute for the SRU. Its 0.7-degree field of view would enable it to lock onto a star, but it would not be able to issue attitude determinations to the computer as rapidly as the SRU because it would be able to provide images at a rate no greater than once every 30 seconds. The Autonav would process each image and deliver the location of the star to the attitude control system, which would make a determination of the spacecraft's attitude. Using a star tracker disadvantaged by what amounted to 'tunnel vision' would involve substantial programming. To overcome the severe light-scattering effect in the MICAS, a 'dark' image was taken prior to a star-imaging sequence to allow the scattered light to be 'subtracted out'. The system recorded 3 by 3 frame mosaics of images in its 'search' mode (using the gyros to control attitude during the modest interval required to do so), and took two images per position in order to reject any cosmic ray strikes on the detector that might masquerade as stars. The short time available to implement a rescue limited the sophistication of the onboard software. Operating in this mode posed some interesting mission design challenges. The ion propulsion system, which had to operate for thousands of hours to drive the spacecraft to its next encounter, had to be pointed in roughly the right direction continuously. That direction had also to point the MICAS at a sufficiently bright star for guidance (dubbed a 'thrustar'). An approximation to the ideal thrusting direction was achieved by breaking the thrust arc into about 10 segments, each with its own thrustar. Another complication was that whenever the main antenna needed to be aimed Earthward, the reference star had to be selected to simultaneously permit the camera to point at it, the solar arrays to face the Sun, and the antenna to point at the Earth. In practice, the star had to be changed every few weeks as the spacecraft and Earth progressed in their separate orbits around the Sun. The greatest challenge, of course, was to conduct the comet encounter with the camera serving 'double duty', because the spacecraft had to orient itself to sight on a star for attitude reference and then reorient itself to aim the

MICAS at the target.[35,36] The only occasions on which the spacecraft lost its stellar reference without being able to recover autonomously were during periods of unusually high solar activity. By the time that Comet Borrelly loomed, Deep Space 1 had been in space for three times its nominal lifetime, and had almost used up its hydrazine attitude control gas. Several contingencies were devised in case the supply ran out, one of which involved placing the spacecraft on a ballistic trajectory to its target and then keeping the ion engine operating to enable the attitude control system to use it to control the attitude, but this required alternating the direction by approximately 180 degrees every few weeks in order to cancel out the effect of this thrusting. The successful flyby of the nucleus of the comet on 22 September 2001 provided a second bonus for what started out as a technology demonstration mission.[37,38]

TOPEX–Poseidon

Launched on 10 August 1992 by an Ariane 42P, the TOPEX–Poseidon satellite was a Fairchild multipurpose bus (similar to that used by SolarMax) with a US–French instrument package that included a radar altimeter that was to measure the topography of the oceans with exceptional precision in order to infer the currents running beneath, and thereby investigate the El Niño phenomenon.[39,40]

An artist's impression of the TOPEX–Poseidon satellite.

The satellite used a suite of sensors for attitude control, including two advanced star trackers with CCD arrays. For high sensitivity, each incorporated a thermoelectric cooler to reduce its dark-current background. To preclude damaging the CCD, a shutter closed each time a bright object moved within 20 degrees of the boresight. As the shutter had a few pinholes to admit a small amount of light when closed, it reopened when the light level fell below a predefined value. After four months of nominal operations, one of the star trackers erroneously closed its shutter when a single-event upset latched the state of the analogue-to-digital converter in the CCD readout electronics to indicate a twos complement binary output instead of the true value, giving rise to a 'reverse video' effect that the protection logic interpreted as a bright object. The Attitude Determination and Control System adapted by using the remaining star tracker and the Sun sensor. This worked well until mid-1997, when the background level in the operational star tracker began to increase, most likely due to the degradation of its readout electronics. By April 1998, the level had increased sufficiently to introduce hysteresis in the control logic: if the shutter were to close, the logic would *always* see a 'bright' level irrespective of what was in the field of view, and therefore prevent the shutter from reopening. The engineers set out to predict the occasions when the Sun or Moon would impinge on the star tracker's field of view, at which times the satellite would be ordered to rotate on its yaw axis to face the other way. Unfortunately, during one such turn, an unexpected reflection of the Sun in the interior surface of the star tracker's light baffle prompted the shutter to close, and it has remained in that position to this day. With both star trackers closed, engineers commanded a power cycle, which cleared the latch-up in the first unit to fail, enabling it to take over, with the added advantage that its background level was still low. Although designed for only three years of operation, the satellite is still monitoring the oceans.[41]

NOTES

1.   http://www.jpl.nasa.gov/missions/past/explorer.html
2.   http://nssdc.gsfc.nasa.gov/database/MasterCatalog?sc = 1958-001A
3.   http://leonardo.jpl.nasa.gov/msl/QuickLooks/tomsQL.html
4.   http://www.spacedaily.com/spacenet/text/adeos-j.html
5.   'The importance and difficulty of Attitude Control System (ACS) phase and polarity testing: Lessons learned from TOMS-EP early operations', R. Lundquist, P. Sabelhaus, S. Scott and E.B. Holmes, AAS 98-075 in Guidance and Control 1998, *Advances in the Astronautical Sciences*, vol. 98, p. 633.
6.   http://www.timed.jhuapl.edu/
7.   http://stp.gsfc.nasa.gov/missions/timed/timed.htm
8.   'Recent G&C experiences of the TIMED spacecraft', W. Dellinger, H.S. Shapiro, C. Ray and T. Strikwerda, AAS 03-074, Guidance and Control 2003, *Advances in the Astronautical Sciences*, vol. 113, p. 569.
9.   http://www.sop.usra.edu/terriers.html
10.   http://nssdc.gsfc.nasa.gov/database/MasterCatalog?sc = 1999-026A
11.   http://www.bu.edu/satellite/mission/missionops.html
12.   http://www.itc.nl/~bakker/earsel/9709.html
13.   *Lewis Spacecraft Mission Failure Investigation Board: Final Report*, 12 February 1998.
14.   http://www.astronautix.com/craft/clark.htm
15.   *Spacecraft System Failures and Anomalies Attributed to the Natural Space Environment*, K.L. Bedingfield, R.D. Leach and M.B. Alexander, NASA RP-1390, August 1996.
16.   http://heasarc.gsfc.nasa.gov/docs/asca/newsletters/mission_overview1.html
17.   'In-orbit performance of ASCA satellite attitude control system', K. Ninomiya *et al.*, AAS 94-065, *Advances in the Astronautical Sciences*, p. 555.
18.   http://heasarc.gsfc.nasa.gov/docs/asca/newsletters/gof_status4.html
19.   'Angular momentum-based safing attitude control: An approach and experiences of the ISAS spacecraft', K. Ninomiya *et al.*, AAS 02-015, Guidance and Control 2002, *Advances in the Astronautical Sciences*, vol. 111, p. 149.
20.   http://heasarc.gsfc.nasa.gov/docs/asca/safemode.html
21.   http://heasarc.gsfc.nasa.gov/mail_archive/ascanews/msg00497.html
22.   'Gyroless attitude estimation for Hipparcos', S. Val Serra *et al.*, 2nd International Conference on Guidance, Navigation and Control, ESTEC 12-15 April 1994, ESA, WPP-071, p. 545.

23. 'ROSAT in-orbit attitude measurement recovery', L. Kaffer, A. Boeinghoff, E. Bruderle, W. Schrempp and P. Wullstein, AAS 92-075 in Guidance and Control 1992, *Advances in the Astronautical Sciences*, vol. 78, p. 565.
24. http://www.asdc.asi.it/bepposax/sax_orbit.html
25. http://www.spacedaily.com/news/salvage-01a.html
26. This club included Rosat, Soho and ERS-2.
27. www.sat-index.com/failures/bepposax.html
28. 'Magellan attitude control mission operations', E.M. Dukes, AAS 93-045, Guidance and Control 1993, *Advances in the Astronautical Sciences*, vol. 81 p. 375.
29. *Aviation Week & Space Technology*, 25 March 1985, p. 22.
30. *Aviation Week & Space Technology*, 3 August 1998, p. 31.
31. http://www.gsfc.nasa.gov/gsfc/service/gallery/fact_sheets/spacesci/hst3-01/gyros.htm
32. 'The Deep Space 1 extended mission', M.D. Rayman and P. Varghese, *Acta Astronautica*, vol. 48, 2001, p. 693.
33. http://nmp.jpl.nasa.gov/ds1/arch/mrlogP.html
34. http://www.oasis-nss.org/articles/1999/99Oct.html
35. 'The Deep Space 1 extended mission', M.D. Rayman and P. Varghese, *Acta Astronautica*, vol. 48, p. 693, 2001.
36. 'The Deep Space 1 extended mission: challenges in preparing for an encounter with Comet Borrelly', M.D. Rayman, *Acta Astronautica*, vol. 51, 2002, p. 507.
37. http://www.jpl.nasa.gov/images/ds1/ds1_borrelly.html
38. 'The successful completion of the Deep Space 1 mission: important results without a flashy title', M.D. Rayman, *Space Technology*, vol. 23, 2003, p. 185.
39. http://www.deos.tudelft.nl/signal/topex.shtml
40. http://www.tsgc.utexas.edu/spacecraft/topex/intro.html
41. 'TOPEX–POSEIDON: Controlling the attitude of a 10-year old', P. Sanneman, B. Lee and P. Vanderham, AAS 03-077, Guidance and Control 2003, *Advances in the Astronautical Sciences*, vol. 113, p. 597.

12

Electrical failures

POWER SYSTEM FAILURES

With the exception of passive reflectors such as the Echo 1 mylar balloon (used as a communications relay prior to the development of active transponders) and 'mirrorball' laser-reflecting satellites like LAGEOS for geophysical studies, spacecraft tend to be full of electrical systems. In fact, even passive satellites require electrical systems to ensure their correct deployment. Electrical systems, of course, also need power. When a modern electronic system such as a calculator or a cordless telephone fails, it usually fails in its power system. A single short or open circuit in the power source can render an entire system completely dead, whereas a single-point failure elsewhere may only degrade it. On 8 April 1966 an Atlas Agena-D deployed NASA's first Orbiting Astronomical Observatory to open a new ultraviolet 'window on the sky', but two days into its startup routine it suffered high-voltage arcing in its star trackers and was crippled by the failure of its power supply, depleting the battery.[1] The Seasat satellite, which was integrated into an Agena upper stage, demonstrated synthetic-aperture radar imaging from orbit in 1978. However, after 105 days the bus lost electrical power as a result of a "massive and progressive" short circuit in one of the slip-ring assemblies that fed power from the rotating solar arrays to the main body of the spacecraft. The investigation concluded that this must have been caused by an electrical arc triggered by a contaminant bridging two rings of opposite polarity, or perhaps wires contacting the brushes.[2] Open or short circuits on solar arrays are not uncommon. Soon after a Japanese H-1 launch vehicle inserted the Martin Marietta

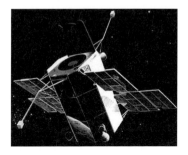

An artist's impression of the 1st Orbiting Astronomical Observatory.

3000 series satellite, Yuri 3A, the first in the BS 3 series, into geostationary orbit on 28 August 1990 to provide a DBS television service, it suffered a partial solar array failure that reduced its power output by 25 per cent.[3,4] On its interplanetary cruise, the European Space Agency's Mars Express suffered a problem in the solar array bearing and power transfer assembly that resulted in the loss of 30 per cent of its power.[5]

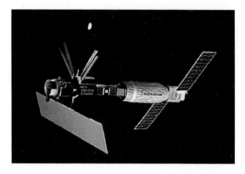

An artist's impression of the Seasat satellite.

SolarMax

The Solar Maximum Mission (SMM or SolarMax) that was launched by a Delta on 14 February 1980 blew its fuses. It may seem absurd that a satellite should have fuses at all, since it is generally impractical to replace them. However, fuses can protect a satellite during ground processing, when all manner of mishandling can occur. Unfortunately, the fuses on SolarMax were undersized. They handled the excess current for some months in orbit, but they were not hermetically sealed, and slow outgassing sustained a residual gas pressure in the material that helped to remove heat from the glowing wire. Once this pressure had fallen, the filament burned out. On 13 November 1980 the first fuse to fail disabled the roll reaction wheel; identical failures claimed the yaw wheel nine days later and the pitch wheel three weeks after that. When operational, the satellite had been 3-axis stabilised to point within a few seconds of arc of the Sun. With three of its four wheels out of action, this was no longer possible. The flight controllers devised a new, spin-stabilised mode using magnetorquers to adjust the attitude, but this meant that the instruments that required a fixed perspective could no longer be used, and even the data from the other instruments was degraded by the fact that mispointings could be as great as 15 degrees.

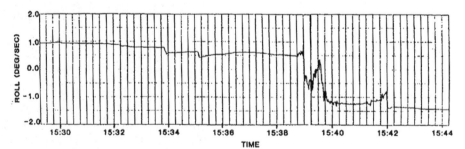

A plot showing how an astronaut's efforts at capturing SolarMax perturbed the satellite. ('The dynamics of the Solar Maximum Mission spacecraft capture and redeployment on STS-41C', K.J. Grady at the Guidance and Control Conference in 1985)

Fortunately, SolarMax had a modular bus that had been designed to be serviced in orbit. It was therefore decided that in April 1984 STS-41C should rendezvous with it, and send out two astronauts to attempt a repair. However, a device that was intended to enable a spacewalker to 'dock' with the satellite failed to engage, and further endeavours left it tumbling, prompting the astronauts to withdraw. It would take some time to stabilise the satellite using only its magnetorquers, during which time it would be reliant on its battery. To give it a fighting chance, the controllers at the Goddard Space Flight Center switched off non-essential loads, such as the radio transmitter. They got a lucky break, in that the arrays finally received sufficient illumination to enable the battery to recharge. After nearly 46 hours, SolarMax was declared ready for the Shuttle to return to make another attempt, this time by capturing it using the robotic arm. Once this had been achieved and the satellite was berthed in the payload bay, the two spacewalkers replaced the module containing the faulty attitude control system. When the retrieved module was examined after the flight, it was discovered that both of the star trackers had been destroyed during the rescue. Their protective covers required power to be actuated, but the attitude control system had been switched off during proximity operations in order to preclude a malfunction endangering either the Shuttle or the spacewalkers, and the tumbling resulting from the mishandling had aimed the trackers at the Sun with the covers open, burning them out. However, as these were on the module that was replaced there was no impact on post-repair operations.[6] The module was returned to Fairchild, refurbished, and incorporated into the Upper Atmosphere Research Satellite, which was deployed by STS-48 in 1991.

Oscar 13's battery charger

A good example of a perceived, but nonexistent failure is that of Oscar 13, a satellite for ham radio enthusiasts. It was launched on 15 June 1988 on the maiden Ariane 4 and manoeuvred into a highly elliptical orbit inclined at 58 degrees in order to spend most of its time over the northern hemisphere. Five years later, it was found that the battery would not hold charge. Further investigation revealed that this was simply a case of the battery *not receiving* any charge. The satellite's repeated transits of the van Allen belts had aged the solar arrays, altering their current–voltage characteristic such that the originally set point of the battery charge regulator no longer allowed a charging current. When this was adjusted by a fraction of a volt, all was well.[7]

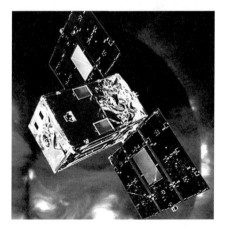

Yohkoh in the dark

The Japanese Yohkoh (Sunbeam) satellite launched on 31 August 1991 monitored the solar corona at X-ray wavelengths for a

An artist's impression of the Yohkoh satellite.

decade.[8] However, a few months later, on 14 December 2001, it flew into the Moon's shadow, interpreted the onset of darkness as an unscheduled sunset and switched to its batteries. An undervoltage condition switched off the scientific instruments – a contingency procedure called Disconnect Non-Essential Loads (DNEL) – but as the power level fell the attitude control system shut down and the satellite began to tumble. As the satellite was not in view of its ground station at the time, there was no way to intervene to save it, and full control was never regained.[9,10]

BSS-702 solar array problems

Modern communication satellites have prodigious power requirements to blast high-bandwidth television signals directly to small dishes mounted on roofs. A satellite that feeds millions of consumers is therefore a substantial money-making asset. Its revenue-generating capacity is determined by the amount of propellant available for north–south station-keeping in geostationary orbit, and by the rate at which its solar arrays are degraded by radiation – typically about 1 per cent per annum. One series developed by Hughes incorporated mirrors along the sides of the arrays to concentrate the sunlight on the transducer cells, which were high-efficiency gallium arsenide with an initial output of 16 kilowatts that was expected to decrease by only 1 kilowatt during a satellite's 15-year operating life.[11] It therefore came as a shock to find that the power was declining so rapidly that by the end of a satellite's operational life the output would be 12 kilowatts. It was

An onboard camera confirms the deployment of the 'solar concentrator' panels of an HS-702 satellite.

suspected that the arrays were fogging over, and an investigation revealed that the sealant was decomposing and, in outgassing from the arrays, was depositing material on the cells and (especially) the mirrors which, being highly reflective, and therefore chilly, served as excellent cold-traps for condensible vapour. While similar effects occur on all spacecraft surfaces, the heating of the arrays by the concentrators, and the 'trough' geometry that presented condensation surfaces to outgassed vapour, made the problem particularly severe in this case. It had not been detected in ground tests because correct simulation of the radiative environment around a large structure is difficult.[12,13] By the time that this flaw became evident, PanAmSat's Galaxy 11 and PanAmSat 1R, Telesat Canada's Anik F1 and XM Satellite Radio's XM 1 and XM 2 (nicknamed 'Rock' and 'Roll') were already in space, and their operators faced the prospect of a significant reduction in the likely operating life.[14]

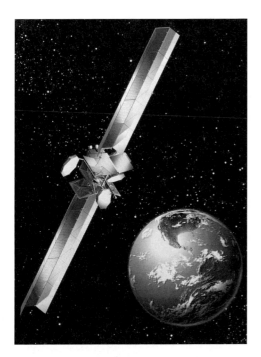

An artist's impression of Anik F1.

BATTERY FAILURES

Batteries are tricky devices. Firstly, they are chemical systems whose performance is strongly temperature dependent, and can be irreversibly damaged by mistreatment such as overcharging. Worse, if they fail, they can do so catastrophically by leaking corrosive electrolyte or even exploding.

Mariner 7

A few days before Mariner 7 reached Mars in 1969, the Johannesburg ground station lost contact with it for a period of seven hours. It responded when commanded to use its low-gain antenna, but 15 telemetry channels were missing and others were garbled. Tracking established that the trajectory had been deflected, shifting the point of closest approach on 5 August by 130 kilometres. The spacecraft had evidently suffered a major trauma. The early speculation was that it had been hit by a micrometeoroid, but it was discovered that the silver–zinc battery had exploded. As the casing burst, the electrolyte was sprayed into space and this had acted like a thruster. The explosion (or perhaps the corrosive electrolyte) caused the various electronic units to short out.[15] The only serious impact on the flyby was that the loss of the camera scan platform's calibration meant that a new set of orientation references had to be devised at short notice.

A model of Mariner 7.

NOAA 8 disabled

The first of the Advanced-TIROS-N meteorological satellites, NOAA 8, was placed into a polar orbit at 800 kilometres on 28 March 1983 by an Atlas-E from Vandenberg.[16,17] In June 1984 a battery charge regulator failed. NOAA 6, which it had superseded, was reactivated to provide cover while a workaround was devised. The controllers treated the battery with care, manually setting a low charge rate, and the satellite resumed service. A year later, the power supply for the attitude control

An artist's impression of NOAA 8.

system's clock oscillator suffered an intermittent problem that resulted in the loss of attitude control, which aggravated the battery-management. On 30 December 1985, while the satellite was out of contact, the oscillator failed momentarily, causing the computer to adopt a safe mode that included a default mid-range charge setting for battery number 1. Nine hours later, the telemetry indicated that this battery was inoperative, and the satellite was tumbling. On 3 January 1986, NORAD warned the National Oceanic and Atmospheric Administration that it was tracking several fragments close to the satellite. Nevertheless, contact was regained four days later. The telemetry history indicated that five hours after the battery charging had been set to the mid-range value the battery had overcharged, undergone a thermal runaway, and exploded, possibly as a result of two of its cells having been weakened by a previous anomaly. The explosion had blown off the thermal blankets. The satellite could be controlled, but since the loss of the battery and the likelihood of further oscillator failures made continued operations risky, it was deactivated. Prior to the introduction of the NAVSTAR Global Positioning System, polar-orbiting weather satellites carried a search-and-rescue system for locating lost yachts and crashed aircraft, and there was concern that if NOAA 8's oscillator were to fail again and cause the satellite to tumble while this transmitter was on, it might disrupt the operations of other such satellites.[18]

Sunstroke for SUNSAT
An interesting example of thermal-electrical difficulties is the Oscar 35 satellite, also known as SUNSAT (Stellenbosch University Satellite) because it was developed by the University of Stellenbosch as South Africa's first satellite.[19,20,21] The 64-kilogram microsatellite hitched a ride on a Delta II on 23 February 1999, and was released into the same 400 by 840-kilometre polar orbit as the Oersted magnetic survey satellite developed by Denmark. For the first year, the plane of the satellite's orbit enabled it to cool down while it was in the Earth's shadow. However, as the orbital plane precessed to become normal to the Sun direction, the satellite was in continuous illumination for five months. In a larger, more sophisticated satellite, this would be welcomed because it would provide a higher orbit-averaged power level. However, the microsatellite had only modest margins on its thermal performance, and a simple power system. Although the satellite was reoriented in an attempt to mitigate the situation, it suffered high temperatures and overcharged its batteries. Thereafter, the

battery voltage tended to rapidly decrease under load, and while it was in shadow it fell to the point where the processors reset. This suggested that the nickel–cadmium batteries had degraded. The battery's capacity was at least partially restored by a reconditioning procedure (rapid discharge) and operations returned to normal, but in late February 2001 the satellite fell silent.[22] Apart from *deus ex machina* explanations such as an impact by space débris, the most likely failure scenario would appear to be a battery cell failure that prompted some other failure.

Hiten and Hagoromo

With just 18 seconds remaining in the countdown to launch Muses A – the first in a series of Mu Space Engineering Spacecraft – on an M-3SII-5 solid-propellant launcher on 23 January 1990, the hydraulic pump that powered the

The SUNSAT satellite.

nozzle gimbal on one of the strap-on boosters malfunctioned. Tests showed that a thyristor in the pump had ceased to operate in the low winter temperatures at the Kagoshima Space Centre. A hot air blower was hastily set up on the pad, and the vehicle lifted off the next day. Once in space, the spacecraft was renamed Hiten, after an angel in Buddhist lore. It was to test technologies for Japanese deep space exploration by entering cislunar space and releasing an orbiter. Because the launcher underperformed by 50 metres per second, the apogee fell 40 per cent short of the intended 476,000 kilometres, beyond the Moon's orbit. However, as injection errors are not uncommon, the spacecraft was to use accelerometers to monitor the final velocity, and if a shortfall was within specific limits it was to use its hydrazine engine to make up the difference. Unfortunately, in this case the shortfall exceeded the limit that had been imposed to prevent a faulty accelerometer from ruining an acceptable trajectory, with the result that the spacecraft ignored this real error. Nevertheless, several later manoeuvres restored the nominal orbit.[23] The 200-kilogram spacecraft was carrying a 12-kilogram secondary satellite named Hagoromo (the veil

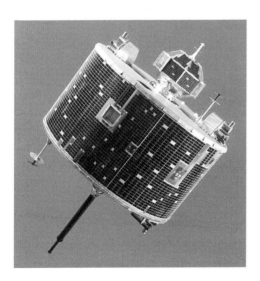

The Hiten spacecraft with Hagoromo on top.

worn by Hiten). Although Hagoromo's S-Band transmitter failed on 21 February, it was decided to go ahead and release it on 18 March, as Hiten flew within 16,475 kilometres of the Moon on its first flyby. Hogoromo was to use a solid rocket to enter an orbit ranging between 7,400 and 20,000 kilometres. The motor is known to have fired, because its plume was spotted by the 1-metre telescope of the Kiso Observatory at the University of Tokyo. The failure of Hogoromo's transmitter was not a serious issue, as it carried no instruments. The satellite was a small 26-faced polyhedron, with 16 of its surfaces covered with a new type of solar array utilising indium phosphide transducers. The performance of these arrays was characterised by measurements in space, but telemetry ceased when Hagoromo switched to Hiten's power supply, and ground tests showed that a transient high voltage could occur during this switch over and prompt a transistor in the subsatellite's transmitter to break down. After a successful mission during which it performed the first demonstration of aerobraking of a spacecraft at near-escape speeds in a planetary atmosphere – in this case, the Earth's atmosphere – Hiten inserted itself into lunar orbit on 15 February 1993, and on 10 April 1993 was deliberately crashed.[24]

Unlucky Indian satellites

India's first satellite, Aryabhata, which was named after a fifth-century scientist, was deployed by a Soviet Cosmos 3M launcher on 19 April 1975, but five days later suffered an electrical transformer failure.[25] The Insat 1A geostationary communications satellite was built by Ford Aerospace and launched by a Delta on 10 April 1982 to provide India with geostationary meteorological and communications services. It took 12 days to coax the C-Band antenna to deploy. When the solar panel failed to fully unfold, the S-Band transponder overheated, which degraded the television and radio service, and the power shortage resulted in the meteorological instruments shutting down on 13 August.[26] On 4 September it expended its propellant attempting to recover after losing Earth-lock caused by an "unanticipated Moon interference", and was deactivated two days later.[27] As Insat 1B was ejected from its cradle in the payload bay of STS-8 on 31 August 1983, the Shuttle's crew heard a "clunk", but the satellite sailed away apparently unscathed riding its solid perigee kick motor. On achieving geostationary orbit, it took several weeks to flatten out its segmented solar panel. At the end of its seven-year nominal mission the satellite was relegated to serving as an in-orbit spare, because one of its two travelling-wave-tube amplifiers had suffered a drastic reduction in gain and its nickel–cadmium batteries had failed.[28] Insat 1C, which was launched by an Ariane 3 on 21 July 1988, also had difficulty in deploying its solar panel.[29] Shortly thereafter, a short between the external portion of the solar array drive and the array itself wiped out one of the two power buses, thereby denying the satellite six of its 12 C-Band transponders and one of two S-Band transponders.[30] It was abandoned on 22 November 1989 when it lost Earth-lock.[31] Launched by an Ariane 4 on 3 June 1997, Insat 2D suffered a short circuit on 1 October that caused it to lose Earth-lock, prompting it to be written off.[32]

Deaf British satellites

On 16 November 2000, a pair of British technology demonstration microsatellites named STRV 1C and 1D hitched a ride with PanAmSat 1R on an Ariane V to study the harsh radiation environment in geosynchronous transfer orbit. Two weeks later, a problem developed in STRV 1C and a few days later 1D showed a similar anomaly.[33] A software design error had denied their command receivers power. Specifically, latching relays needing only a short pulse had been instead driven by a continuous current, and this had heated the relays, degrading their insulation and causing a short circuit that blew the fuses in the main receiver. Ironically, there was a secondary receiver, but this had been isolated by a trip-switch that required a ground command to be reset, and with the primary receiver out of action this could not be done.[34]

Nuclear power

As we have seen, batteries can be prone to failure, and their degradation is often the life-limiting factor on a spacecraft. Similarly, the sudden death of many spacecraft upon loss of attitude control follows from the draining of the battery as a result of the inability of the solar arrays to supply power. Nuclear power sources that provide continuous power without requiring that the spacecraft maintain a particular orientation would be much more reliable. Their initial development was stimulated by the poor lifetime of rechargeable batteries available early in the space programme for use in conjunction with solar arrays. The Department of Defense ordered the System for Nuclear Auxiliary Power (SNAP) for satellites that were required to operate for many years.[35] As the atoms in a material such as plutonium-238 decay by nuclear fission they release energy that thermocouples can convert into electricity. The first use of the SNAP power system was the Transit series of navigational satellites, which had a design life of five years. The first, Transit 4A, was launched on 29 June 1961 by a Thor–Able. Transit 4B followed on 15 November 1961. They carried SNAP 3A systems. The first incident was on 21 April 1964 when Transit 5BN-3 failed to achieve orbit. The faulty Able stage placed the satellite on a ballistic arc with a 1,600-kilometre apogee over the south pole. Because the cell of the SNAP 9A was not designed to survive re-entry, it would have scattered its 1 kilogram of plutonium high in the atmosphere. NASA's first use of such a system was to supplement the solar arrays of its Nimbus weather satellites, but in view of the Transit loss the agency ordered that its cells be modified to survive re-entry. On 18 May 1968 a gyro failure caused a Thor–Agena-D from Vandenberg to veer off course, and it was destroyed by the range safety officer at T + 120 seconds. The wreckage of Nimbus B splashed into the Santa Barbara Channel 5 kilometres north of San Miguel Island. After a six-month search, the Navy retrieved its two SNAP 19 units from 150 metres of water. They were refurbished, installed on Nimbus 3, and relaunched on 14 April 1969.[36]

Each of the Apollo Lunar Surface Experiment Packages (ALSEP) deployed on the Moon by astronauts were powered by SNAP 27 units. When Apollo 13 had to abort *en route*, its lunar module acted as a 'lifeboat' and was discarded just before the mothership re-entered the atmosphere. Although the delicate lunar module burned up, the SNAP that it carried would have fallen into the Pacific in the vicinity of the 20,000-foot-deep Tonga Trench.

The Soviets used thermionic reactors to power their Radar Ocean Reconnaissance Satellites (RORSAT) which were put in 1,000-kilometre orbits at 65 degrees in order to monitor Western navies.[37,38] On 25 April 1973 the failure of a Tsyklon launch vehicle dumped a reactor into the Pacific, north of Japan. American 'sniffer' aircraft flew over the site looking for radioisotope traces, in order to characterise the reactor.[39] A RORSAT was to eject its reactor on the conclusion of its mission and boost this into a higher orbit, but a malfunction prevented Kosmos 954 doing so. The situation was exacerbated by the fact that this satellite was in an orbit that was much lower than usual, and when it re-entered on 24 January 1978 its 50 kilograms of uranium-235 were scattered across the barren wilderness of Canada's northwest territories, prompting an international incident and a major clean-up exercise dubbed 'Morning Light'.[40] After Kosmos 1402 failed to eject its reactor, and re-entered over the Atlantic on 7 February 1983, a backup ejection system was added, and in April 1988 when the primary ejector on Kosmos 1900 failed the backup succeeded. The RORSATs used liquid sodium–potassium as the heat-transfer medium. However, this tended to leak out, and the tiny droplets introduced a new type of space débris.[41] On 19 February 1969 the shroud of a Proton launch vehicle collapsed due to launch vibrations just a few seconds after lift-off, destroying the payload. Forty seconds later, the vehicle's self-destruct system was activated. The wreckage was sifted in search of the radioactive polonium isotope that was to have kept the electronics of the Lunokhod lunar rover warm, but this was not recovered. The Mars '96 spacecraft that suffered an escape-stage failure and re-entered the atmosphere over South America in November 1996 had four small radioisotope thermoelectric generators.[42] It is not known where they landed.

Nuclear power sources are also essential for exploring the outer Solar System, where sunlight is too weak to be practicably converted into electricity. They are, however, expensive, and the political issues pose a formidable programmatic challenge – approval from the President of the United States is required for NASA to launch a nuclear-powered vehicle like Galileo or Cassini. Even just the paperwork to do this takes years and costs millions of dollars.

ELECTROMAGNETIC INTERFERENCE

One man's signal is another man's noise, or so the saying goes. Systems for spacecraft must be protected from the ambient electromagnetic fields of the space environment. In addition, the different subsystems must be protected from each other, because even a wire that is electrically insulated is still susceptible to radiated or induced electromagnetic interference from nearby systems. A related problem is that of electrostatic discharge, in which an electric charge builds up on a surface (or a person) and discharges into an electronic component, often causing it to fail. This a particularly insidious issue in spacecraft assembly and test, because it usually leaves no visible indication of damage: a part may work one moment, then suddenly cease to do so without obvious reason. If they involve power cabling, short and open circuit failures can be crippling, and if they occur on the ground they can start fires – the most tragic example being the loss of the Apollo 1 crew during a ground test on 27 January 1967 in a fire that was started by a spark from a short circuit. However, this section will focus on situations involving signal wires or transducers rather than high-current power cables.

Spurious launches

In the most extreme cases, electromagnetic interference can lead to total disaster. On 14 April 1964 an Orbiting Solar Observatory was being mated with its solid-fuel third stage in an assembly room at Canaveral for prelaunch testing, and a spark of static electricity triggered the motor's igniter, instantly filling the room with searing gases that burned 11 technicians, killing three of them.[43,44,45]

In 1987, a lightning strike activated the ignition systems of three sounding rockets at NASA's Wallops Island facility. The launch was on hold due to the thunderstorm. Two rockets were small radar targets, but the third, which was much larger, was an Orion rocket to study the ionosphere. It was mounted almost horizontally at the time of the incident and shot off on a shallow trajectory for 150 metres before falling into the sea. Fortunately, the personnel were in the command bunker at the time, and therefore out of danger.[46]

Oscar 6

The shoestring-budget 'ham' radio satellite Oscar 6 was launched from Vandenberg by a Delta on 15 October 1972 into low circular orbit. The magnetically stabilised boxy satellite had solar cells covering its faces, and was the first long-service amateur satellite to use a rechargeable nickel–cadmium battery.[47,48] Unfortunately, as the satellite crossed the terminator, the jump in power bus currents as the illumination on the solar panels changed induced spikes in adjacent logic-control lines. This electromagnetic interference caused the satellite to switch modes, not only inconveniencing the users communicating at that time but also risking a spurious command that would completely discharge the battery.[49] A set of ground stations was established to continuously maintain the satellite in the correct mode.[50] With careful management, it was nursed along well beyond its 12-month design lifetime, until the battery finally failed on 21 June 1977.[51,52]

Captain Midnight

What engineers might term electromagnetic interference can be deliberate. Military spacecraft generally are protected against jamming, and satellite control commands are usually encrypted. In January 1986, shortly after the Home Box Office network began to scramble its C-Band transmissions from the Galaxy 1 satellite, a disgruntled satellite TV dealer whose business was suffering decided to send a message, perhaps prompted by the announcement that Showtime and the Movie Channel networks planned to scramble their signals too. He was moonlighting as an operations engineer at a ground station servicing other satellites, and early on the morning of 27

April he transmitted to Galaxy 1 with a greater effective power than that of the HBO uplink and caused it to interrupt a movie for five minutes and display the message:[53,54]

> *Good Evening HBO*
> *From Captain Midnight*
> *$12.95 a month?*
> *No Way!*
> *[Showtime/The Movie Channel beware]*

Mars 3 and DS2

Landing a spacecraft in an unknown environment presents many problems, both to the designer and to the investigators attempting to determine the reason for its failure. It is possible that electromagnetic interference or electrostatic discharge contributed to the loss of several Mars probes.

In 1971 the Soviet Union dispatched the Mars 2 and Mars 3 missions, each of which was to drop a probe into the planet's atmosphere. After entry, the probe was to discard its conical heat shield, deploy a parachute to make a 'semi-hard' landing, and then open four petals to expose its instruments in much the same manner as did Luna 9, the first Soviet probe to land on the Moon. Unlike the Viking missions that NASA was planning to send, on which the landers were to be released after the main spacecraft had entered orbit, the Soviet mission design called for the landers to be released four hours prior to the main spacecraft entering orbit, so there was no possibility of revising either the time of landing or the target site.[55] By sheer bad luck, when they made their landings on 27 November and 2 December, one of the most extensive dust storms ever recorded was raging. Nothing was received from the first one, but the other touched down at 25 metres per second and its parent relayed an enigmatic 14 seconds of contrast-free facsimile television, forming only a few scan lines, not full frames.[56]

For many years the loss of these probes was blamed on the dust storm – perhaps the wind had caught Mars 2's parachute and instead of touching down vertically it had smashed into the ground on a near-horizontal trajectory, and perhaps Mars 3 had been blown over. However, it now appears that the autonomous trajectory correction performed by Mars 2 six days prior to its arrival at the planet set the approach hyperbola too low, with the result that its probe entered the atmosphere at too steep an angle and it hit the surface before the timer could release the parachute. This has been attributed to a lack of time to verify the autonomous

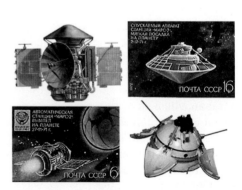

The Mars 3 mission involved the main spacecraft releasing a lander prior to entering planetary orbit.

navigation programs in the case when the actual trajectory was very close to the predicted one.[57] The theory that Mars 3 fell silent because it was rolled over by the wind is difficult to reconcile with the fact that, with its petals deployed, it was a very stable configuration. It has also been suggested that the relay system on the main spacecraft failed.[58] However, during the Second World War, radio operators serving in desert units had problems with their transmitters during dust storms due to 'corona discharge', which occurs because dry dust accumulates an electric charge as it is swept up and collides with other particles. A more likely scenario for Mars 3's silence is therefore that either corona breakdown disrupted the radio transmission, or some kind of discharge zapped a critical system onboard.

When Mars Polar Lander was launched on 3 January 1999, it carried two probes for the Deep Space 2 mission of the New Millennium programme. A few minutes before it penetrated Mars's atmosphere on 3 December, the cruise stage released the main lander and, a few seconds later, the two microprobes. Each of the 4-kilogram microprobes was to passively orient itself during entry with the heat-resistant forebody facing forward. On striking the surface at a speed of no more than 200 metres per second, and at an angle of no more than 12 degrees from vertical, the frangible shell was to shatter and the probe was to embed itself about 1 metre into the ground.[59] The probes had a very robust construction that had been tested by blasting them into the ground using a powerful air gun (tests in which one of us, RL, participated).[60] However, nothing was heard from them. A variety of failures were considered plausible, ranging from the ground being too cold, too rocky, or softer than expected, to component failure at

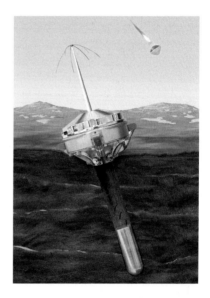

An artist's impression of how a Deep Space 2 penetrator was to embed itself in the Martian surface.

impact, or even the failure of the pyrotechnic ejection system on the cruise stage. Two possibilities related to electrostatic discharge are that the antennas on the probes, which had not been tested in conditions simulating the thin atmosphere of Mars, might have suffered breakdown or corona discharge. Another possibility is that handling at Canaveral could have inadvertently switched on the probes by faking the separation pulse, in which case the probes would have drained their batteries. Such a spurious turn-on had been noticed and corrected during the assembly of the hardware, but the susceptibility to subsequent unintended activations remained. If this occurred on the pad, it would have gone undetected since there was no interface from the piggyback payloads to the spacecraft's telemetry system.[61]

Report on the Loss of
the Mars Polar Lander and
Deep Space 2 Missions

JPL Special Review Board

REPORT ON THE LOSS OF MARS POLAR LANDER / DEEP SPACE 2
JPL SPECIAL REVIEW BOARD
— SIGNATURE PAGE —

(signatures of the board members)

Arden Albee   Charles Leising

Steven Battel   Duncan MacPherson

Richard Brace   Wesley A. Menard

Garry Burdick   Richard Hoff

Peter Burr   Robert L. Sackheim

John Casani, Chair   Al Schallenmuller

Duane Dipprey   Charles Whetsel, Deputy Chair

Jeff Lavell

22 March 2000

JPL
Jet Propulsion Laboratory
California Institute of Technology

JPL D-18709

Mars Polar Lander/Deep Space 2 Loss — JPL Special Review Board Report
JPL D-18709 — page iii

The opening pages from the investigation into the loss of the Deep Space 2 probes.

A catalogue of troubles

A number of failures due to electromagnetic interference were identified in a NASA survey of the issue, both on spacecraft and on other systems.[62] In a circumstance akin to the inadvertent ignition of rockets discussed above, this survey cited the extreme case of an aircraft that "experienced an uncommanded release of munitions" while landing on the *USS Forrestal* during the Vietnam War. The investigation determined that a degraded wire shield on the aircraft had allowed interference from the radar on the ship to issue a spurious release signal to the weapons system. The munitions struck a fuelled and armed aircraft on the deck, and 134 people were killed in the resulting conflagration. A number of aircraft crashes and even the actuation of the brakes on Mercedes cars have been attributed to electromagnetic interference when close to powerful radio transmitters, notably those of the *Voice of America* in Germany. There are several documented instances of interference to systems on commercial aircraft, the worst offenders being laptop computers, cellphones and radios. As happened to Oscar 6, VHF interference from the ground, particularly over Europe, triggered spurious commands in the NOAA 11 and NOAA 12 weather satellites and NASA's Compton Gamma-Ray Observatory. While this was able to be countered, doing so imposed extra vigilance on controllers. In the case of Compton, the situation was complicated by a design flaw that allowed electromagnetic interference to lock-up its command transponder, resulting in two 13-hour losses of communications. The Wake Shield Facility, a free-flyer deployed temporarily by the Shuttle for materials-processing experiments, lost attitude control because the power bus was coupled inductively to the unshielded cables of its attitude sensor; the problem was fixed for a later flight by adding shielding. When a Spacelab was carried on the Shuttle, the experiment data was transferred to the ground using a Remote Acquisition Unit. This had a susceptibility to transient voltage drops and, during a mission in 1985, operation of the vacuum cleaner on the mid-deck caused the unit to shut down. Also, the Spacelab intercom would encounter noise problems if used close to the Payload General Support Computer. The great variety of equipment flown on the Shuttle, much of it Commercial Off-The Shelf (COTS) apparatus, made it a challenging environment from the point of view of 'electromagnetic cleanliness and control'.

The Magellan spacecraft was deployed by STS-30 on 4 May 1989, and dispatched to Venus by a two-stage IUS. A solid rocket motor inserted the spacecraft into orbit around the planet on 10 August 1990 and was jettisoned two days later. Seven seconds after the pyrotechnics were activated, the spacecraft's computer received spurious signals. Part of the addressing circuitry for memory B had failed. The investigation constructed a failure model that reproduced the symptoms. It seemed that when one of the separation squibs was fired, a voltage transient was conducted through the spacecraft's chassis and caused a latch-up failure in which bit number 4 of a 2-kilobyte (TCC244) random access memory chip became stuck at 'high'. The conclusion was that the manner in which electro-explosive devices were initiated required some attention. Fortunately, the analysis predicted that the problem would heal itself in six months and in February 1991 the memory was restored to function.[63,64]

Transient signals produced by débris shorting across the slip-rings between the spun and de-spun parts of the Galileo spacecraft were a major cause of anomalies. The issue was particularly problematic because the intense radiation around Jupiter was prompting single-event upsets that were causing other problems. Although robust software helped to limit such anomalies, the fact that they were most likely to occur while deep inside the magnetosphere, and could last for hours or even several days, meant that observations of the inner moons Io and Europa were often interrupted. For example, on 20 July 1998 a string of three spurious signals hit both Command and Data Subsystems. CDS A handed authority to CDS B which, since the primary processor was already off-line, disregarded the spurious signals.[65] However, in November a pair of bus resets occurred within an interval of 66 milliseconds and forced the spacecraft into safe mode, resulting in the loss of most of the science data from a flyby of Europa.[66] Nevertheless, by the conclusion of its repeatedly extended mission, the spacecraft had endured *five times* the cumulative radiation dosage for which it had been designed, and was still in fairly good condition.[67]

The WIRE mishap
On 4 March 1999 a Pegasus put NASA's Wide-field Infrared Explorer (WIRE) into the planned 540-kilometre polar orbit to conduct a 'deep' infrared extragalactic survey for the Origins Program.[68] Ten seconds after the satellite was released, its attitude control electronics were turned on and its solar panels deployed.[69] However, the satellite was in trouble. The solid hydrogen that was to chill the detector of the telescope had absorbed a small amount of heat after active cooling was terminated prior to launch, creating gas, and to maximise the mission duration it was important to open the vent in the cryogen tank as soon as possible upon achieving orbit. The

STS-30 deploys an IUS with the Magellan spacecraft.

Preparing the WIRE satellite.

command to the vent pyro was to be issued by the McMurdo Ground Station on the first pass over Antarctica, and if this failed there was an onboard sequence that would be executed later in the first orbit. It was expected that as the gaseous cryogen was vented it would produce a certain increase in the satellite's spin rate, and this was to be countered by the attitude control system using magnetorquers. When Poker Flats in Alaska acquired the satellite, it was evident that even though the attitude control system was working, the spin rate was still increasing. The coolant was boiling off at a much higher rate than expected, and the torque from the venting was beyond what the magnetorquers could handle. In fact, the hydrogen vent rate was consistent with Sun and Earthlight falling on and warming the instrument, which indicated that the cover on the telescope must have been ejected prematurely. The investigation traced this to an initialisation error in the electronics of the pyrotechnics for the cover, which allowed one of the programmable gate-array chips to start up in a non-deterministic state that allowed a false signal to be sent to fire the pyro to release the cover. This had not been detected in ground testing because the external power supplies had not permitted this ambiguous initial state to occur.[70] At its worst, the satellite was spinning at a rate of 60 revolutions per minute.[71] Although attitude control was regained, the telescope was useless without the coolant, so on 8 March the mission was declared a write-off. On 6 October 2004 NASA announced the selection of a new Medium Explorer mission named WISE (Wide-field Infrared Survey Explorer) which, although not stated as such, would recover the science that was intended for the WIRE satellite.[72]

The NICMOS instrument that was installed on the Hubble Space Telescope by STS-82 in 1997 was also cooled by solid hydrogen, but in late 1998 the instrument was shut down after a heat leak enabled this to boil off more rapidly than planned. However, the instrument was restored to operation in 2002 when STS-109 fitted it with a closed-cycle radiative cooler.

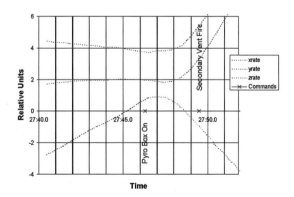

Telemetry from the first communications session with the WIRE satellite.

COMMUNICATIONS FAILURES

By their very nature, spacecraft must be remote from their users and radio links are used to control them and receive their data; but unlike terrestrial radio systems, a radio on a spacecraft must operate in a vacuum or a near-vacuum, and often endure large temperature excursions. Furthermore the remarkably high speed of objects in space also introduces the issue of frequency variations due to the Doppler effect. All these challenges – together with the more conventional problems of broken cabling, bent antennas and blocked or reflected (multipath) signals – have resulted in the failure of space missions.

The early UoSATs

UoSAT 1.

In the late 1970s the University of Surrey in England set out to design and construct low-cost satellites, thereby setting the trend for the 'microsatellites' that were to become popular in later years. The UoSAT 1 project was completed within 30 months at a cost of only £250,000. The 40-kilogram package hitched a ride into polar orbit with NASA's Solar Mesosphere Explorer on a Delta from Vandenberg on 6 October 1981. Because its payload was for use by 'ham' radio enthusiasts, it was named Oscar 9. It had uplinks and downlinks that shared an antenna – a phased-monopole canted turnstile operating at the fundamental of the VHF frequency and in a harmonic overtone mode at UHF. Although the system was designed to permit full-duplex operation (i.e. simultaneous transmit and receive) at either, or both, frequencies, the impedance of the antenna changed somewhat in space from that determined on the ground, with the result that the command receivers for both frequencies became 'desensitised' by the telemetry transmissions. This meant that communications had to be either half-duplex (i.e. turn the transmitter off while receiving) or cross-band duplex (e.g. uplink on UHF and downlink on VHF, or *vice versa*). Compounding difficulties, the antenna's UHF radiation pattern was rather ragged, making its performance sporadic. During a software test in April 1982 both of the downlinks were inadvertently activated, and as the spacecraft was 'shouting' at both of the available frequencies it was unable to hear commands from the ground. A simulation reproduced the problem and showed that a power level of +64 dBW would be required to blast commands into the desensitised receiver. This exceeded the power of the Surrey ground station, but in September the Stanford Research Institute in America employed a 50-metre-diameter dish to generate +71 dBW at the UHF uplink frequency, and once the VHF downlink had been turned off communications were restored.[73],[74] For later satellites the software was written to prevent this problem. Apart from this hiccup, UoSAT 1 was a great success, and it was still operational when its orbit decayed on 13 October 1989.[75]

UoSAT 2 (Oscar 11) was of the same design, and was launched with Landsat 5 on 1 March 1984. Three communications sessions were successfully held with the satellite on its first day in space, but the next day there was no response to ground commands. Tests conducted over subsequent weeks failed to make progress. There was even doubt that the object being tracked by NORAD was the satellite, and not a piece of débris from the launch vehicle – the satellite was very small, after all. UoSAT 2 was equipped with VHF, UHF and SHF uplinks and downlinks. A review of the measurements taken prior to launch to determine the satellite's electro-magnetic characteristics indicated that the local oscillator of its SHF uplink command receiver 'leaked' about 1 microwatt of radio-frequency power at 1.2 gigahertz. (Leakage from local oscillators is the means by which unlicensed televisions are tracked down in Britain.) This oscillator was connected directly from the spacecraft's battery rather than via its power conditioning module, and hence relied only on the spacecraft having power and not on the state of the computer, which had switched off the downlinks. Ten weeks into the mission, a ground station in Greenland detected the 1.2-gigahertz radiation, confirming that the object being tracked was indeed the satellite, and that it was receiving power. With the orbital position and status of the satellite confirmed, controllers were able to use the UHF uplink to regain control, and once the telemetry system was activated the fault was located: a component (either a logic gate, or its associated diode, resistor or capacitor) in the VHF command receiver had failed on the first day. This was an example of 'infant mortality'. Statistics show that most failures occur in the first few days of a flight. The UHF uplink had been working, but the antennas were more directional at this higher frequency and, because the satellite was still tumbling following its deployment, the antenna had evidently been pointing away from Earth on the few occasions on which an uplink had been attempted – the early recovery effort had focused on the usually more reliable, but in this case non-operational, VHF uplink. The system was reprogrammed to bypass the failed circuit and to re-route its command receiver through a digital communications experiment. Like its predecessor, UoSAT 2 went on to operate successfully for many years.[76]

After the successes of UoSATs 1 and 2, the University of Surrey began to develop a more advanced satellite for launch on a Delta in 1988, but Arianespace offered the team the use of the new Ariane Structure for Auxiliary Payloads (ASAP) that could carry six microsatellites. Because this offer allowed only a little over six months for development, the payloads intended for the 60-kilogram satellite had to be split across a pair of smaller (40-kilogram) satellites, which became UoSAT 3 and 4.[77] The satellites were to provide demonstrations of several new technologies (CPUs, memory chips, solar cells, etc.), a small CCD camera for Earth imaging, radiation monitors and a digital store-and-forward communications experiment. (These projects were particularly dear to one of the present authors, RL, who had the opportunity as an undergraduate in the summer of 1989 to build and

UoSAT 3.

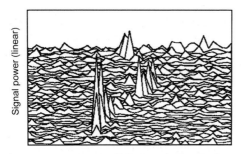

A power spectrum of 'leakage' signals from UoSAT 4 proved that the satellite was still operating. ('A possible explanation for the brief life span of UoSAT 4', K. Clark *et al* at the 1996 AMSAT conference.)

calibrate UoSAT 4's Earth-horizon sensor.) The Ariane 4 that launched Spot 2 on 22 January 1990 successfully ejected the two UoSATs and four other microsatellites. UoSAT 3 was used for six years, but sadly UoSAT 4 fell silent after only 30 hours. The early suspicion fell on the power system, in which there were several known potential single-point-failure modes,[78] but when the 35-metre-diameter antenna of the Stanford Research Institute later detected the radio-frequency leakage from the local oscillator on the 144 megahertz receiver it was realised that the satellite was powered up and that it must have suffered a telecommunications and/or control system failure. At a late stage in the design, an output filter had been introduced into the UHF transmitter both to reduce the desensitisation that the VHF receiver suffered on previous projects and to prevent interference with the DORIS precision radio-ranging experiment on the Spot 2 satellite. This filter took the form of a tuned cavity containing helical transmission lines, and large voltages could be generated in this resonant cavity when full radio-frequency power was applied. The practicalities of the satellite thermal vacuum test meant that the radio system could not be tested end-to-end: the transmitter was monitored by connecting a cable to its output, bypassing the filter and antenna. Thus the filter was never tested in a vacuum. It has been suggested that a corona discharge may have occurred as air slowly leaked from this cavity, ablated metal from the coils and deposited it on the filter wall, resulting in the filter becoming 'lossy' and sapping the signal strength.[79] This scenario has been demonstrated in ground tests. Such corona would not have occurred in true vacuum conditions, nor at sea-level pressure conditions, only in partial vacuum.

Leasat 4

The Leasat 4 communications satellite was deployed by STS-51I on 29 August 1985 and successfully achieved geostationary orbit, but a week later, as it was being checked out, it lost its UHF downlink.[80,81] This failure was traced to a malfunctioning cable linking the downlink transmitter to the deployable antenna. Unlike its predecessor, Leasat 3, which was repaired by astronauts, Leasat 4 was beyond the Shuttle's reach.

Shuttle Ku-Band link

One communications problem encountered by a Shuttle merits discussion here. The Shuttle has several times carried an imaging radar in its payload bay, and such devices generate prodigious quantities of data. In October 1984 STS-41G flew the second such mission. It was to downlink its data in real-time using the Ku-Band link, but when the Shuttle adopted the mapping attitude with its bay facing the Earth, the Ku-Band antenna failed to track the geostationary Tracking and Data Relay Satellite. As a workaround, the crew disassembled the system's electronics box in order to lock its antenna in a specific attitude, and then revised the procedure so that a swath of radar data was recorded on a tape recorder, then the vehicle rotated to aim the antenna at the satellite in order to download the tape, after which the vehicle would resume Earth-pointing to tape another mapping swath. This inspired solution permitted only 20–40 per cent of the planned targets to be covered, but this was better than nothing. A separate electrical problem in the radar unit caused an 8 to 10-decibel drop in signal-to-noise ratio.[82] This mission saw another example of human intervention: as the robotic manipulator prepared to deploy the Earth Radiation Budget Satellite, the solar panel failed to deploy, so the satellite was shaken to coax it out.[83]

Multipath issues

The inability of Galileo to unfurl its high-gain antenna on 11 April 1991 ranks as one of space exploration's historic failures. It is less well known that the spacecraft later had a temporary problem with its low-gain antenna. Shortly after the second gravitational slingshot by the Earth in December 1992, the signal strength of the downlink fluctuated unexpectedly and fell significantly. The changing geometry during the flyby required the signal to pass through the sunshade at the tip of the still-furled high-gain antenna. This structure comprised nine electrically conductive ribs with a radio-transparent shade and a circumferential wire. Reflection of the signal from the ribs interfered with the direct signal – so-called multipath loss – causing a signal drop that varied with the spacecraft's rotation. The telecommunications functional requirements stated that the tip sunshade must be transparent to radio frequencies, but because ground tests had used a transparent mockup that did not represent the conductive ribs, the multipath losses were not detected.[84]

In February 1994, as STS-60 prepared to deploy the 1,800-kilogram Wake Shield Facility developed by the Space Vacuum Epitaxy Center at the University of Houston, the robotic manipulator held the free-flyer above the payload bay to check it out, but the communications were intermittent as a result of multipath problems that corrupted 80 per cent of the signals.[85] After a horizon sensor failed and made deployment impractical, the experiment to produce semiconductor was performed *in situ*, but the package

The Wake Shield Facility

became contaminated by ice crystals from the outgassing of other apparatus in the bay.

IUS anomaly

On STS-51J in October 1985, the Shuttle released an Inertial Upper Stage carrying a pair of DSCS communications satellites for the Department of Defense. Although the autonomous vehicle successfully delivered its payload to geostationary orbit, it had ceased to transmit telemetry one-third of the way through its mission.[86]

Yuri 2A

A component in communications systems that is occasionally viewed with the same suspicion as gyros in the attitude control realm, is the travelling-wave-tube amplifier which enhances the downlink signal by passing it through a tuned cavity along which an electron beam flows in the same manner as an old-fashioned thermionic vacuum tube. The failure of the 12-gigahertz transponders on the Yuri 2A (BS 2A) communications satellite that was launched by Japan on 23 January 1984 was attributed to the failure of the three 100-watt TH3579 amplifiers supplied by Thomson–CSF of France. (The same company, at about the same time, had to delay the delivery of other amplifiers, including the 200-watt TH3619A model to be used on Olympus and TVSat 1 due to problems that included uncommanded shut downs due to discharges in the units containing the tubes and their power supplies.) The Japanese complained that the units were received in a sealed condition without technical documentation, which prevented quality-control inspection in Japan.[87] In fact, the 100-watt 12-gigahertz amplifiers on Japan's experimental Yuri 1 (BSE) satellite launched in 1978 tripped off after 14, 25 and 26 months in orbit.[88] The investigation by Thomson–CSF found that arcing in the high-voltage power supply due to degradation of the insulation had tripped protective circuits and disabled the power.[89]

Mars 2 and 3

On 19 May and 21 May 1971 the Soviet Union dispatched Mars 2 and Mars 3 to the Red Planet. On 24 June the decimetre-wavelength transmitters on *both* probes fell silent, as did their backups. It was decided to switch to the centimetre-wavelength transmitters, but they failed. After several days of trouble-shooting, the backup decimetre-wavelength transmitters were able to be reactivated, and they remained operational, but the failure of the centimetre-wavelength transmitters was a puzzle. One suggestion was that the high-gain antennas had been inadvertently pointed at the Sun for about 10 seconds, during which time the focused sunlight had damaged the transmitters, but calculations did not support this idea because the melting temperature of the silver solder used on the transmitter was 700°C. Nevertheless, as a precaution, future spacecraft were fitted with a cloth tent over the high-gain antenna in order to prevent sunlight being concentrated this way.[90,91]

In the days before encryption

While communication might be defined as the successful conveyance of a desired message, this definition would be incomplete, in that undesired conveyances should not occur. After 'Captain Midnight' struck the HBO channel, there was some initial fear that the failure of Galaxy 4 might have been due to hackers, but this was due to a manufacturing flaw. To safeguard operations, telemetry – and in particular, telecommands – are frequently encrypted. However, in the earliest days of spaceflight, the data-handling electronics were sufficiently primitive that the downlink signals were readily interpreted by skilled eavesdroppers. On 3 February 1966, astronomers using the 250-foot-diameter Jodrell Bank radio telescope in England were not only able to measure the Doppler shift on the transmission by the Soviet Union's Luna 9 probe as it fired its retro-rocket in an attempt to become the first probe to land on the Moon, but were also able to capture the subsequent telemetry and recognise this as a type of facsimile transmission such as used by newspapers to transmit images between sites. The astronomers therefore invited the *Daily Express* to feed the signal into one of its machines, with the result that the first picture from the surface of the Moon was published in the West before it was released by the Soviets. Unfortunately, since the calibration data was not so readily interpreted, the panoramic vista was depicted with the wrong aspect ratio, which made the lunar surface appear much rougher than was the case.[92,93]

Venera 7
When the Soviet Union realised that its early probes to penetrate the atmosphere of Venus had been overwhelmed by the tremendous pressure, they had decided to play safe by designing one to withstand a pressure of 180 bars and a temperature of 540°C. When the first such probe, Venera 7, arrived at the planet on 15 December 1970 it deployed its parachute at an altitude of 60 kilometres, and for the next 35 minutes transmitted data as it descended, then contact was abruptly lost. Had it been disabled by striking the surface, or had it, too, succumbed to the pressure? Several weeks later, an analysis of what had been thought to be 'noise' on the frequency used by the probe revealed that the signal had persisted for 23 minutes at a strength only 1 per cent of nominal. The probe had indeed reached the surface and rolled over, unfavourably positioning its antenna. However, the commutator that alternated the telemetry signal between various sensors had 'stuck' on the temperature sensor, and no other instrument data was returned. Nevertheless, the surface pressure could be inferred from the way in which the Doppler shift on the signal had changed.[94,95] By revealing that the temperature at the surface was a hostile 475°C, at a pressure of 90 bars, this mission paved the way for the future exploration of Venus.

Huygens relay
The Huygens probe was to transmit its data during a 2.25-hour parachute descent through Titan's atmosphere in November 2004 as the Cassini orbiter flew towards it at around 6 kilometres per second. Testing the link between the probe's transmitter and the receiver on the main spacecraft would require complex apparatus to simulate the Doppler shift and the signal drop on the ground between the two systems, and a revalidation of the connections after this apparatus was removed. As no requirement was identified in the specification, such an elaborate test was not performed.[96]

Nevertheless, at the request of some of the project team for the Huygens mission, a relay-link test was performed as the spacecraft flew by the Earth in August 1999 in which the 34-metre-diameter antenna of the Deep Space Communications Complex at Goldstone in California simulated a transmission from Huygens to validate the receiver-side of the relay link under realistic conditions.[97] There were objections, not only because such a test cost money, but also because anything that proved to be amiss would be beyond fixing. When the results of the test were analysed, it was found that the receiver had strongly detected the carrier signal, but no data could be recovered from the subcarrier on which the telemetry had been modulated.[98,99] The initial suspicion was that the test had been performed incorrectly, but continued investigation, which included simulating the digital receiver's data-transition tracking loop, established that while the receiver had worked correctly in the ground tests on a signal that was not Doppler-shifted, it had failed to synchronise adequately on the bit stream that simulated this shift. The receiver's design had been optimised for low signal levels, and had a very low tolerance for changes in the received symbol rate, and when this rate was changed slightly the receiver lost synchronisation and rejected the data. In a general sense, this was a failure of specification, in that the document did not explicitly state that the Doppler shift must be applied to the data-rate and subcarrier frequencies. A diagram in the document did show the relationship between the Doppler-shifted frequencies, and the radio system engineers ought to have thought through the implications, but in large projects the overall picture does not always trickle down to subassembly designers, and this omission in the wording was never noticed or queried. Had the flaw been noted prior to launch, it could easily have been rectified. The data-rate parameter was adjustable in software, but was specified in a table that could not be updated in flight (there was no reason it could not have been made updateable but, on the other hand, no requirement was stated for it). Similarly, some software revisions to the receiver might have allowed partial data recovery from unsynchronised data, but the receiver simply rejected this instead of doing the best it could – again an arbitrary design choice. As things stood, the Huygens team faced the prospect of a significant fraction of the probe's data being lost. The challenge for the Huygens Recovery Task Force was to identify how, with little direct influence on the receiver's behaviour, the data could be saved.

Since the Doppler shift was in a sense responsible for the failure, an obvious solution was to reduce this by either slowing the Cassini spacecraft (which would be costly in propellant) or increasing the flyby distance so that only a fraction of the flyby speed was projected on the line of sight. Either option would, however, disrupt the carefully planned flyby geometry, during which Titan would modify the spacecraft's orbital path. A redesign would have to establish a different gravity assist, to arrange the correct arrival time for the *next* flyby. While revising the trajectory was a major effort for planners, and resulted in the probe's arrival being slipped to January 2005, it enhanced the scientific return from the early part of Cassini's orbital tour by introducing an extra flyby of Titan, close observations of Enceladus, and an opportunity to observe Iapetus. Manipulating the received signal strength was also an option, since the tracking loop's performance was dependent on the signal, but the situation (although captured accurately in models of the receiver's behaviour which were validated by subsequent special relay-link tests on Huygens as Cassini cruised past Jupiter) was tricky because strong signals would trigger an automatic gain control to attenuate the signal, with the result that a small increase in signal strength might make the synchronisation *worse*.

A late decision was whether to 'pre-heat' the probe. Huygens was designed to coast in a dormant state for 22 days after its release by the Cassini spacecraft, and a timer was to awaken it about 15 minutes prior to penetrating Titan's atmosphere. This preliminary period was to allow time for the ultrastable oscillator for Doppler tracking to settle onto a precise frequency. However, with some software changes (requiring separate software for five of the six instruments, as well as a rewrite of the probe's command and data system) Huygens could be activated several hours earlier. With the systems powered up, the quartz oscillators driving the telemetry transmitters would warm up, causing their frequency to shift and slightly reduce the frequency difference. The batteries had sufficient margin to permit 4 hours of pre-heating without jeopardising the primary mission during the atmospheric descent.

Some important lessons can be learned from this story. The solution (which will be put to the test as this book goes to press) was facilitated first by executing the test that discovered the flaw. There was ample time to devise a solution – were the spacecraft *en route* to Mars or Venus it would have been challenging indeed to fix it prior to arrival. The flexibility in mission design made possible the principal corrective measure (i.e. the trajectory change) but other tools, such as the pre-heating, were enabled by the probe having a healthy battery energy margin.

Mars Pathfinder

Spacecraft communications links must often operate with very weak signals, and crucial to their receipt is that the receiver and transmitter be very accurately tuned. Since oscillator frequencies are usually temperature sensitive (a fact that was used to positive effect in recovering the Huygens relay), poorly known or poorly controlled temperatures can detune radio systems such that communications break down. Given only 26 months to develop the system by which the Mars Pathfinder lander and its Sojourner rover were to communicate on the Martian surface,[100] JPL's engineers used off-the-shelf Motorola 9,600-baud radio modems. Environmental testing showed that they would function on Mars, and that radiation effects such as latch-up could be fixed by power-cycling the modems. The system worked impressively well, considering its rapid development, but there were frequent outages that appear to have been due, at least in part, to temperature differences between the lander and the rover that caused a frequency mismatch – the bit-error rate jumped from 10^{-7} when both radios were at 20°C, to some 10^{-3} when the lander's modem was at 0°C and the rover's was at –20°C.[101]

COMPONENT FAILURES

A failure is not just an event, it is a process. Somehow the system must change from being in a non-failed state, to a failed one. A non-failed system in complete isolation should remain happy. In a purely physical sense, failure is a process that may require time and energy. In some cases, an instantaneous delivery of a large amount of energy causes the failure, such as a cosmic-ray strike, or a large electrical current or discharge. In other cases, the process is slow, and only after some time does the gradual change of the system cross a threshold after which it no longer functions. Chemical reactions such as corrosion come in this category. All physical processes, including chemical reactions, are governed by thermodynamic laws, and therefore their rate can depend on the ambient temperature – so-called Arrhenius kinetics, where the rate of the process R is proportional to a function of the form $\exp(-E/kT)$ where E is the activation energy of the process, k is the Boltzmann constant, and T is the absolute temperature. Inserting some typical values indicates that raising the temperature by 10 or 20 K can increase the rate of a reaction by orders of magnitude. Electronic engineers are familiar with the fact that a part may grow hot prior to failing, and usually the part fails *because* it is too hot. The high temperature allows chemical reactions – usually the growth of intermetallic compounds, or the break down of insulation to make short circuits – to occur in minutes or seconds, whereas it might take centuries to produce such ageing in normal conditions. However, poorly designed parts may well have reactions occurring within them at normal operating temperatures that take years to reach failure. These are perhaps the most dangerous types of failure, because if a part takes years to fail, the failure mode is unlikely to be discovered before the part is in widespread use.

Boeing's 601-type satellites

One of the most popular series of geostationary communications satellite was the 3-axis stabilised HS-601, later the BSS-601 after Boeing purchased Hughes Space and Communications. It was afflicted by failures of the above type. The first was PanAmSat's Galaxy 4, whose Satellite Control Processor (SCP) failed on 19 May 1998 after five years. As the redundant processor that should have taken over was inoperable, the satellite lost attitude control and tumbled.[102] The next failure was PanAmSat's Galaxy 7, with a brief loss of service when its primary SCP failed in June 1998.[103] It was prudently demoted to serve as a backup, and provided occasional service until its second SCP failed in November 2000.[104] After PanAmSat 4 lost its primary SCP in November 1998, its transponders were leased out at a discounted rate due to the risk of the satellite 'going dark' without warning.[105] The primary SCP of Solidaridad 1, operated by the Mexican company Satmex, failed in May 1999, and its secondary failed in April 2000.[106] The primary SCP of DBS 1 operated by DirecTV failed in July 1998.[107,108] After DirecTV's third satellite lost its primary SCP in May 2002, the satellite was moved to a graveyard orbit because its operators deemed the risk of its secondary unit failing too great. It emerged that the cause of these failures was related to a coating applied to *prevent* short circuits in the processors.[109,110] The tin plating on relay switches was causing electrical shorts when

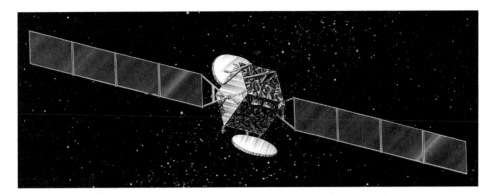

An artist's impression of Galaxy 4.

several factors were concurrently present. The telemetry from the secondary SCP on Galaxy 4, both SCPs on Galaxy 7, and the primary SCP on DBS 1 showed the unit's 50-volt current from the power distribution module increasing until a 10-amp fuse blew. In the case of the primary SCP on Galaxy 4, the failure was at its output to the solar array drive motor. A different coating method had been applied to satellites launched after August 1997, and these have shown no such problems.

This story highlights not only the absolute risk of failure, but also that *perceived risk* can end a mission. While scientists might make do with whatever they can get, a reputation for reliability is essential in the telecommunications industry, and even although a satellite running on its sole SCP might operate for years, the risk of a sudden interruption of the TV-feed to millions of customers is too great a business risk. Another point of this story is that a generic risk places a significant burden on insurance companies, which operate by spreading risk. An insurer might diligently spread risk by booking satellites on different launch vehicles, but if a generic flaw is found to affect a whole series of satellites then the insurer can be left heavily exposed.

Voyager 2 was tone deaf
In preparation for their epic exploration of the outer Solar System, the two Voyager spacecraft were fitted with many redundant systems, one example being the command receiver. Voyager 2 was launched on 20 August 1977. After the initial period of intense communication to check its systems, the level of commanding declined as the spacecraft entered its interplanetary cruise. Sensibly, the commanding system was preprogrammed to switch to the backup receiver in the event that the primary received no commands for a week. When the operators inadvertently allowed a week to pass without issuing any commands, the spacecraft responded by switching to its backup receiver. It was told to switch back to the primary, but to no effect. It was then discovered that a capacitor in the frequency tracking loop had failed in the backup receiver. This meant that a command had to be received at *precisely* the right frequency, or it would be rejected. By issuing the command to resume using the primary receiver at a range of frequencies, the switch was effected,

but this blew both its redundant fuses, which disabled it forever! After another week without commands the spacecraft reverted to its backup receiver.

The bandwidth of this receiver should have been 200 kilohertz, but with the failed capacitor the bandwidth of the backup receiver was a thousandth of this. To put this into context, the Doppler shift of the transmission due to the Earth's axial rotation is of the order of 3 kilohertz. Most awkwardly, the frequency at which the receiver operated depended on an oscillator whose frequency was temperature dependent, and a change of only 0.25°C altered the central frequency by 96 hertz – almost half the bandwidth. The temperature of the receiver depended on the attitude of the spacecraft, and on the power dissipated by various spacecraft components, so elaborate models of the temperature of the receiver had to be developed. Furthermore, each time the electrical load used by the spacecraft changed by as little as 2 watts a command moratorium of 24 to 72 hours had to be imposed, as commanding could not be relied upon until the 'best-lock' frequency of the receiver had stabilised. In case the backup receiver failed, or was disabled during its encounters, the spacecraft was programmed with the so-called Backup Mission Load in the hope that even though it could no longer be commanded it would go on to execute a minimum set of scientific observations and transmit its data to Earth. Happily, despite being 'tone deaf', Voyager 2 delivered a formidable science return from Jupiter, Saturn, Uranus and Neptune.[111],[112]

Magellan's transmitter failure

Among the Magellan spacecraft's problems, one of its downlinks was lost – albeit in the third extension of the mission. Although the X-Band carrier on chain A was present, the data subcarriers were missing when the spacecraft turned back to the Earth after a routine star calibration. The failure was judged to have most likely occurred in a Harris 2520 operational amplifier chip in the signal conditioner for the telemetry. When one of these amplifiers had failed prior to launch, it was determined that the contractor had employed air rather than dry gas when repackaging the components, which may have trapped moisture inside. In general terms, water is bad for electronics. In this instance, it was suspected that the water was combining with the phosphorus dopant in the glass passivation layer, forming phosphoric acid that was attacking the nichrome resistors on the chip. Although it was felt that additional screening had eliminated potentially faulty chips, in view of the prelaunch failure close attention was paid to these devices. The in-flight failure occurred after 20,600 hours of flight operation, and more particularly after 2,600 power cycles (with the temperature varying between 10 and 15°C each time) and it was concluded that the part probably failed due to these thermal stresses. Magellan had a second telemetry transmitter, but this was less effective, and the failure resulted in a lower data rate for the remainder of the mission.[113]

Purple plague

A well-known materials problem is the growth of brittle intermetallic compounds at, for example, junctions between gold and aluminium. Because joins between these two metals are susceptible to failure, they are best avoided. Since intermetallic compounds show up purple under microscopic inspection, this problem is often referred to as 'purple plague'. The Soviet Union sent four spacecraft to Mars in 1973. Mars 4 and 5 were to enter orbit, and Mars 6 and 7 were to release landers before flying past the planet, with the landers' signals being forwarded by the orbiters. Unfortunately, all had a congenital defect that proved fatal. In an effort to reduce gold consumption, an engineer at the Voronezhskiy factory revised the fabrication procedure of a commonly used type of transistor. Substituting aluminium leads for the gold ones made the transistors susceptible to lead-failure by intercrystalline corrosion – a failure that tended to occur 18 to 24 months after production. This problem was discovered prior to launch, but there was no time to replace the hundreds of transistors in all of the spacecraft if they were to meet the launch window. It was decided to proceed, and accept the 50 per cent probability of suffering a failure at about the time that the flotilla arrived at Mars.[114]

As Mars 4 approach the planet on 10 February 1974, a propellant leak precluded the orbit-insertion manoeuvre, but it snapped a few pictures as it flew by at a range of 2,200 kilometres. Two days later, Mars 5 entered the desired synchronous orbit. When Mars 7 arrived on 9 March it released its lander slightly early, prompting it to miss the planet by 1,300 kilometres. Telemetry from Mars 6 had ceased after only two months. It could be commanded, but no health information on its systems could be issued. However, Soviet interplanetary spacecraft were rigidly programmed, and it released its probe with Mars 5 standing by to relay the transmission. This was of the same configuration as the previous probes and, as they had done, it fell silent immediately on reaching the surface. By the time that Mars 5 succumbed to its instrument compartment losing pressure, it had returned fewer than 60 pictures, which was a meagre return on the tremendous investment.[115] The ministerial report ordered that future missions use specially produced electronic components costing almost 10 times as much as the standard components.

A similar problem befell the laser-ranging payload of NASA's IceSat in 2003. This polar-orbiting satellite was to measure the thickness of the polar ice using a lidar. The solid-state laser diodes had a predicted lifetime of 18 months, but the first one shut down after only 36 days of operations.[116] It was discovered that all three lasers had a manufacturing defect: indium solder used on the laser's circuit boards had contacted a maze of tiny gold wires, and the indium reacted to grow a layer of gold indide. Because the rate of growth was slow, the units had passed their acceptance and performance tests. As the other two lasers were expected to have similar lifetimes, it was decided to take data only every few months, accepting a discontinuous dataset in order to monitor the ice for the planned duration.

An artist's impression of IceSat.

Ranger frustrations

The Ranger missions were flown by the Jet Propulsion Laboratory in Pasadena to reconnoitre the Moon in preparation for the Apollo landings. The first two spacecraft, launched in 1961, were to have conducted engineering tests in Earth orbit to assess the basic systems, but in both cases the Agena stage malfunctioned. The Agena that was to have sent Ranger 3 to the Moon released it on a trajectory that missed the target by some 32,000 kilometres, a divergence that was beyond the capability of the spacecraft's own propulsion system to eliminate. Just before it crashed, the spacecraft was to have ejected a small spherical hard-landing capsule protected by a balsa wood shock absorber with a payload of instruments to determine the mechanical properties of the lunar surface. A contingency plan to snap pictures as the spacecraft sailed by the Moon was foiled when its computer and sequencer failed when the camera system was activated. On 26 April 1962, Ranger 4 became the first NASA probe to hit the Moon, but it was inert because the clock in its sequencer had failed. A malfunction robbed Ranger 5 of solar power, and it drained its battery before it could make the mid-course correction, with the result that it missed the Moon by 725 kilometres. This very disheartening series of failures prompted a halt to review the design and manufacture of the spacecraft and the management of the project. One issue was the baking of the spacecraft at 125°C for 24 hours to sterilise it. In fact, Ranger was in some respects overdesigned, as it was being used to test systems and procedures for planetary missions. While sterilisation was affirmed for probes to the planets, it was decided to be unnecessary for a lunar mission. Tests showed that immersing components in ethylene oxide for this treatment had made wires brittle, and therefore susceptible to damage by the vibration at launch.[117] The reviewers also called for more reliable solar panels, redundant attitude thrusters, and a more powerful engine for the mid-course correction in order to overcome less than nominal performance by the Agena. It was also decided to cancel the surface science and focus on imaging during the final approach. The hope of resuming flights towards the end of 1963 was dashed by the discovery that tiny gold flakes in diodes could float freely in space and compromise their performance, and these had all to be replaced. When Ranger 6 was finally launched on 30 January 1964, it hit the Moon but failed to transmit pictures because the power supply to the television system had been shorted out by an electrical arc that occurred as the Agena separated from its Atlas booster. After the umbilical hatch through which the hot gas had penetrated the aerodynamic shroud was hermetically sealed, Ranger 7 flew a perfect mission, transmitting 4,316 pictures prior to striking the lunar surface on 31 July 1964, and this success was repeated by Rangers 8 and 9.[118,119]

A model of Ranger 7.

INSTRUMENT FAILURES

Frequently the most delicate parts of a spacecraft are its instruments. In the case of communications satellites, the only instrumentation is the suite of attitude sensors, but Earth-observation satellites and scientific probes usually have a wide variety of payloads.

Spying fog

A major factor in the Cold War was the ability to peer over the Iron Curtain to assess the strategic potential of the Soviet Union. When the surface-to-air missile threat made U-2 overflights too risky, satellite imaging became the primary method of reconnaissance. The satellites were named CORONA (after the Smith-Corona typewriter) and incorporated a major advance in camera technology. However, the name proved prescient for another reason. If the film emulsion dried out, the system suffered electrostatic build-up – a plastic tape on a set of rollers has much in common with a Van de Graaf generator, after all – and the discharges fogged the film, ruining the pictures. When the acetate-based film that was initially used dried out it became brittle in space conditions (even though it had been used satisfactorily on balloons up to 80,000 feet) and tended to crack or even break altogether. Although switching to polyester-based film helped with the drying-out issue, and permitted a thinner film, thereby increasing the length of film that could be carried, the corona discharge issue remained. Ultimately, this was resolved by a process of trial and error. It was found that the fogging was more or less eliminated if the rubber film transport rollers were treated for several hours in a pressure cooker and thereafter maintained in a scrupulously clean state.[120] Related problems were suffered by the Lunar Orbiter cameras, which used a similar design.

Cassini

In early 2001, the Cassini spacecraft took a series of pictures of bright stars to enable the interplanetary navigators at JPL to verify that the gravitational slingshot of the recent Jupiter flyby had put the spacecraft on course for Saturn. However, after having returned excellent pictures of Jupiter, the camera was found to have become fogged, and was now diffusing 70 per cent of the starlight, evidently due to the deposition of a contaminant, possibly as a result of using the thrusters for attitude control after a problem temporarily disabled the gyroscopic system. The camera system normally operated at –90°C, but since it was periodically warmed as a maintenance function it was decided to try to 'bake' the contaminant off. This had to be attempted with great care, however, since if the material was organic strong heating might polymerise it and make it permanent. A series of cycles in which the camera was gently heated to about 5°C for four weeks brought the diffusion down to about 5 per cent.[121]

Stardust

The camera on the Stardust spacecraft also developed a severe fog. This mission was launched on 7 February 1999 to make a close flyby of Comet Wild 2 on 2

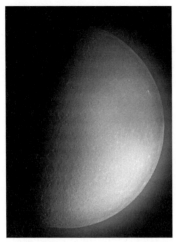

A 'foggy' image of the Moon taken by the Stardust spacecraft.

January 2004, catch some cometary dust in aerogel, and return this to Earth in early 2005 in a capsule. On the interplanetary cruise, scientists became concerned when a blurring fog developed on the camera in late 2000 as a result of an unidentified oily substance contaminating its main mirror. Pictures taken of the Moon from a range of 100,000 kilometres shortly after the Earth flyby on 15 January 2001 confirmed this. Activating the heaters to warm the optics from –35°C to +8°C had removed the fog by the end of the year,[122] but it had reappeared by next March and the decontamination procedure had to be repeated. Long-range images taken of asteroid 5335 Annefrank on 2 November 2002 were excellent, as were those from the comet encounter.

Landsat 7

The Earth sciences community was traumatised in May 2003 when a fault developed in the principal Earth-observing instrument on Landsat 7, which had operated flawlessly since launch on a Delta II in April 1999. A scan-line corrector that compensated for the spacecraft's motion as an image was built up by the Enhanced Thematic Mapper ceased to operate correctly, leaving 'gores' of missing data. No other satellite had a comparable instrument. Although 78 per cent of each scene was imaged correctly, scientists that had grown accustomed to receiving contiguous data sets were frustrated by the gaps.[123] The company who marketed Landsat data began to issue products where the missing data was substituted with data from previous images, but in such cases was obliged to reduce the fee that it charged.

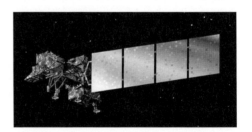

An artist's impression of Landsat 7.

GOES woes

Even the large scanner used to generate weather pictures can fail. When GOES 6 was launched on 28 April 1983, it had four lamps for the optical encoder that measured the position of the mirror of its scanner; by October 1988 three had failed, and the failure of the last one was expected at any moment – it blew on 21 January 1989.[124] Although by that time the satellite had far exceeded its nominal lifetime, the absence of a replacement for the GOES-WEST operating station seriously degraded the ability of the National Oceanic and Atmospheric Administration to issue storm warnings.[125] GOES 7, which had taken the GOES-EAST station in February 1987, had been launched with two of its four lamps superseded by more reliable light-emitting diodes. As a compromise until the next-generation of this type of satellite could to be launched, GOES 7 was moved to enable it to cover both the Atlantic and the Pacific coasts of the Americas, although with less coverage of each ocean area. It was relocated again when Eumetsat 3 (which had exceeded its service life and been placed into orbital storage) was leased from the European Space Agency.[126] Unfortunately, Space Systems/Loral had many problems in developing GOES 8, involving both the bus and the instruments, and it was unable to be launched until 13 April 1994, and even then a problem with its propulsion system meant that it was only barely able to reach geostationary orbit.

NOAA 11's radiometer
The polar-orbiting NOAA 11 satellite launched in 1988 had an Advanced Very-High-Resolution Radiometer that was expected to operate for two years, but it lasted much longer, developing an erratic scan synchronisation in 1994, and finally failing the following year.[127]

Viking seismometers
No data was acquired from the seismometer on the Viking 1 lander. A seismometer uses a mass against which the motion of the ground and the spacecraft can be measured. To make the instrument as sensitive as possible, the mounting of this mass is often delicate, so the mass is supported by a cage during the shock of landing. Unfortunately, in this case the cage jammed.[128] The seismometer on Viking 2 uncaged, but since it was on an arm rather than on the ground it was susceptible to being shaken by the wind. Nevertheless, when there was no wind the instrument was able to operate at its maximum sensitivity. In addition to a tremor rated at 2 on the Richter scale, it detected a magnitude 6 Marsquake that showed that the planet is not yet geologically dead.

Galileo's instruments
Galileo took a beating from the harsh Jovian radiation environment, as well as the simple toll from being in space for over a decade. Generally, the faults it suffered were manifested as transient faults that could subsequently be worked around. Some losses of science data occurred because observations used areas of memory on the spacecraft's computer that had been damaged,[129] or because the spacecraft entered safe mode during the observation, but some anomalies affected the instruments. After the filter wheel on the photopolarimeter radiometer jammed, most of the subsequent data was taken in just one wavelength band. After the grating on the near-infrared mapping spectrometer stuck, this was able to measure at only 13 wavelengths instead of hundreds. On the first Ganymede flyby on 27 June 1996, the energetic-particle detector entered safe mode and recorded no data. After the main camera system, the solid-state imager, began to suffer problems,[130] planners chose a 2 by 2 pixel-summing mode for the particularly harsh Io encounters, in the hope that the pixel summing would minimise the effect of the radiation hits on the CCD array

 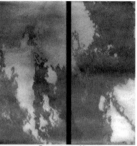

Original scrambled image   Reconstructed image

Unscrambling a radiation-corrupted image of Io taken by the Galileo spacecraft.

of the camera, but on one occasion this process *scrambled* the data. Luckily, engineers at JPL developed a hypothesis to explain this fault. Rather than two pulses of the CCD serial clocks occurring per pixel-sampling period in the summation mode, only one serial clock cycle was occurring. In most cases, the result was that the left and right halves of each frame were summed together, with a 7-pixel horizonal offset from the right-hand side of alternate rows of data. Although some pixels were irretrievably lost, it proved possible to disentangle most of the pixels, and make decent guesses at those that were lost in order to recover most of the imagery.

Hubble's STIS

In August 2004, as this book was being compiled, an observation planned by one of the authors, RL, with the Hubble Space Telescope was thwarted by the failure of the 5-volt power supply on the Space Telescope Imaging Spectrograph. Because the redundant unit had already suffered a short circuit, the instrument was made inoperable by the second failure. If a Shuttle crew is permitted to service the spacecraft, this failed instrument will be removed and replaced by a more advanced instrument to restore the telescope to full use.

NOTES

1. http://www.skyrocket.de/space/doc_sdat/oao-1.htm
2. 'Report of the Seasat failure review board', *Readings in Systems Engineering*, p. 201, 1993.
3. *The Japanese and Indian Space Programmes: Two Roads into Space*, B. Harvey, Springer–Praxis, 2000, p. 25.
4. http://www.tbs-satellite.com/tse/online/sat_yuri_3a.html
5. 'Glitch cuts Mars Express power to 70 per cent', *New Scientist*, 3 July 2003.
6. 'The dynamics of the Solar Maximum Mission spacecraft capture and redeployment on STS-41C', K.J. Grady, AAS 85-060 Guidance and Control 1985, *Advances in the Astronautical Sciences*, vol. 57, p. 495.
7. 'Oscar-13's life and death', J. Miller, *Amsat Journal*, vol. 20, no. 2, March–April 1997.
8. http://www.gsfc.nasa.gov/topstory/20010917yohkoh.html
9. http://science.nasa.gov/ssl/pad/solar/yohkoh.htm
10. http://isass1.solar.isas.ac.jp/sxt_co/011221/press.txt
11. http://www.skyrocket.de/space/doc_sat/hs-702.htm
12. www.sat-index.com/failures/index.html

13. Interview with Dave Ryan, President of Boeing Satellite Systems, *Space News*, 22 September 2003, p. 22.
14. http://www.spaceandtech.com/digest/flash2001/flash2001-082.shtml
15. *Solar System Log*, A. Wilson, Jane's, 1987, p. 58.
16. http://www.earth.nasa.gov/history/noaa/noaa8.html
17. http://www.met.fsu.edu/explores/Guide/Noaa_Html/noaa8.html
18. *Aviation Week & Space Technology*, 13 January 1986, p. 21.
19. http://esl.ee.sun.ac.za/projects/sunsat/
20. http://www.cellular.co.za/sunsat.htm
21. http://space.skyrocket.de/doc_sdat/sunsat.htm
22. http://sunsat.ee.sun.ac.za/news/20010201.html
23. http://nssdc.gsfc.nasa.gov/database/MasterCatalog?sc = 1990-007A
24. 'Space odyssey of an angel: summary of the Hiten's three-year mission', K. Uesugi, AAS 93-292, Spaceflight Dynamics 1993, *Advances in the Astronautical Sciences*, vol. 84, p. 607.
25. *The Japanese and Indian Space Programmes: Two Roads into Space*, B. Harvey, Springer–Praxis, 2000, p. 134.
26. http://www.tbs-satellite.com/tse/online/sat_insat_1a.html
27. http://www.flonnet.com/fl1421/14210520.htm
28. 'Novel experiences with Insat-1B operations', S. Rangarajan, AAS 93-318, Space Flight Dynamics 1993, *Advances in Astronautical Sciences*, vol. 84, p. 989.
29. http://www.tbs-satellite.com/tse/online/sat_insat_1c.html
30. *Aviation Week & Space Technology*, 24 October 1998, p. 47.
31. http://www.skyrocket.de/space/doc_sdat/insat-1a.htm
32. http://www.tbs-satellite.com/tse/online/sat_insat_2d.html
33. http://klabs.org/mapld04/tutorials/mishaps/strv1c_d.htm
34. http://www.sat-index.com/failures/strv.html
35. *Spaceflight*, April 1998, p. 136.
36. http://space.skyrocket.de/doc_sdat/nimbus-1.htm
37. http://fti.neep.wisc.edu/neep602/SPRING00/lecture35.pdf
38. http://www.rssi.ru/IPPE/General/spacer.html
39. http://www.astronautix.com/craft/usa.htm
40. *Operation Morning Light: Terror in our Skies – The True Story of Cosmos 954*, Leo Heaps, Paddington Press, 1978.
41. http://apollo.cnuce.cnr.it/~rossi/publications/iaf97/iaf97_html.html
42. http://www.iki.rssi.ru/mars96/08_mars_e.htm
43. http://www.wordiq.com/definition/Space_disaster
44. *Beyond the Atmosphere: Early Years of Space Science*, H.E. Newell, SP-4211, NASA, 1980, p. 164.
45. *Astronautics and Aeronautics*, 1964, SP-4005, NASA, 1965, p. 135.
46. *Aviation Week & Space Technology*, 15 June 1987, p. 66.
47. 'The sixth amateur satellite', *QST*, July-August 1973.
48. http://space.skyrocket.de/doc_sdat/amsat-p2a.htm
49. 'The amateur space programme', *Journal of the British Interplanetary Society*, 1979, p. 378.
50. http://www.spacetoday.org/Satellites/Hamsats/Hamsats1970s.html
51. http://www.sat-net.com/winorbit/help/satao6.html
52. 'Oscar 6: Gone but not forgotten', *QST*, November 1977, p. 31.
53. *Aviation Week & Space Technology*, 5 May 1986, p. 28.

54. http://www.signaltonoise.net/library/captmidn.htm
55. This meant that the main spacecraft required less propellant for the insertion manoeuvre, this economy being necessary despite the fact that 1971 was a particularly favourable time to send a mission to Mars in terms of energy requirements.
56. *Solar System Log*, A. Wilson, Jane's, 1987.
57. *The Difficult Road to Mars*, V.G. Perminov, NASA Monographs in Aerospace History, no. 15, 1999.
58. *Exploring Space: Voyages in the Solar System and Beyond*, W.E. Burrows, Random House, 1990, p. 199.
59. http://www.solarviews.com/eng/deepspace2.htm#mission
60. http://mars.jpl.nasa.gov/msp98/ds2/fact.html
61. *JPL Special Review Board, Report on the Loss of the Mars Polar Lander and Deep Space 2 Missions*, JPL D-18709, 22 March 2000.
62. *Electronic Systems Failures and Anomalies Attributed to Electromagnetic Interference*, R.D. Leach and M.B. Alexander, NASA Reference Publication 1374, Marshall Space Flight Center, July 1995.
63. klabs.org/DEI/References/pyros/magellan_pyro_memory_upset.pdf
64. 'Magellan attitude control mission operations', E.M. Dukes, AAS 93-045, Guidance and Control 1993, *Advances in the Astronautical Sciences*, vol. 81, p. 375.
65. *Aviation Week & Space Technology*, 3 August 1998, p. 31.
66. *Aviation Week & Space Technology*, 30 November 1998, p. 30.
67. http://www.jpl.nasa.gov/releases/2002/release_2002_205.cfm
68. http://spacelink.nasa.gov/NASA.Projects/Space.Science/Origins/WIRE.Mission/.index.html
69. www.aoe.vt.edu/~cdhall/courses/aoe4065/NASADesignSPs/wire.pdf
70. *WIRE Mishap Investigation Board Report*, NASA, 8 June 1999.
71. http://edition.cnn.com/TECH/space/9903/05/wire.01/
72. http://wise.ssl.berkeley.edu/
73. Special Issue on 'UoSAT – The University of Surrey's Satellite', *Journal of the Institution of Electronic and Radio Engineer*, vol. 52, no. 8–9, August/September 1982.
74. 'UoSAT 1: A Review of Orbital Operations and Results', M.N. Sweeting, *Journal of the Institution of Electronic and Radio Engineers*, vol. 57, no. 5 (supplement), September–October 1987, p. S184.
75. http://www.ee.surrey.ac.uk/SSC/CSER/UOSAT/missions/uosat1.html
76. 'The University of Surrey UoSAT 2 spacecraft mission', M.N. Sweeting, *Journal of the Institution of Electronic and Radio Engineers*, vol. 57, no. 5 (supplement), September–October 1987, p. S99.
77. 'The UoSAT D and UoSAT E Technology Demonstration Spacecraft', M.N. Sweeting and J.W. Ward, IAF-88-045.
78. J.E. Haines, unpublished ESA note, 1990.
79. 'A possible explanation for the brief life span of UoSAT 4', K. Clark and C. Thamviriyakul, Paper at the 1996 AMSAT conference. http://www.ee.surrey.ac.uk/SSC/CSER/UOSAT/papers/amsat96/kcc/kcc.html
80. *Aviation Week & Space Technology*, 23 September 1985, p. 21.
81. http://www.skyrocket.de/space/doc_sdat/leasat.htm
82. *Space Shuttle Log*, T. Furniss, Jane's, 1986, p. 62.
83. http://samadhi.jpl.nasa.gov/msl/QuickLooks/erbsQL.html
84. http://llis.nasa.gov/llis/plls/ (entry 0350).
85. *Aviation Week & Space Technology*, 14 February 1994.

86. *Aviation Week & Space Technology*, 15 November 1985, p. 16.

87. *Aviation Week & Space Technology*, 21 May 1984, p. 18.

88. 'Satellite broadcasting experiments and in-orbit performance of BSE', S. Shimoseko, M. Yamamoto, M. Kajikawa and K. Arai, *Acta Astronautica*, vol. 9, 1982, p. 499.

89. *Aviation Week & Space Technology*, 18 June 1984, p. 23.

90. *The Difficult Road to Mars*, V.G. Perminov, NASA Monographs in Aerospace History, no.15, 1999, p. 55.

91. klabs.org/richcontent/Reports/mars/difficult_road_to_mars.pdf

92. *Solar System Log*, A. Wilson, Jane's, 1987, p. 33.

93. 'Observations of the Russian moon probe Luna 9', J.G. Davies, A.C.B. Lovell, R.S. Pritchard and F.G. Smith, *Nature*, 26 February 1966.

94. *Solar System Log*, A. Wilson, Jane's, 1987 p. 59.

95. http://www.space.com/news/spacehistory/venera7_000817.html

96. Huygens communications link enquiry board report, 20th December 2000.

97. 'Resolving the Huygens communication anomaly', L.A. Deutsch, IEEE Aerospace Conference, 2002.

98. 'Titan calling', J. Oberg, *IEEE Spectrum*, October 2004.

99. http://www.spectrum.ieee.org/WEBONLY/publicfeature/oct04/1004titan.html

100. http://mars.jpl.nasa.gov/MPF/rovercom/radio.html

101. http://mars.jpl.nasa.gov/MPF/rovercom/images/ber_plot.jpg

102. http://www.boeing.com/defense-space/space/bss/factsheets/601/galaxy_iv/galaxy_iv.html

103. http://www.boeing.com/defense-space/space/bss/factsheets/601/galaxy_vii/galaxy_vii.html

104. http://www.spaceref.com/news/viewpr.html?pid = 3144

105. http://www.panamsat.com/global_network/satellites.asp

106. http://www.spaceandtech.com/digest/sd2000-25/sd2000-25-002.shtml

107. http://www.adec.edu/user/skyreport/1998/sky07-09.html

108. *Space News*, 11 November 2002, p. 1.

109. *Aviation Week & Space Technology*, 17 August 1998, p. 31.

110. http://nepp.nasa.gov/whisker/failures/

111. *Uranus*, E.D. Miner, Springer–Praxis, 2nd edn, 1998.

112. 'Engineering the Voyager Uranus Mission', R.P. Laeser, *Acta Astronautica*, vol. 16, 1987, p. 75.

113. http://llis.nasa.gov/llis/plls/ (entry 0362)

114. *The Difficult Road to Mars*, V.G. Perminov, NASA Monographs in Aerospace History, no.15, 1999.

115. *Solar System Log*, A. Wilson, Jane's, 1987.

116. *Space News*, 13 October 2003, p. 6.

117. *Exploring Space: Voyages in the Solar System and Beyond*, W.E. Burrows, Random House, 1990, p. 122.

118. *Solar System Log*, A. Wilson, Jane's. 1987.

119. *The Eagle Has Wings*, A. Wilson, British Interplanetary Society, 1982.

120. 'The Corona camera system: ITEK's contribution to world security', F. Madden, *Journal of the British Interplanetary Society*, vol. 52, 1999, p. 379.

121. *Aviation Week & Space Technology*, 5 August 2002, p. 17.

122. www.planetary.org/html/news/articlearchive/headlines/2001/foggylens2.html

123. Landsat Monthly Update June 2004, USGS (landsat.usgs.gov)

124. *Aviation Week & Space Technology*, 24 October 1988, p. 36.

125. *Aviation Week & Space Technology*, 30 January 1989, p. 30.
126. *Aviation Week & Space Technology*, 4 April 1994, p. 73.
127. *Aviation Week & Space Technology*, 17 October 1994, p. 60.
128. *Solar System Log*, A. Wilson, Jane's, 1987.
129. http://www2.jpl.nasa.gov/galileo/today991019.html
130. 'High-resolution images of Io from SSI', A. McEwen *et al.*, Lunar and Planetary Science Conference XXXI (abstract 1995).

13

Environmental failures

THE SPACE ENVIRONMENT

Space is a dangerous environment. In addition to the challenges of heat transfer by radiation, an object in space is bathed in strong ultraviolet light from the Sun, and by nuclear radiation. Further, as a result of the large relative velocities involved in orbital and interplanetary flight, even specks of dust in space can have the kinetic energy of a rifle bullet. While our understanding of this environment has improved over the years, it continues to take its toll, often unavoidably.

RADIATION FAILURES

Nuclear radiation is an oft-misunderstood hazard, not least because there are many types. Radiation affects living tissue most usually by creating damaging chemicals (free radicals) in a cell which then damage its DNA. In a spacecraft, the damage is usually manifested in the crystalline structure of semiconductors in electronics. Some radiation effects are 'soft' – transients or memory errors that can be repaired. Other failures can be permanent – sudden death by 'latch-up' or some other process. Still other failures are progressive total-dose effects, in a way resembling radiation-induced ageing. Dosage is usually expressed in units of rads: 100 rad corresponding to an energy dose of 1 joule per kilogram. By way of a scale, a prompt dose of ~ 500 rad would typically be lethal to a human.

Radiation sources
There are four principal natural sources of radiation to be considered:

- Any radiation source on board the spacecraft, such as a radioisotope thermo-electric generator power system, radioisotope heaters, or sources associated with instruments such as an X-ray fluorescence spectrometer.

- High-energy particles from astrophysical sources, usually nuclei of large atomic number ('heavy-Z' or 'high-Z' particles). Such galactic cosmic rays provide an essentially constant background of very penetrating particles that are difficult to shield against – and indeed, shielding can actually exacerbate the problem, because a single strike can cause the shield to spray out a shower of less energetic, but still dangerous particles.
- Solar wind particles which, although usually less energetic and of lower atomic mass, and therefore easier to shield against, are more numerous, with their flux varying with time and being particularly associated with solar flares.
- The greatest risk usually comes from the charged particles circulating in a planet's magnetosphere. Although ultimately of solar origin, the energies of such particles are increased by being accelerated in the local magnetic field, which also confines their spatial distribution, producing intense belts. The fact that there are intense belts around the Earth was discovered on 31 January 1958 by America's first satellite. As Explorer 1 ascended towards its first apogee, it's radiation detector reported a rapidly increasing count. When the count suddenly ceased, the instrument's designer, James van Allen of the University of Iowa, correctly inferred that the Geiger–Muller tube was taking so many 'hits' that it had become saturated and rendered incapable of generating pulses – in a sense, this condition was a specification design error in the instrumentation!

In addition, human activity can augment the magnetospheric radiation environment.

Starfish
On 9 July 1962, in a test named Starfish, the US Air Force launched a Thor missile from Johnston Island in the Pacific which detonated a 1.4-megaton nuclear warhead at an altitude of 325 kilometres.[1] The flash was visible for a radius of some 2,500 kilometres, and the effects were more profound than expected. An artificial aurora lasted seven minutes, and extended over Hawaii. The unforeseen and most militarily significant effect was the electromagnetic pulse, which caused power mains surges in Oahu, knocking out street lights, blowing fuses and circuit breakers, and triggering burglar alarms – and recall that this predated the introduction of microelectronics.[2] It seems the intense X-rays from the explosion (which are usually absorbed within metres by the dense air at sea level to drive the rapid expansion of the fireball) caused ionisation over a large region of the upper atmosphere. This released vast numbers of electrons into the magnetosphere – in essence creating a new radiation belt that lasted for a decade. Over the next few months, seven satellites fell victim to total-dose effects, usually on their solar arrays. One of these was the Telstar satellite that was launched on 10 July to relay television across the Atlantic. Its electronics had been carefully screened to counter the radiation that had been measured in orbit prior to the test, but several transistors in the command system of its repeater succumbed to total-dose effects within four months; the controllers implemented a workaround, but this only kept it going for another two months. Between the Starfish test and a similar test by the

Soviets that October, satellites were receiving about 100 times the expected radiation dose.[3,4,5,6]

Total-dose effects can be subtle. For example, one common comparator chip (the National Semiconductor LM139) used for a variety of applications is sold as a 100,000-rad part, a specification it consistently exceeded while it was manufactured at a semiconductor plant in Greenock, near Glasgow, in Scotland. The company transferred manufacture to a plant in Texas, expanded the 4-inch silicon wafer to 6 inches, and laid the circuit out slightly differently in order to eliminate unused space and thereby pack on more devices. Although to all intents and purposes the chip remained the same, testing established that the new parts tended to fail at only 30,000 rad. When analysis showed that a change in the distribution of metal around one of the transistors was the cause of this increased susceptibility, it was redesigned. It is this type of subtlety that leads major space operators to rigorously test new components and maintain databases of their radiation tolerances, one important criterion being whether they are manufactured using 'known good dies'. Diligent manufacturers issue 'process change notifications' if suppliers, materials, or manufacturing changes occur that may prompt retesting.[7] Note also that in testing it is insufficient simply to reproduce the total radiation dose, as many effects are rate-dependent, and some semiconductors are particularly susceptible to low dose rates. As most space projects cannot wait for years for a test dose to slowly accumulate, testing at a high dose rate may have to do, but can give a false sense of security.

Reactor emissions

The emissions from reactors on satellites, while not a major radiation dose source to spacecraft other than those in which they are incorporated, can pose a nuisance to certain types of scientific instrument. In the 1980s, the reactors on the Radar Ocean Reconaissance Satellites (RORSAT) operated by the Soviet Union disrupted instruments on a number of NASA satellites, notably the gamma-ray spectrometer on the SolarMax satellite and the Burst and Transient Source Experiment on the Compton Gamma-Ray Observatory. The reactors caused two types of disruption. When SolarMax's orbit brought it within about 500 kilometres of a RORSAT, the satellite directly detected the gamma rays emitted by the reactor. When the two satellites were travelling along the same geomagnetic field line, clouds of positrons and electrons from the reactor caused disruptions lasting from a few seconds up to several tens of minutes. In both cases, SolarMax's spectrometer was blinded. The operations of the Compton Gamma-Ray Observatory were modified to turn off its instrument when a close encounter with a RORSAT was predicted – a significant complication to instrument operations.[8]

Shielding

The modelling of radiation effects is challenging, because not only are different types of component susceptible to differing degrees, their exposure depends in a complex manner on the distribution of mass, and so on masking or shielding effects. Shielding is usually expressed as an equivalent thickness of aluminium, and the optimum shielding material depends on the expected radiation source: for example, tantalum is particularly effective against stray neutrons from radioisotope thermoelectric generator power systems. More shielding is not necessarily better, since a cosmic ray striking a shield may give rise to even more damaging Bremmstrahlung radiation. Hence, whereas shielding might reduce the total dose in a radiation belt, it may increase the damage suffered on a cruise through deep space. Various simulation codes have been developed to assist in modelling these effects.

Damage Mechanisms

Radiation damage usually manifests itself by disrupting semiconductor devices, but a high dosage can also affect non-electronic systems: structural materials can weaken due to changes in the arrangement of the crystal lattice and lattice defects, and opacification of certain glasses can degrade optical instruments and cut the efficiency of solar arrays.

On 3 December 1973 Pioneer 10 became the first spacecraft to venture into the Jovian magnetosphere, and by flying through the inner radiation belt it received an exceptionally high dose. The outer surface of the spacecraft suffered 1.5 million rad, one-third of which were due to electrons and the remainder due to protons.[9] One of its non-imaging optical sensors was so darkened as to be rendered useless. Pioneer 11 passed though a year later, and the radiation disrupted its primary imaging instrument, a photopolarimeter, making its drive mechanism skip backward and forward in a random manner instead of stepping progressively across the planet. Its distorted images had to be reconstructed by sophisticated software after receipt.[10]

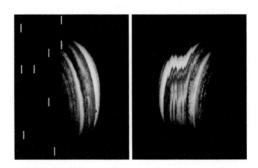

When the operation of Pioneer 11's imaging spectrometer was disrupted by radiation, the pictures of Jupiter had to be reconstructed on Earth.

A related total-dose effect is seen in some electronic components wherein the gate voltage or leakage current steadily increases with time sufficiently to exceed the levels at which the circuit will function correctly. A similar effect is seen in certain detectors such as a CCD wherein the 'dark current' progressively increases to produce 'hot pixels'; an example being a sensor in the attitude control system of the TOPEX–Poseidon satellite. Sometimes, this effect can be reversed by running the device at a high temperature in order to allow the charge-pairs to recombine.

There are also 'sudden death' radiation damage mechanisms. One of these is the single-event upset (SEU), wherein the passage of an energetic particle through a digital component such as a memory chip randomly flips a bit from '0' to '1' or *vice versa*. If the bit represents data (such as an image pixel) this is unlikely to be catastrophic, but if it is in a computer instruction then the effects may be impossible to predict and potentially irrecoverable. The main protection against such effects is to use memory and processors that have been 'radiation hardened'. The most critical functions are made less vulnerable by, for example, the use of alternative substrates (e.g. silicon-on-sapphire) and the use of larger gates (the energy required to flip a bit depends on the operating voltage and the capacitance of the memory cell, and since high-density memories use lower voltages and smaller cells they are more vulnerable; on the other hand, they use less power, and using them represents a trade-off between performance and risk). Another approach is to use 'coded memory', whereby an 8-bit word is represented by 12 physical bits that are arranged in a code such that two distinct words differ by more than one bit-change in order that an inconsistent single bit can point to the fact that a memory cell has been flipped, and the incorrect bit identified and corrected. A failure involving two bit flips in the same word can be noted as having taken place, but cannot be corrected directly. Other, more elaborate codes can be constructed with different levels of redundancy and protection, but at the expense of extra memory and processing needs. This Error Detection and Correction (EDAC) relies on bit errors occurring in a sparse manner, without many occurring in the same word. Because all words will steadily accumulate errors if left in a radiation environment, a process must be introduced (either in software or in hardware) to correct the errors periodically in order that an error in one word is corrected before it suffers another error. This procedure has been termed 'washing' the RAM (random-access memory). If possible, memory chips should be arranged such that multiple bits from the same word do not occur on the same chip, to make it unlikely that a single particle or a shower of particles will corrupt more bits per word than the code can recover. EDAC has become particularly important now that solid-state memories have superseded tape recorders. The 2.5-gigabit units on the Cassini spacecraft proved to have an EDAC-correctable single-event upset rate of typically 200 per day, but the uncorrectable multiple-bit error of about 0.7 per cent was orders of magnitude greater than predicted.[11] The issue was a design flaw that meant that although the bits in each 39-bit word were spread across five different chips, the bits in each chip were adjacent pairs, which meant that a single particle could often cause a multiple-bit error.

Most of the single-event upsets suffered by Earth-orbiting satellites derive from the van Allen belts. In the case of low orbit, these occur predominantly while the satellite is crossing the auroral ovals (i.e. where the closed field lines funnel towards the magnetic poles) and in the South Atlantic Anomaly off the coast of Brazil where the net field is rather weak, and the inner radiation belt dips down somewhat towards the ionosphere, leading to increased interaction with low-orbiting satellites. Sensitive operations, Hubble Space Telescope observations in particular, are often scheduled to avoid the heightened probability of single-event upsets during the several minutes that it takes to transit the South Atlantic Anomaly. During normal operations, between 1.5 and 3 per cent of the pixels in a Hubble Space Telescope image will be influenced by cosmic rays on a 1,000-second exposure. Longer exposures are often built up by adding together shorter exposures, in order to help to discriminate real targets from cosmic ray hits. The radiation hazard in low orbit (which, for radiation exposure, can reasonably be defined as being altitudes lower than 1,000 kilometres) is relatively modest. The hazard is much more intense at higher altitudes and modest orbital inclinations, since these penetrate the doughnut-like van Allen belts. The orbits most affected are those at altitudes of several thousand kilometres. Although these are generally avoided, satellites destined for even higher orbits and spacecraft leaving on deep space missions must pass through the belts, and their transits are designed to minimise radiation exposure, in particular to avoid loss of solar array capacity. This hazard is often the major cause of urgency in the recovery of satellites that are stranded in geosynchronous transfer orbit by the failure of either the upper stages of their launchers or their apogee motors (ETS 6, Hipparcos, etc.) Another example is the 'ham' radio satellite, Oscar 10. This was released into geosynchronous transfer orbit on 16 June 1983 by the Ariane 1 that deployed Eutelsat 1F1. Oscar 10 was the first such satellite to have its own propulsion, and this bipropellant motor was to steepen the inclination of the orbit to about 60 degrees. However, the motor became damaged and its underperformance left the orbit at 27 degrees, which complicated the task of using it. At this intermediate inclination, the satellite passed through a more intense part of the van Allen belts than planned. The memory chips in its control computer failed in 1986, presumably due to total-dose effects, leaving the satellite operational but without control.[12] Its continued use as a radio relay depended on the evolution of its uncontrolled, but spin-stabilised, attitude to aim its antennas towards the Earth. As the harsh environment caused the battery to 'fail open', the satellite had to draw power directly from its solar panels, which required that these be illuminated.[13] Nevertheless, Oscar 10 was still in use as late as 2000.

Some spacecraft – for example, Explorer 50, which was launched on 26 October 1973 into an elliptical orbit with a very high apogee and a 12-day period to serve as the eighth and final Interplanetary Monitoring Platform, and the International Ultraviolet Explorer that was launched on 26 January 1978 and placed into geostationary orbit – have reportedly never suffered a single-event upset.[14,15] On the other hand, the memories required to support the large data volumes from modern missions lead to high rates of single-event upsets. For example, the fifth satellite in the Meteor 3 series that was launched by Russia on 15 August 1991 had a Total Ozone Mapping Spectrometer from NASA's Goddard Space Flight Center.[16] Because the Hitachi 256-kilobit chips from which the instrument's 128-megabit memory was built were known to be sensitive to single-event upsets, they were arranged so that only one-bit-per-word errors would be likely to arise, and RAM washing was employed. This appears to have worked well: the system fixed an average of 350 errors per day, almost all of which were associated with the South Atlantic Anomaly, with the zone matching the contours where protons in the energy range 75–100 MeV were likely to reach the satellite. When the rates for a similar instrument on the TOMS–EP satellte are compared with the memory single-event upsets of UoSAT 2 and UoSAT 3,[17] it is interesting that whereas all three suffered in the South Atlantic Anomaly and the UoSATs were upset in the auroral ovals, the TOMS–EP was largely unaffected in the polar regions.

In addition to memory devices, single-event upsets influence photodiodes, (e.g. those experienced by the Magellan star mapper). They can even occur where such

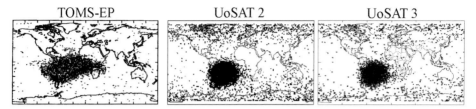

The memory single-event upsets suffered by UoSAT 2, UoSAT 3 and the TOMS–EP satellites.

devices are not being used in particularly sensitive applications (e.g. in opto-isolators). Such events affected the fibre-optic data bus on the SAMPEX satellite,[18] and also the power-handling electronics of the SMART 1 spacecraft *en route* to the Moon, causing its ion propulsion system to shut down. The infrared scanwheels on the TOMS–EP satellite proved to be susceptible to spurious pulses from single-event upsets in the South Atlantic Anomaly. Oddly, slightly different scanwheel failures occur in the zone 20 degrees each side of the equator between longitudes 80 and 170 degrees east – far from the South Atlantic Anomaly – as a result of ground-based radio interference that in some way couples into the sensitive pyroelectic detectors of the Earth sensor.[19]

Although most single-event upsets are 'soft' errors, in that the affected bit can be rewritten correctly, some permanent failures can occur where bits 'stick' and cannot be changed. These errors often require substantial software modifications to work around the affected memory locations.

The 'latch-up' damage mechanism is not reversible, but it is preventable. The passage of a charged particle through a semiconductor creates a parasitic transistor that enables a large current to flow if the device is powered up, and the heat that results may destroy the device. The protection measure is a current-sensing logic that rapidly detects when this occurs and deactivates the circuit before it can be damaged. Although, this interrupts the function of the circuit until it is powered on again, it is preferable to outright destruction. One well-documented case of latch-up failure was the loss, five days after launch, of the German-built Precision Range and Range-Rate Equipment on the first European Remote Sensing satellite. Apparently, as a cost-saving measure, not all the devices used were screened for susceptibility to single-event upset or latch-up. After extensive ground testing, this failure was attributed to a proton impact on a random access memory chip (supplied by NEC) that had a low damage threshold, and this impact occurred in the centre of the South Atlantic Anomaly. The instrument suffered a high-current transient and shut down, but could not be restarted.[20,21]

IMPACT FAILURES

Impact by a micrometeoroid or a piece of artificial space débris is frequently cited as a possible mechanism of spacecraft failure, although very few known instances actually exist. Because it may be expected to lead to 'sudden death', it is a difficult mechanism to eliminate entirely, and it has the appealing *deus ex machina* characteristic of leaving the spacecraft manufacturer and operator plausibly blameless. On the other hand, near-Earth space has become progressively cluttered with defunct satellites, upper stages, fragments from exploded spacecraft, items lost by astronauts, etc. The windows of the Shuttle are routinely replaced after a mission because of cratering by débris, and whenever NORAD predicts that a significant piece of débris will approach within several kilometres of the International Space Station, this is manoeuvred to open the encounter range.

Cerise

The first credited collision between two catalogued space objects occurred on 24 July 1996. The Cerise microsatellite was built by Surrey Satellite Technology in England to characterise terrestrial radio emitters for the French military. Based on the UoSAT series, it had magnetorquers for attitude acquisition and adjustment, and used a 6-metre-long boom and the gradient in the Earth's gravitational field to passively maintain itself Earth-pointing. It was in a polar orbit at an altitude of 700 kilometres, and pointed within a few degrees of nadir, gently

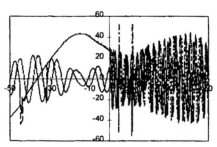

Elapsed Time Since Collision [minute]

A plot of the magnetometer signal from the Cerise satellite at the time of its collision.

spinning around its long axis, as intended, when suddenly it started to tumble. The investigation determined that the boom had been completely severed almost at its base when a fragment of débris from an exploded third stage of an Ariane launcher struck it at a closing speed of 14 kilometres per second.[22] NORAD radar tracking later discovered an object in a very similar orbit to Cerise – the end of the boom with its cylindrical tip-mass. Fortunately, Cerise remained under ground control, and its telemetry showed that the angular momentum had jumped instantaneously, but only by 0.005 newton-metre-second, which implied that the boom had been locally vaporised by the impact, without conveying much momentum to the satellite. The control logic of the magnetorquer was reprogrammed to enable much of the mission to continue, but the loss of gravity-gradient torque complicated maintaining the Earth-pointing orientation.

An artist's sketch of the Cerise collision.

Kosmos 539 and 1275

In several cases, NORAD radar tracking has shown an abrupt change in the periods of satellites suggestive of an impulsive change in their orbital energy. The fact that often there were several objects travelling in slightly different orbits that could be projected back to intersect a satellite's position at the key moment suggested that the satellite had suffered an energetic disassembly in a collision with (unknown) space débris.

A recent example is Kosmos 539, a Soviet geodesic satellite launched in 1972 into a long-lived orbit at an altitude of 1,370 kilometres. Many aged satellites shed fragments, presumably as a result of thermal cycling and the degradation of materials, but these are usually released at very low velocity. On 21 April 2002, a fragment calculated to have been roughly 50 centimetres in diameter was released from Kosmos 539 at a speed of 19 metres per second, which suggests that it was blown off by some sort of small explosion or impact. This event was accompanied by a change in the orbital period of the satellite by 1 second, consistent with a collision with a piece of orbital space débris of around 1 centimetre in size. (At that altitude, the flux of débris particles of this size is 10 times greater than natural meteoroids, although the collision speeds would be lower.) The orbit of the fragment rapidly decayed as a result of solar radiation pressure and re-entered the atmosphere after only 43 days, implying that it had a very high ratio of area to mass, like a sheet of foil from a thermal blanket.[23] A similar event befell NOAA 7 in August 1997, long after it had been decommissioned, when the orbital period jumped by 1 second and it shed three fragments, one of which entered a higher orbit.

Some space débris is instrument covers, separation bolts, yo-yo de-spin masses, etc., but 95 per cent of the 100,000 known centimetre-sized objects are due to fragmentation. Some of these break-up events were deliberate, as in anti-satellite weapon tests, but most appear to be due to the spontaneous ignition of the residual propellant in the spent stages of launchers. The first of these, Ablestar, blew up an hour after it released the Transit 4A navigational satellite on 29 June 1961 and two-thirds of the 300 or so fragments are still in orbit.[24] Indeed, 4 per cent of all space missions have been linked to break-up events. Six cryogenic upper stages of Ariane 4 launch vehicles have exploded in orbit, in most cases after several years. Once this was realised, Arianespace introduced measures to vent excess propellant following the deployment of the payloads.[25]

The Kosmos 1275 navigational satellite launched on 4 June 1981 into a polar orbit at an altitude of 1,000 kilometres suffered a catastrophic event on 24 July that ejected several hundred fragments.[26,27] This was initially believed to have been an impact, but after several other satellites suffered similarly it was surmised that their batteries had exploded.[28]

Tethered satellites

The Tethered Satellite System (TSS) built by Italy was a 518-kilogram experimental satellite some 1.6 metres in diameter that was to be unreeled by a Shuttle on the end of a 20-kilometre cable, to study the electrodynamics of tethers

STS-75 starts to unreel the TSS.

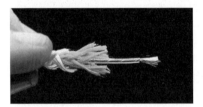

The conducting cable of the TSS.

that might be used to generate power in space (because it was an electrical conductor moving across the lines of the Earth's magnetic field).[29] On its first test in June 1992, on STS-46, a protruding bolt prevented the tether from extending more than 267 metres, at which length the electrical potential on the tether was just 60 volts. It was reeled back in. On 25 February 1996, on STS-75, the deployment proceeded smoothly for 5 hours, until the tether snapped several hundred metres short of its full length. The 2.54-millimetre-diameter tether was constructed using kevlar and nomex with 10 strands of 34 AWG copper wire and a teflon sheath. It was designed with a breaking strength of 1,780 newtons, greatly exceeding the expected load of 100 newtons, but it snapped after some of the strands were eroded by an electrical discharge into the ambient plasma. It was initially thought that a microscopic débris impact might have damaged the tether, but it seems more likely that a pre-existing flaw in the electrical insulation had facilitated the discharge. The situation may have been exacerbated by the fact that the power generated was greater than predicted. The current expected was 0.5 amp (based on a nominal 5-kilovolt potential across the tether) but it had reached almost twice this when it snapped – possibly as a result of the Earth's radial electric field adding to the dynamo voltage because the system was oriented vertically in the gravity gradient.[30] The TSS remained in orbit until 19 March 1996. Remarkably, the tether was visible to terrestrial observers moving across the night sky as a short vertical line capped by a little star (the satellite).

The Small Expendable Deployment System was to demonstrate the use of a tether to de-orbit a small payload. In two tests in March 1993 and 1994, after the second stage of a Delta launcher had released its primary payload, it successfully unreeled a small satellite on a 20-kilometre-long non-conducting tether to show that the tether acted like a thruster and caused the lower object to 'hover' at an altitude below that which its orbital angular rate would allow, thus causing its orbit to decay.[31] On the second test, the tether snapped at the 7-kilometre point after 3.7 days. This was believed to be the result of débris impact.[32] Models of the débris population implied that it ought to have lasted about 11 days (with an uncertainty of a factor of 2). To reduce the possibility that a débris impact would sever its entire width, the Tether Physics and Survivability (TiPS) experiment, launched on 12 May 1996, used a pair of 53-kilogram masses linked by a yarn composed of several fibre strands that were puffed out rather than a narrowly spun tether. The predicted lifetime for the 4-kilometre-long TiPS tether was at least a year, and when funding for tracking was terminated after two years it was still intact.[33,34]

Perseid strikes

The hazard from micrometeoroids is generally less than that from space débris, but the flux is highly variable. The Perseid meteors derive from Comet Swift–Tuttle. This was discovered in 1862, and travels in a 135-year orbit. The fact that the comet had made

its perihelion passage in December 1992 suggested that the display in 1993 would be very spectacular, with a brief but very intense peak in the early hours of 12 August.[35,36] The crew of the Mir space station reported audible meteoroid strikes.[37,38] One object punched a 10-centimetre-diameter hole in a solar panel.[39] The European Olympus geostationary communications satellite appears to have met its demise indirectly as a result of a Perseid impact. After six months of trouble-free service the satellite lost its Earth-pointing orientation and began to spin at 23:55 Zulu on 11 August, just five minutes short of the time set in an ominous warning that its manager had received earlier in the day saying 'Olympus dies at midnight'. Perseid meteors have a relative speed of 60 kilometres per

The Mir space station.

second, which is somewhat greater than a random sporadic meteor. The impact ionisation caused by a meteor strike varies strongly with impact speed, and even small Perseids can produce disproportionate ionisation, so a 1-milligram Perseid will generate the same amount of plasma as a 50-milligram sporadic. While the Olympus anomaly could not be *proved* to have been due to an impact, the most likely scenario was that a small impact on its southern solar array had generated a plasma near the opening where an umbilical had given access prior to launch to the gyro signals, and this connector provided a current path into the interior that shut off the roll gyro and caused a capacitor in the safing control circuit to fail. Although control of the satellite was recovered, the propellant consumed during the anomalous condition, and in the subsequent de-spinning, was such that by the end of the month the satellite had to be manoeuvred off-station into a graveyard orbit, safed, and shut down.[40] This experience prompted geostationary satellite operators to predict periods when such bombardment was likely to be intense, and to take protective measures.[41,42] The ability of astrodynamicists to predict the intensity and timing of micrometeoroid streams has dramatically improved in recent years.[43,44]

Direct encounters with comets

Launched by an Ariane 1 on 2 July 1985, the Giotto probe really invited trouble from collisions by penetrating the inner coma of Halley's comet to make a flyby of its nucleus on 13 March 1986, at a range of 600 kilometres and the tremendous relative speed of 72 kilometres per second.[45] Developed by the European Space Agency, the spacecraft was a spin-stabilised drum with a conformal array of solar cells. The 'rear' had a de-spun high-gain antenna that was maintained aimed at the Earth. There was a shield on the 'front' in which impacting dust grains would be broken up by an outer sheet of thin alloy and left to splash near-harmlessly onto a thick kevlar-reinforced layer. Also at the front, was the nozzle of the solid rocket motor used to accelerate the spacecraft away from the Earth, now covered by two clamshell doors. The shield was instrumented with microphones to count the number and energy of the impacts, in order to measure the distribution of dust in the coma. A camera built by the Max Planck Institut für Aeronomie in Katlenburg–Lindau (and in fact the first CCD camera to fly on a planetary mission) used a periscope to peer around the edge of the shield.[46] At the very least, it was expected that the dust would disturb the attitude of the spacecraft. As the encounter proceeded, the crescendo increased and then, 7.6 seconds prior to the point of closest approach, the signal abruptly ceased. On reacquiring the signal later, it became apparent that an impact had caused an electrical discharge that desynchronised the motor of the high-gain antenna, resulting in its beam missing the Earth, and either at that same time or as the spacecraft continued to close in on the comet an impact induced a wobble that lasted until the nutation dampers could restabilise the spacecraft.[47] A small change in the moment of inertia and the failure of the camera to produce a signal except when the spin pointed it near the Sun indicated that the camera's mirror had been torn off! Apart from this, Giotto was in good shape: it was a little warm because some of its thermal coatings had been degraded, and some of the high-voltage parts had lost their function, but the solar arrays had lost only about 1 per cent of their generating capacity. It was put into hibernation until a 200-kilometre flyby of Comet Grigg–Skjellerup on 10 July 1992. Although there could be no pictures, the other instruments returned fascinating results, and since this nucleus was inactive the shield reported just three dust impacts.[48]

An artist's impression of the Giotto spacecraft.

Each of the two VeGa (Venus–Halley) spacecraft dispatched by the Soviet Union in 1984 dropped off probes at Venus during flybys *en route* to Halley's comet, which they met several days ahead of Giotto.[49] Despite approaching no closer than 8,000 kilometres, they suffered rather heavy impact damage, presumably as a result of flying through a jet of dust emanating from the nucleus that produced (in the case of VeGa 1) 1,000 impacts per square centimetre by motes of 1.5×10^{-16} kilograms travelling at a relative speed of 78 kilometres per second. Both spacecraft lost 40 per cent of their solar array capacity;[50] both lost their low-frequency plasma analysers; and Vega 1 lost its high-frequency plasma analyser and Vega 2 lost its 3-axis magnetometer.

Unlucky Oscar 10

When Arianespace offered rides to micro-satellites as secondary payloads, a third generation of 'ham' radio satellites was developed by the amateur satellite community, AMSAT.[51] These 70-kilogram spin-stabilised satellites were fabricated in the shape of a three-pronged star, and were to be released into a transfer orbit inclined at 17 degrees to the equator. Whereas on reaching apogee the primary payload would cancel this inclination as it entered geostationary orbit, the amateur satellite would *steepen* it to 57

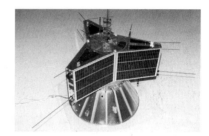

Oscar 10.

degrees to spend most of its time high above the northern hemisphere – an orbit pioneered by the Molniya communications satellites operated by the Soviet Union. AMSAT Phase IIIA was to use a Star 13C motor for the 1,600-metres-per-second orbit change, but this satellite was lost on the second test of the Ariane 1 launch vehicle on 23 May 1980.[52] When Arianespace decided to deploy satellites heading for geostationary orbit into a transfer orbit at a lower inclination in order to ease their apogee manoeuvre, this meant that AMSAT had to seek a more powerful motor for its *increased* plane change. Messerschmitt–Bölkow–Blohm donated a 400-newton bipropellant motor that, while more complex and hazardous for ground operations, was capable of delivering the required delta-V.[53] The team designed an innovative tankage to carry the monomethyl hydrazine and nitrogen tetroxide, but did not have the resources for a commercial high-pressure helium bottle for the pressurant. They instead used a mass-produced item manufactured for fire extinguishers which, with the addition of carbon fibre windings, gave the required pressure rating. The Phase IIIB satellite was launched on an Ariane 1 on 16 June 1983 and named Oscar 10. However, shortly after releasing the satellite, the third stage collided with it, changing its attitude and spin rate, and damaging its antennas, degrading their radiation pattern.[54] When the satellite returned to perigee, the controllers used its magnetorquers to stabilise it in the proper attitude and spun it up for the first burn. This deviated slightly from the planned duration, but otherwise seemed normal and, as planned, steepened the angle to 26 degrees. A reorientation was required in preparation for the second burn, and because magnetorquers do not permit rapid turns this took some time, during which the telemetry indicated that the helium pressure was falling. It is thought that the seal on the bottle had developed a leak, due either to the collision or to rapid cooling associated with the first burn. In any case, by the time the spacecraft was lined up for the second burn the pressure had fallen below the minimum required to operate the motor valves. Nevertheless, the satellite was able to be used in its intermediate orbit (albeit only by users with the patience and skill to track the satellite in this less convenient orbit) until it eventually succumbed to radiation damage.

SOLAR FAILURES

Spacecraft in deep space sail in the 'solar wind', and are often threatened by intense surges of plasma emitted by solar flares and coronal mass ejections.

Magellan's star scanner
The Magellan spacecraft, which suffered more than its fair share of troubles, proved to have several interesting susceptibilities in orbit around Venus. When near periapsis at an altitude of 250 kilometres, the spacecraft aimed its large high-gain antenna at Venus to operate as a synthetic-aperture radar, to map a long, narrow strip of the surface, and as it climbed towards the 8,750-kilometre apoapsis it swung around to aim this antenna at the Earth in order to transmit the taped data. These turns were made primarily under control of rate-integrating gyros. Once per orbit, at apoapsis, the craft used a simple star scanner to obtain a stellar attitude reference to update the gyros. This scanner focused the scene onto a mask comprising two slits, one vertical and the other inclined, behind which were two silicon photodiodes. The amplified and filtered output from these diodes was passed to a threshold detector to produce a digital pulse when a star's image crossed one of the slits. As the spacecraft turned during a scan, a star would produce two pulses, one from each slit, yielding information on the position of the star in the field of view. A month after launch, *en route* to Venus, spurious pulses caused some concern. The reference voltage of the threshold detector could be set by ground command, so it was changed to eliminate the smaller pulses, but the problem persisted. Finally, it was realised that the failed scans were correlated with solar activity, more specifically with the flux of high-energy protons. Alpha Carinae, also known as Canopus, is widely used as the reference star for spacecraft in deep space because it is bright and almost orthogonal to the plane of the ecliptic. Over a period of several months in orbit around Venus, the star scanner's response to Canopus decreased by about 6 per cent due to solar proton damage, as did the output from the solar arrays, which were also silicon photodiodes.[55]

The star sightings caused another problem. A week after arriving at Venus, Magellan fell silent for a period of 14 hours. Several days later, it was silent for 18 hours. If these blackouts were to occur often, they would jeopardise the mapping. A flaw was found in the software that made the star sighting: in essence, a sequence of instructions that were meant not to be interrupted was *sometimes* being interrupted and causing the spacecraft to safe itself. The intermittent nature of this asynchronous fault delayed its diagnosis, but once this was done the software was patched to speed up the recovery.[56]

Another issue that showed up while Magellan was *en route* to Venus, and after being resolved reappeared later, was spurious sightings by the star tracker that correlated with the spacecraft's attitude motion. When high-rate telemetry from the star tracker showed a profile suggestive of unfocused objects passing in front of the optics, it became apparent that the thermal blankets were shedding dust which was reflecting sunlight. These were made of Astroquartz, a material woven from 35-micron-diameter fibres of quartz–glass. This was a mixed blessing, because when tests suggested that the intensity of sunlight at Venus might darken the material and reduce its effectiveness as thermal protection, the chemical binders that it contained to prevent flaking of the fibres were baked out. Now the material was shedding particles, which were repelled from the surface by electrical charging due to ultraviolet photoelectric emission. (Later tests of blankets in a vacuum chamber revealed that the lack of grounding for the aluminium backing of the blankets was a contributing factor.) The solution was to take sightings on stars that were on the shaded side of the spacecraft. When the problem recurred in orbit around Venus, it was realised that the dust motes in the shadow of the spacecraft were being illuminated by the sunlight reflecting off the planet, which has a very high albedo as a result of being completely enshrouded by white cloud. Interestingly, it was also noted that these events tended to occur at a specific part of the spacecraft's orbit, near the ionopause, where plasma from Venus's atmosphere interacts with the solar wind.

Particles shed from spacecraft were of course noticed by astronauts – 'fireflies' John Glenn called them – and have been an occasional problem for star trackers in particular. Mariner 6 successfully locked onto Canopus *en route* to Mars in 1969, only to lose it when bright débris was shaken loose by the firing of the pyro squib to unlock the scan platform.[57] Ironically, given the contemporary concern over the long-term reliability of gyros, it was decided that attitude control during the critical close-encounter phase of the mission would be conducted under gyro control rather than star tracker control.

The case of Marie

The Mars Odyssey spacecraft suffered a rather ironical radiation failure. Its Martian Radiation Environment Experiment (Marie) failed as a result of radiation damage shortly prior to arrival at Mars. (In a curious coincidence, the instrument's principal investigator died unexpectedly around the same time.) The failure is believed to have been due to a single-event upset in the instrument's memory which prompted the instrument to cease science operations in order to initiate an error detection and correction procedure, during which time the instrument ceased to communicate with the spacecraft's computer. The spacecraft interpreted this lack of communication as a failure, and attempted to reset the instrument, which was unable to respond while its memory wash was in progress. Since the spacecraft had been programmed in the expectation that this process would require 8 hours, and it actually took 10 hours, by the time the instrument was ready to respond the computer had written it off. This incident occurred in August 2001. In late February 2002 it was realised that the instrument was actually operational (although it was being ignored) and the flight controllers intervened to bring it back on-line.[58] Unfortunately, on 28 October 2003, during a period of intense solar activity, it was overwhelmed and finally ceased to function.[59],[60]

ELECTROSTATIC DISCHARGE

Among the accoutrements of the spacecraft technician – cotton gloves to prevent the deposition of skin oils on components, hats, overalls, masks and little booties to prevent the shedding of particulates such as hair or skin flakes – probably the most important is the antistatic wrist band. This is a conductive band worn usually on the left wrist, which a grounding cable connects to Earth by way of a convenient conductive working surface, the objective being to prevent charges that have built up in the body by movement from discharging into electronic components. It is all too easy, especially in dry conditions, to generate potentials of many kilovolts simply by walking – with carpets made of artificial (insulating) fibres such as nylon being particularly dangerous. If a discharge, which may or may not be accompanied by a visible or audible spark, connects with the pins of an electronic device it can readily burn it out. Electrostatic discharge is an insidious hazard that (without leaving any visible indication of damage) can cause partial and intermittent failures in components, as well as total failures. In passing, it may be noted that part of the function of clean rooms, hats, booties and electrostatic discharge bracelets is actually psychological, because by the act of donning these items, which are neither glamorous nor comfortable, a worker acknowledges that he or she is about to work on delicate hardware and must pay particular attention.

Electrostatic discharge can also occur in space.[61] A particular problem is that of deep dielectric charging. Satellites in geostationary orbit are prone to suffer it. Although they are beyond the van Allen belts, such satellites are bathed in a flux of moderate-energy electrons which, while insufficiently energetic to cause latch-ups or single-event upsets, travel fast enough to penetrate insulating materials like plastics and accumulate a charge that will cause an arc when the potential difference exceeds a breakdown threshold. This effect introduces unpredictable, and often large, currents and voltages on nearby wires. Open wires, such as uncovered connectors used in ground handling, can be particularly vulnerable. As a result of launching spacecraft such as P78-2 (also known as SCATHA, for Spacecraft Charging AT High Altitude) to study the phenomenon, it is now fairly well understood, and the 'space weather' circumstances that lead to it are known.

Pagers go dark
In the middle of the business day in the USA on 19 May 1998, Galaxy 4, an HS-601 in geostationary orbit off the coast of Ecuador run by PanAmSat, suddenly failed. In

addition to serving 45 million pagers – recall that as cell phones had not yet become ubiquitous in the United States, pagers were commonplace – it provided many other services such as linkage for the credit card terminals on petrol ('gas') pumps, and so its loss considerably disrupted people's daily lives.[62] The fear that hackers had somehow caused it to fail was soon dismissed. It may be that the failure was connected with the tin 'whisker' issue in the HS-601's control processors, but it is notable that the failure occurred at a time when the magnetosphere had been infused with energetic electrons.[63,64] Several other satellites (Equator-S, Polar, and Japan's GMS) had experienced anomalies in the previous two weeks, although in these cases the correlation of failure time with the electron fluxes is not entirely persuasive.

Anik E glitches

Some other geostationary satellite failures are more convincingly shown to be due to environmental effects. The two Anik satellites, E1 and E2, operated by Telesat Canada, suffered near-simultaneous failures. Anik E1 lost its primary momentum wheel system on 20 January 1994. This same system failed 70 minutes later on Anik E2, which was 2,800 kilometres away.[65] Although controllers stabilised Anik E1 by using its secondary wheel after seven hours, Anik E2's backup failed to function.[66] Nevertheless, over the next five months a scheme was devised whereby the satellite maintained an Earth-pointing orientation by firing its thrusters for several milliseconds every few minutes. Although the pitch and roll data was available on board from the sensors, the yaw information had to be developed by equipment on Earth – in essence, part of the attitude control system was spliced in from the ground.[67] The satellite was operable, but at the expense of having a dedicated ground facility. The investigation traced the problem to a specific chip, and suggested that it was due to bulk charging associated with a magnetic disturbance that enhanced the energetic electrons in that environment. This hypothesis is supported by the near-simultaneity of the failures, together with an anomaly on the Intelsat K satellite, which started to wobble several hours before the first Anik E failure.[68] A few years later, Telstar 401 suffered a failure that has been associated with a coronal mass ejection that similarly perturbed the magnetosphere.[69] Plasmas and electrostatic discharges disproportionately affect satellites in geostationary orbit. Most of these incidents trigger glitches or failures in attitude control systems, or flip bits that cause 'phantom commands' in command or telemetry logic.[70] An occasional occurrence (e.g. MARECS A) is degradation of the power output of solar panels due to arcing across their surfaces while in eclipse during times of enhanced solar activity. Telesat Canada's Anik B1 experienced a rise in the operating temperatures of some of its components as a result of the degradation of the thermal coating attributed to local electrostatic discharges while in eclipse.

The case of Pioneer 10

When Pioneer 10 passed through the most intense part of Jupiter's radiation belts, it suffered spurious signals (which, for example, made its imaging photopolarimeter skip, resulting in the loss of close-up images of Jupiter and Io) as well as some permanent damage.[71] Recall that this was December 1973, some years before electronics were miniaturised to the point that individual energetic particles could cause a bit to flip, and thus the damage was probably due to electrostatic discharge – the first change in state of a computer memory attributed to a cosmic ray did not occur until 1975.[72]

Acts of God?

A note of cynical caution is warranted – an objective analysis of these failures is not easy. A satellite maker or operator might find it convenient to attribute a sudden failure to an 'act of God' such as a solar storm, rather than accept responsibility for a design or construction flaw. Similarly, scientists studying obscure aspects of magnetospheric physics may materially benefit from the association of 'real-world' failures with their discipline. Nevertheless, one study identified no less than 40 plasma-related (and mostly electrostatic discharge) spacecraft failures, a number comparable with the number attributed in the same study to radiation effects.[73]

NOTES

1. http://www.astronautix.com/lvs/thor.htm
2. http://www.astronautix.com/lvs/thodsv2e.htm
3. http://roland.lerc.nasa.gov/~dglover/sat/telstar.html (quotes NASA SP-32 extensively)
4. http://www.eas.asu.edu/~holbert/eee460/tiondose.html
5. http://www.smecc.org/james_early___telstar.htm
6. *Telstar 1: Dawn of a New Age*, J.M. Early, *SMEC Vintage Electrics*, vol. 2, 1990.
7. www.national.com/appinfo/space/files/Bipolar_Voltage_Comparator.pdf
8. *Aviation Week & Space Technology*, 15 May 1989, p. 52.
9. 'Pioneer 10 and 11 Jovian encounters: radiation dose and biological lethality', M.W. Miller, G.E. Kaufman and H.D. Maillie, *Life Sciences and Space Research*, 1976, vol. 14, p. 195.
10. *Pioneer: First to Jupiter, Saturn and Beyond*, R.O. Fimmel, J. van Allen and E. Burgess, SP-446, NASA, 1980, p. 171.
11. 'In-flight observations of multiple-bit upset in DRAMS', G.M. Swift and S.M. Guertin, *IEEE Transactions in Nuclear Science*, vol. 47 (6) in press.
12. http://www.amsat.org/amsat/sats/n7hpr/ao10.html
13. http://www.cstone.net/~w4sm/AO-10.html
14. http://pwg.gsfc.nasa.gov/istp/collaborating/imp8.html
15. http://www.vilspa.esa.es/iue/iue.html
16. http://toms.gsfc.nasa.gov/m3toms/m3sat.html
17. 'Instrument and spacecraft faults associated with nuclear radiation in space', J.H. Trainor, *Advances in Space Research*, vol. 14, no. 10, 1994, p. 685.
18. 'Instrument and spacecraft faults associated with nuclear radiation in space', J.H. Trainor, *Advances in Space Research*, vol. 14, no. 10, 1994, p. 685.
19. 'On-orbit anomalies in the TOMS–EP Earth sensors', B.P. Robertson, J.K. San and V.H. Selby, AAS 97-073 in Guidance and Control 1997, *Advances in the Astronautical Sciences*, vol. 94, p. 331.
20. 'PRARE 2: Building on the lessons learnt from ERS 1', W. Schaefer and W. Schumann, *ESA Bulletin*, no. 83, August 1995.
21. http://esapub.esrin.esa.it/bulletin/bullet83.htm

22. http://www.ee.surrey.ac.uk/SSC/CSER/UOSAT/press/cerisepr1.html
23. www.orbitaldebris.jsc.nasa.gov/newsletter/pdfs/ODQNv7i3.pdf
24. www.wws.princeton.edu/cgi-bin/byteserv.prl/~ota/disk2/1990/9033/903306.PDF
25. www.orbitaldebris.jsc.nasa.gov/library/SatelliteFragHistory/13thEditionofBreakup-Book.pdf
26. 'Determining the cause of a satellite break-up: A case study of the Kosmos 1275 break-up', D.S. McKnight, IAF Congress, IAA-87-573, 1987.
27. *Aviation Week & Space Technology*, 9 March 1992, p. 19.
28. www.orbitaldebris.jsc.nasa.gov/newsletter/pdfs/ODQNv3i2.pdf
29. http://www.aleniaspazio.it/program/sci/tss/tss.htm
30. http://liftoff.msfc.nasa.gov/Shuttle/sts-75/tss-1r/brochure/page_13.html
31. http://www.daviddarling.info/encyclopedia/S/SEDS.html
32. http://www.tetherapplications.com/seds2.htm
33. http://ilrs.gsfc.nasa.gov/satellite_missions/list_of_satellites/tips_atex.html
34. 'The risk to satellite tethers from meteoroid and débris impacts,' N. McBride and E.A. Taylor, in *Second European Conference on Space Debris*, B. Kaldeich-Schurmann and B. Harris (eds), ESA SP-393, European Space Agency, 1997, p. 643.
35. http://science.nasa.gov/newhome/headlines/ast09aug99_1.htm
36. http://www.oarval.org/perseids.htm
37. 'The potential danger to space platforms from meteor storm activity', M. Beech, P. Brown and J. Jones, Quarterly Journal of the Royal Astronomical Society, 1995, p. 127.
38. http://www.fas.org/spp/guide/russia/piloted/1993.htm
39. *The Story of Space Station Mir*, D.M. Harland, Springer–Praxis, 2004, p. 225.
40. http://www.selkirkshire.demon.co.uk/analoguesat/olympus.html
41. http://www.zetatalk.com/info/tinfo14y.htm
42. http://www.house.gov/science/ailor_05-21.htm
43. 'Olympus and the 1993 Perseids: Lessons for the Leonids', D. Caswell, Leonids Threat Conference Manhattan Beach, 26–27 April 1998.
44. http://sci2.esa.int/leonids/leonids98/OLYMPUS_and_the_1993_Perseids/
45. http://nssdc.gsfc.nasa.gov/planetary/giotto.html
46. *Giotto to the Comets*, N. Calder, Presswork, 1992.
47. 'Giotto–Halley encounter: When was the large nutation generated?' M. Paetzold, M.K. Bird and H. Volland, *Astronomy and Astrophysics*, vol. 244, 1991, L17.
48. http://www.spacedaily.com/news/comet-99a.html
49. http://stardust.jpl.nasa.gov/comets/vega.html
50. 'Solar array degradation by dust impacts during cometary encounters', R.D. Lorenz, *Journal of Spacecraft and Rockets*, vol. 35, 1998, p. 579.
51. 'The propulsion systems of the Phase-III series satellites', R. Daniels, *AMSAT-NA Technical Journal*, vol. 1, no. 2, 1987–1988, p. 9.
52. http://www.skyrocket.de/space/doc_sdat/amsat-p3a.htm
53. http://www.ralfzimmermann.de/phase3d/400n.html
54. http://www.tbs-satellite.com/tse/online/sat_oscar_10.html
55. 'Magellan star scanner experiences: what a long, strange trip it's been', E.H. Seale, AAS 91-072, Guidance and Control 1991, *Advances in the Astronautical Sciences*, vol. 74, 1992, p. 513.
56. *The Evening Star: Venus Observed*, H.S.F. Cooper, Johns Hopkins Press, 1994, p. 46.
57. *Solar System Log*, A. Wilson, Jane's, 1987, p. 58.
58. Mars Odyssey Mission Status, 13 March 2002.
59. http://astrobiology.arc.nasa.gov/news/expandnews.cfm?id=9830

60. http://SkyandTelescope.com/news/article_1116_1.asp
61. *Failures and Anomalies Attributed to Spacecraft Charging*, R.D. Leach and M.B. Alexander, NASA RP-1375, Marshall Space Flight Center, August 1995.
62. http://www.southcoasttoday.com/daily/05-98/05-20-98/a02wn043.htm
63. 'Pager satellite failure may have been related to disturbed space environment', D.N. Baker, J.R. Allen, S.G. Kanekal and G.D. Reeves, *EOS*, 6 October 1998, p. 477.
64. http://www.agu.org/sci_soc/articles/eisbaker.html
65. http://satjournal.tcom.ohiou.edu/Issue4/historal_cfseries.html
66. *Aviation Week & Space Technology*, 31 January 1994, p. 28.
67. *Aviation Week & Space Technology*, 7 February 1994, p. 58.
68. *Aviation Week & Space Technology*, 31 January 1994, p. 28.
69. http://www-istp.gsfc.nasa.gov/istp/cloud_jan97/event.html
70. *Spacecraft System Failures and Anomalies Attributed to the Natural Space Environment*, K.L. Bedingfield, R.D. Leach and M.B. Alexander, NASA RP-1390, August 1996.
71. *Solar System Log*, A. Wilson, Jane's, 1987, p. 72.
72. 'Radiation-induced anomalies in satellites', E.G. Stassiniopoulos, G.J. Brucker, J.N. Adolphsen and J. Barth, *Journal of Spacecraft and Rockets*, vol. 33, no. 6, 1996, p. 877.
73. *Spacecraft System Failures and Anomalies Attributed to the Natural Space Environment*, K.L. Bedingfield, R.D. Leach and M.B. Alexander, NASA RP-1390, August 1996.

14

Structural failures

STRUCTURAL FAILURES

Without a structure, a spacecraft is just a set of circuit boards, wires and propulsion pipework. Although, in some senses, a strong structure is unnecessary in microgravity, it is required to hold the subsystems together on the ground, and during launch. In addition to the rigid structure, some elements are designed to move – either because they are too bulky to fit inside the launch vehicle's aerodynamic shroud, or are too flimsy to support themselves, or because in space they require to be pointed in different directions. These mechanisms, some actuated by motors, others by pyrotechnic devices, pose particular challenges since the provision of lubrication in a vacuum is not a trivial task. Many structural parts, and in particular their coatings, serve a thermal function, and thus we consider failures in these areas together. Indeed, in addition to the flight unit that will be launched and an engineering model that is generally a faithful reproduction of the electronic systems, the first item to be produced in a spacecraft development programme is a structural and thermal model, or sometimes a structural, thermal and pyrotechnic model. This is used to verify the structure (which is often a large component of the mass of the spacecraft) early in the programme. It also often provides engineers with their first impression of how the final spacecraft will look. Many end up in museums. In this section we consider cases where the spacecraft structure fails in some manner, in the sense of failing to sustain loads and actually breaking, or where the rigidity of the structure has proven inadequate.

Alexis failure and recovery
The Alexis satellite is an unusual example of a structural failure that was survivable, albeit at the cost of attitude control and power problems. This 113-kilogram satellite was launched on 25 April 1993 by a Pegasus and was successfully placed into a near-circular orbit at an altitude of 800 kilometres, but, to the consternation of the controllers, there was no signal from the satellite. A video downlink from the launch

vehicle's second stage indicated that one of the four solar panels had prematurely deployed because the bracket that held its hinge assembly was insufficiently rigid. Another likely factor was additional stress on the mating plates of the hot-wax actuator that released the panel. The initial attempts at contact were unsuccessful, in part because the tracking data supplied turned out to be that of the third stage, not the satellite, and the position and Doppler-shift differences were sufficient to inhibit communication. When a satellite is in distress, a variety of assets can be called upon to provide clues. In this case it was confirmed (possibly by the US Space Command using an imaging radar) that all four solar panels had deployed. This was good news, as it meant that the satellite was at least partially functional. Nevertheless, attempts to command it using a 7-kilowatt radio transmitter and a 20-metre-diameter dish were unsuccessful. Tantalisingly, on 2 June, a 15-second transmission was received, indicating that the batteries, processor, and transmitter and receiver were functional. This was followed by a renewal of silence until the end of the month, when a 4-minute transmission enabled the problem finally to be diagnosed. With the exception of the magnetometer, all the systems were functional. The magnetometer, which was required for attitude control, was mounted on the prematurely deployed solar panel, and had evidently been damaged. Lacking attitude control, Alexis had entered a slow spin in an orientation that provided little power, hence the failure to communicate.

After removing some power loads (such as the magnetic torque coils which, without input from the magnetometer, were unable to control the attitude) it became possible to recharge the batteries. However, the attitude dynamics were complicated by the balky solar panel, which was not rigidly deployed, but was attached only by its umbilical cable, and flopped about. Data from Sun sensors indicated that the spin axis was skewed by 18 degrees from the nominal symmetry axis of the satellite as a result of this change in the mass distribution. When Alexis had been released by the third stage of the launch vehicle, the dynamics had put it into a flat spin in which there was an even chance that its solar panels would face the Sun or face away from the Sun, and by sheer bad luck it had ended up in the unfavourable orientation. Of course, this would not have posed a problem if the attitude control system had been functional, as the satellite would have flipped itself around. It is believed that the trickle of power that permitted the brief contact on 2 June was actually due to the illumination of the solar panels by sunlight that was reflected off the Earth.[1] With control restored, the attitude determination and control system had to be redesigned: instead of a closed-loop using primarily the magnetometer, a sophisticated attitude-dynamics model (including the flopping panel) had to be driven by the remaining attitude sensors. As the Sun sensors were unavailable while Alexis was in the Earth's shadow, a Kalman filter in the attitude dynamics model made estimations sufficiently accurate to predict and control the spacecraft, which then went on to perform 4 years of measurements of astrophysical X-ray sources.[2]

Mars Global Surveyor

Another case of recovering from failure is Mars Global Surveyor, which was the first of a series of 'faster-better-cheaper' missions to be launched to reinvigorate the exploration of Mars. It was conceived in the aftermath of the loss of Mars Observer

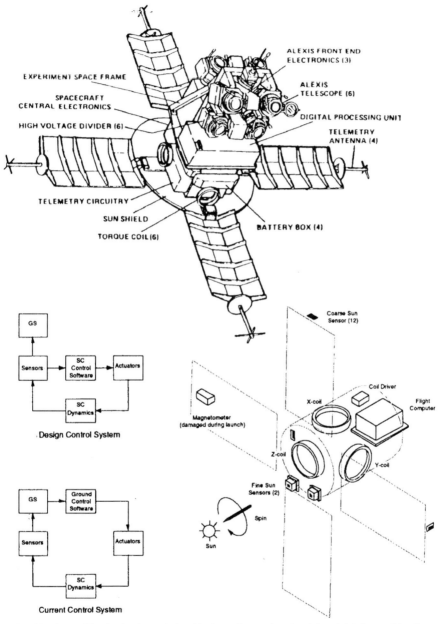

The Alexis satellite in the intended orbital configuration (top) in which it would spin on
its axis at about 2 revolutions per minute, with the cells on the underside of its panels (as
shown) facing the Sun. The major elements of the attitude control system (right). Block
diagrams (left) of the attitude control system as designed, and as modified in use to cope
with the loss of the magnetometer. ('The Alexis mission recovery', J. Bloch *et al.*, AAS-
94-062.)

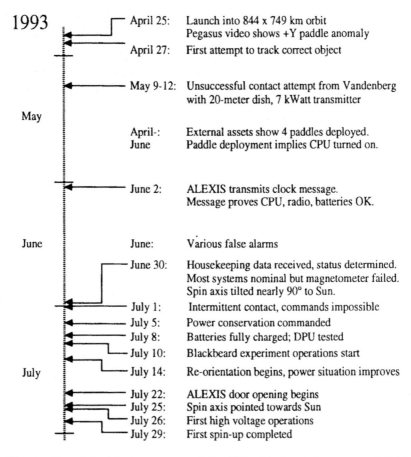

1993		
	April 25:	Launch into 844 x 749 km orbit Pegasus video shows +Y paddle anomaly
	April 27:	First attempt to track correct object
	May 9-12:	Unsuccessful contact attempt from Vandenberg with 20-meter dish, 7 kWatt transmitter
May	April-: June	External assets show 4 paddles deployed. Paddle deployment implies CPU turned on.
	June 2:	ALEXIS transmits clock message. Message proves CPU, radio, batteries OK.
June	June:	Various false alarms
	June 30:	Housekeeping data received, status determined. Most systems nominal but magnetometer failed. Spin axis tilted nearly 90° to Sun.
	July 1:	Intermittent contact, commands impossible
	July 5:	Power conservation commanded
	July 8:	Batteries fully charged; DPU tested
	July 10:	Blackbeard experiment operations start
July	July 14:	Re-orientation begins, power situation improves
	July 22:	ALEXIS door opening begins
	July 25:	Spin axis pointed towards Sun
	July 26:	First high voltage operations
	July 29:	First spin-up completed

The timeline of the Alexis recovery activity. ('The Alexis mission recovery', J. Bloch *et al.*, AAS-94-062.)

in 1993, and was to carry some of the instruments that had been built for that mission (some of the others were to be flown on later spacecraft in the series). To minimise the amount of propellant that had to be carried, and hence the mass of the spacecraft, the size of the launcher and the cost, these new spacecraft were to exploit a technique dubbed 'aerobraking' in which, after using its engine to enter an initial capture orbit with a very high apoapsis, the spacecraft would perform a second burn at apoapsis to lower the periapsis into the planet's atmosphere so that, over successive orbits, the aerodynamic drag would lower the apoapsis. Finally, once the apoapsis was at the altitude required for the mapping mission, a final burn would lift the periapsis above the atmosphere. Aerobraking was first demonstrated in 1991 by Japan's Hiten spacecraft in Earth orbit,[3] and was used by Magellan in 1993 in Venus orbit.[4,5] However, this technique has to be used cautiously, because the drag force depends on atmospheric density, which at high altitudes can vary appreciably with time, location and solar activity. The altitude must be a compromise between dipping into the denser atmosphere in order to aerobrake quickly and the ability of the spacecraft to tolerate the resulting structural and thermal loads. An additional concern is that aerodynamic drag on a spacecraft that is not symmetrical will introduce torques that might result in the loss of attitude control. In contrast to Magellan, Mars Global Surveyor was designed for aerobraking, and its pair of solar arrays were to cant at an angle of 30 degrees, with their rear surfaces facing forward, in order to provide 'weathercock' aerodynamic stability.

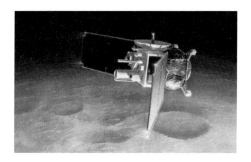

An artist's impression of Mars Global Surveyor in its aerobraking configuration (in this view, travelling to the right).

Unfortunately, when the solar arrays deployed early in the interplanetary cruise, one failed to latch into place. Each array had two panels, and it was suspected that the input shaft of the viscous damper that was meant to prevent the hinge overshooting had sheared when the array was deployed, possibly because the inboard panel was still moving when the outboard panel locked into position. A Sun sensor indicated that whereas the panel had initially been at an angle of 20.5 degrees, it drifted to 19 degrees during the cruise, suggesting that the yoke, which was a triangular epoxy–aluminium honeycomb assembly that connected the panel to its gimbal, was tearing away from the panel to which it was attached. Because the aerobraking loads would exceed the torque of the spring that held the array in position, the unlatched panel was rotated by 180 degrees in the hope that the air pressure during the first aerobraking pass would force it out and latch it into position. Although, this would expose the solar cells to the free molecular flow heating, tests had shown that they could tolerate hundreds of cycles of heating to 190°C, which was greater than the temperature they would have to endure. Of course, once the panel was locked, it would be rotated to perform the subsequent aerobraking as planned.[6]

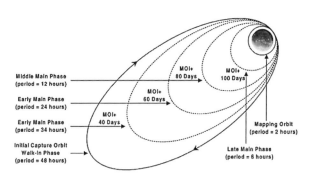

The planned timescale for Mars Global Surveyor aerobraking.

On 17 September 1997 Mars Global Surveyor fired its engine to enter a capture orbit ranging between 110 and 54,000 kilometres, and the plan was to transform this to a near-circular orbit at 400 kilometres by the new year. However, on an aerobraking pass on 6 October, deceleration loads were found to be 50

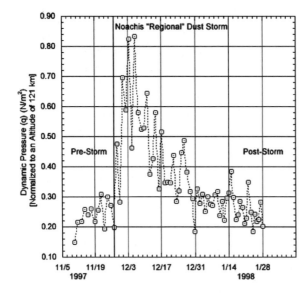

A plot of the dynamic pressure encountered by Mars Global Surveyor during periapsis passages through the upper part of the Martian atmosphere, before, during and after a major dust storm.

per cent higher than expected, and there were indications that the balky solar array had been bent *beyond* its deployed position. Controllers raised the periapsis to keep the spacecraft out of the atmosphere until the problem could be understood.[7] It was eventually decided that the damaged panel would tolerate a dynamic pressure of 0.2 newton per square metre, compared to the planned stresses of about 0.6 newton per square metre and the 0.9 newton per square metre that was unexpectedly encountered. The only option was to perform the aerobraking at a higher altitude than intended in order to reduce the drag, which would make the process much more protracted.[8] This problem is often attributed to unexpected variation in the density of the Martian atmosphere, but the day-to-day variations would have posed only modest difficulties if the structure had not already failed. Deployment tests had been performed on the ground, but not in a vacuum chamber, and the air may well have helped to damp the motion of the panel. When the periapsis was lowered once again on 7 November, aerobraking proceeded without incident, with the spacecraft finally achieving the desired mapping orbit in March 1999.[9] Although the onset of the mapping mission had been delayed by a year, there was a consolation prize as the extra time spent with the periapsis at a relatively low altitude meant that the magnetometer was able to collect high-resolution data which revealed that the planet's surface is magnetised in an intriguing manner.[10] Although the original schedule would have produced such data, it would not have done so on a global scale, and Mars Observer, by entering the higher mapping orbit using a brief sequence of engine firings, would scarcely have revealed anything at all. In the end, Mars Global Surveyor became a classic case of a spacecraft that simply kept on flying, and its mission was repeatedly extended (and, indeed, was still functioning at the time of writing).

Terminator jitters

During the early operations of the Hubble Space Telescope, it was realised that the spacecraft's precise pointing was degraded at certain times. By most spacecraft standards, the angular deviations were tiny, but they resulted in an unacceptable degradation of the imagery, whose exceptional resolution (even with the manufacturing flaw in the primary mirror) required commensurate spacecraft stability. It was quickly noted that this jittering motion occurred when the satellite entered or left the Earth's shadow, which implied it was due to thermal stress on the solar arrays. Developed by British Aerospace as part of the European contribution to the project, these combined very high performance in terms of specific power (i.e. power per unit mass) in a compact package. They were the largest flexible-blanket solar arrays to date, and were unfurled using extensible masts (although when the spacecraft was deployed by STS-31 in April 1990 one of the arrays jammed until controllers overrode the software torque limits on the deployment motor.) However, their flexibility enabled the thermal shock during terminator crossings to propagate into the main body. The arrays were held taut by a pair of bistem booms, which were pre-stressed metal tapes that warped into cylindrical tubes as they were unrolled. In sunlight, one side of a tube warmed faster than the other, and the differential expansion caused the boom to flex. On crossing the terminator, the arrays flapped and imparted a vibration to the craft with a period of approximately 10 seconds. As a workaround, the control law on the attitude control system was revised to damp the vibrations more effectively, and observations were scheduled to avoid terminator crossings.[11]

Fortunately, the Hubble Space Telescope had been designed to be serviced in space, although whether doing so is more efficient than replacing it with a new telescope in a world of full-cost accounting is another question. On the first such servicing mission in December 1993, the arrays were replaced by others on which the extension masts had a flexible aluminised thermal shield to prevent direct sunlight falling on the bistems, to reduce the thermal stress and the attitude disturbance. In fact, one of the original arrays had become so badly distorted that it failed to roll back up, and had to be jettisoned. The other array was returned to Earth for analysis, both for its structural defects and for the accumulated micrometeoroid and space débris impacts.

Similar attitude disturbances afflicted other spaceborne observatories, such as Uhuru (SAS 1) and Copernicus (OAO 3), but this was not a major problem since their telescopes were not capable of such high angular resolution. By warping a structure, differential heating can even cause semi-permanent thermal distortions, but again these are usually a nuisance only to astronomical satellites due to their pointing accuracy. For example, the Japanese ASCA X-ray satellite tended to develop a 0.015-degree mismatch between its estimated attitude and that measured using its star trackers. The fact that this offset was correlated with the day/night cycle suggested that solar heating had distorted its structure in such a manner as to introduce a small rotation between the gyro mounting and the star tracker.[12] Also, Landsat 4, Landsat 5 and the Upper Atmosphere Reseach Satellite, all of which were Earth-observation satellites using a single large solar panel, suffered attitude disturbances at terminator crossings due to thermally induced distortions.[13]

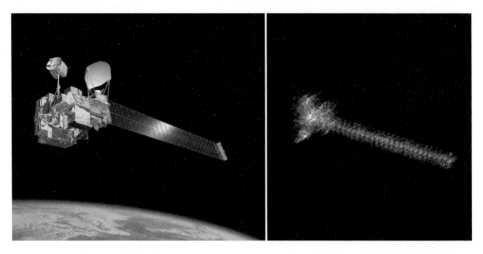

A ground-based radar image of the ADEOS 1 satellite.

The loss of ADEOS 1

The Advanced Earth Observation Satellite (ADEOS 1, also known as Midori 1) developed by Japan, was launched by an H-2 on 17 August 1996 and declared operational on 26 November. However, on 30 June 1997, at some time between a pass over a NASA ground station at Fairbanks in Alaska and a pass over Tokyo an hour later, it suffered a disabling failure and was never heard from again.[14] In diagnosing the problem, the investigators were helped by ground-based images of the satellite provided by the department of High Frequency Physics at the Research Establishment for Applied Science, a primarily defence-related organization in Germany.[15] This imagery undermined the first theory, that the solar panel had failed due to an impact with a major piece of space débris. However, a close analysis of the earlier telemetry showed that the solar panel power output had started to decline on 27 June and its temperature started to fluctuate the following day. It was finally concluded that when the satellite reversed its attitude on 28 August 1996 the rear of the solar array had been illuminated, and at this time the mast that held the array out had expanded more than expected, while simultaneously the array itself had contracted. This thermally induced flexure stressed the solder joint at the base of the array, and while this did not fracture the joint immediately, the thermal cycling of the panel in the following months caused a fatigue crack to grow, prompting a solder joint to fail and thereby deny power to the satellite. The design was also criticised for an inadequate tolerance on the tension spring on the array, and for underestimation of the expansion and contraction of the array at low temperatures. A subtle point contributed to this last factor: the thermal expansivity of most materials increases with temperature, but the layered structure of the solar array blanket (which comprised thin membranes of polymer, adhesive and copper) turned out to have a much higher thermal expansivity at low temperatures than might be predicted on the basis of a value measured at more typical high temperatures.[16]

The ADEOS satellite had suffered several significant difficulties prior to its sudden demise: one of its four thrusters had failed soon after launch, and there had been some difficulties with its onboard computer and several of its instruments. Nevertheless, it had returned some impressive data and it was sorely missed.

Landsat 4
Landsat 4 (which was based on the same Fairchild Multimission Modular Satellite bus as SolarMax) was launched in 1982; by 1984 stresses in power cables resulting from thermal cycling had caused power from two of its four solar arrays to fail and a third to suffer intermittent problems.[17]

Mariner 3
Launched on 5 November 1964, NASA's first mission to Mars was foiled by no fault of the spacecraft. After making the escape burn, the launch vehicle's Agena stage was to release its payload, Mariner 3, which was to deploy its solar panels and report its status. However, the telemetry showed that the panels had not deployed, leaving the spacecraft on its battery, which expired several hours later. The problem was identified as structural failure of the honeycomb material of the payload shroud. This was in part caused by the high temperatures produced by aerodynamic heating during the ascent, combined with pressure loads resulting from inadequate venting. As the Atlas launch vehicle ascended, the ambient pressure dropped rapidly, but the pressure inside the shroud did not. The resultant pressure differential caused the honeycomb cells to crack the inner wall of the shroud, and the heated material bonded with the spacecraft inside, preventing the shroud from jettisoning. It was discovered that Lockheed, the manufacturer, had not perfomed a thermal vacuum test that included the appropriate pressure and thermal profiles.[18,19,20] Even if the shroud had released when commands were sent from the ground, the mission was lost because the Agena, weighed down with the additional mass for the escape burn, had ended up in a heliocentric orbit that fell far short of Mars.

Ulysses
The Ulysses spacecraft for the International Solar Polar Mission was ferried into low orbit on 6 October 1990 by STS-41, accelerated to escape velocity by an IUS, and then boosted by a solid rocket motor in order to make a fast passage to Jupiter for a slingshot that would deflect it back into the inner Solar System in an orbit inclined at a high angle to the ecliptic to enable it to study the solar wind emerging from the Sun at high latitudes. The following day, the 370-kilogram spacecraft deployed its 5.6-metre-long radial boom, which reduced the spacecraft's spin rate from 7 to 5 revolutions per minute, which was to be maintained for stability, and to even out the thermal loads. After a trajectory correction on 15 October, the instruments were checked out one by one, and on 4 November the 7.5-metre axial bistem boom was unreeled by an electrical motor opposite the 1.65-metre high-gain antenna, which was maintained pointed at the Earth. After the deployment, the spacecraft's spin developed a nutation that rapidly built up to 6.5 degrees, and then slowly decayed to 3 degrees. However, as the spacecraft receded from the Earth, even 3 degrees would

An artist's impression of the Ulysses spacecraft.

disrupt communications via the tight X-Band beam. Fortunately, Ulysses had an active closed-loop control system, and when this was activated it *just* managed to overcome the nutation. Thermal stress was causing the boom to flex, and the axial offset was inducing the wobble. It was realised, however, that as Ulysses receded from the Sun the entire length of the boom would fall within the spacecraft's shadow, and the problem would go away of its own accord, which is what happened on 17 December. Although the nutation recurred when Ulysses returned to the inner Solar System in 1994–1995, the closed-loop attitude control was able to maintain it at an acceptably low level.[21],[22],[23]

Crash!

Structural damage can occur in hard landings. A few hours after Neil Armstrong and Buzz Aldrin landed on the surface of the Moon on 20 July 1969, Luna 15, the automated spacecraft dispatched by the Soviet Union to land, scoop up a sample of moondust and return it to the Earth to steal a march on Apollo 11, suffered an attitude control problem in the final phase of its descent and crashed. In 1970 Luna 16 managed to land, retrieve a sample, and return it to Earth, but in 1971 contact with Luna 18 was lost at an altitude of 100 metres when it was orienting itself to land. A sample from this site

was returned by Luna 20 in 1972. When Luna 23 was sent in 1973, it was to employ a coring drill to obtain a sample from a depth of 2 metres. However, a hard landing (attributed, as often seems to be the case with failing landers, to overly rough terrain) disabled the arm on which the drill was mounted. The fact that the ascent stage was not launched simply for engineering tests or for operator training (which one might reasonably have expected as a contingency mission) suggested that damage to the vehicle was severe. It

Artist's impressions of Luna 16 (left) with its swing-down drill and Luna 24 with its rail-mounted drill designed to extract a 2-metre-long core sample.

returned engineering data for three days, until its battery was drained. In 1976 Luna 24 succeeded in drilling at this site and returned a sample of 170 grams.[24] Since then, no spacecraft has attempted to land on the Moon.

MECHANISM FAILURES

Moving parts are the spacecraft engineer's nightmare. The most trivial cable snag or the interference of a gossamer thermal blanket can be crippling, and although it could be fixed by a gentle nudge by a technician, this is little consolation when the nearest one is 1,000,000 kilometres away. Mechanisms must be designed to survive severe vibration and dynamic loads during launch, and then run smoothly in a weightless environment. The vacuum of space poses a particular challenge to lubrication. As many conventional oils would evaporate, it is necessary to employ special low-vapour-pressure greases or exotic solid lubricants like molybdenum disulphide. In vacuum, graphite acts as an abrasive. In the absence of air, clean metal surfaces can cold-weld to each other, seizing completely. Many of these solutions have been developed from bitter experience: wear, friction and lubrication are matters of such import that there is even a facility devoted to their study – the European Space Tribology Lab. However, although these solutions usually work, as with other fields of engineering design for spacecraft, sometimes things still fail.

Voyager 2's scan platform

The two Voyager probes sent to explore the outer planets of the Solar System were each equipped with a scan platform, a motorised turntable that was capable of slewing in azimuth and elevation in order to maintain cameras and other remote-sensing instruments aimed at a specific target while the spacecraft maintained its fixed high-gain antenna pointing at the Earth. There were initial fears that the boom on which Voyager 2's scan platform was mounted had failed to swing out and latch into position, but it was determined that the latch sensor had simply given an improper signal. It operated well at Jupiter, and for a while at Saturn, but shortly after the spacecraft made its closest approach to the planet the azimuthal actuator seized, locking the platform, resulting in the loss of a significant fraction of the scientific observations from the outbound part of the encounter. Worse, if the scan platform was out of service, the observations that could be made at Uranus and Neptune would be very limited. However, the fact that it would take 3 years to reach Uranus gave the engineers time to understand and work around the problem. Intensive testing and analysis determined that the fault was at the interface between a shaft

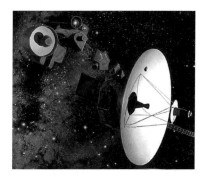

An artist's impression of a Voyager spacecraft, featuring its boom and scan platform.

and its bearing, close to the stepper motor that drove the shaft using a 9,000-to-1 reduction gear. The shaft could be driven at various speeds. During the busy Saturn encounter, the scan platform had been slewed from one target to the next at its maximum speed. However, at that rate lubricant tended to migrate away from the point at issue, enabling the shaft to be scratched and the shards to be redeposited elsewhere, causing the seizure.[25] Temperature cycling ameliorated the situation by crushing the redeposited material. As tests indicated that slewing at a slower rate greatly reduced the likelihood of seizure, it was decided not to slew at the maximum speed of 1 degree per second, but to use either the medium rate of 0.333 degree per second or the low rate of 0.083 degree per second.[26,27] This effort was time well spent, because the scan platform functioned flawlessly at both Uranus and Neptune.

Anik E2's troubled deployment

Built by GE Astro and Canada's Spar Aerospace and launched on 4 April 1991 by an Ariane 4, Anik E2 had separate antennas for its Ku-Band and C-Band systems. Once the satellite had circularised just below geosynchronous altitude, preparatory to drifting onto its assigned station, ground commands were sent to fire pyrotechnics to cut the cables that had been wrapped around the antennas and solar arrays, and their successful firing was inferred from the onboard accelerometers. In their deployed configuration, the solar arrays spanned some 23 metres. Commands were then sent to release the four tie-downs that held the Ku-Band antenna. After these cables (each of which was pre-tensioned to 4,000 newtons) was cut by redundant pyros, a kick-off spring was to initiate the separation. Although there was no telemetry to confirm that the antenna had deployed, the fact that a signal was received from an antenna that it had previously covered indicated that it had done so. Things were not to go so smoothly for the C-Band antenna. This antenna covered a pair of Earth-sensors, and when these were unveiled the satellite was to automatically orient itself to face the Earth, but this did not occur. The pyro commands were re-sent, with the battery at its maximum state of charge in order to prevent a voltage drop inhibiting their firing. The command path was verified using Anik E1, which was identical and still on the ground. A team of engineers studied the logs of Anik E2's assembly in search of clues.

Meanwhile, a signal from Anik E2's gyros registered its Ku-Band antenna swinging into position – two days after its pyros had been fired! Both the duration of the transient and the change in satellite's spin rate as a result of the greater moment-of-inertia confirmed that the Ku-Band antenna was in its fully deployed configuration, but the C-Band antenna was still stuck.

Fifty pyro tests performed over the next three months failed to find the problem. The possibility that the lubricant of the hinge was too cold, inhibiting its action, was studied.

An artist's impression of a satellite in the Anik E series.

The hinge dampers were equipped with a thermostat-controlled heater, but the temperature of the hinge was not available by telemetry. Nevertheless, if this thermostat had failed, the hinge lubricant might be too viscous. The satellite was manoeuvred to face the hinge towards the Sun for a period of 90 minutes (this duration being the time that it would take to warm them to 60°C, a temperature limit imposed to preclude the lubricant from leaking out) but the antenna remained stuck. Attention eventually focused on small thermal blankets that surrounded the tie-downs. These were attached by velcro, reinforced with kapton tape. Using Anik E1 as a testbed, the blanket on the tie-downs of the C-Band antenna was arranged in the worst possible configuration – artificially puffed out – and it was found that interference could occur in which the blanket became snagged on the support. Further tests established that a 5-newton force in an East–West direction was required to release the blanket. While the hinge-springs provided only one-third of this force, the required force could be generated by the action of spinning the satellite at 3 or 4 revolutions per minute about its primary axis. As the satellite was meant to be 3-axis stabilised, putting it into this spin would interfere with the command antenna, but a cross-polarised technique was developed to overcome this. Because the damper on the hinge was designed to cope with the impact of the antenna at a spin rate of only 0.2 revolution per minute, there was concern that once free it might swing out so rapidly as to damage itself. Tests were conducted to determine the safety margins, and the ultimate strength of the hinge.

When the go-ahead was given, the satellite was spun up to 4 revolutions per minute, but the antenna failed to release. Undaunted, the engineers used simulations to develop a thrusting strategy to introduce a nutation to the spin that would place a greater transient force upon the blanket without increasing the spin rate (with correspondingly greater hinge loads). It was also decided to face the hinges away from the Sun for a few hours to 'cold soak' them, and thus increase the viscosity of the damping fluid to absorb the impact more effectively. The manoeuvre for the cold soak posed a challenge, in that it required the satellite to maintain a stable orientation while denied its Earth-sensing capability, but the known relationship between the solar array short-circuit current telemetry and the solar incidence angle was able to be used to infer attitude information. Within hours of this revised procedure being put into effect on 2 July 1991, the C-Band antenna swung out and locked. Gyro response, the change in spin rate, and successful acquisition of the Earth (now that the sensors had been unveiled) all confirmed that the satellite was in its operational configuration.[28] Anik E1 was launched in October of that year with slightly modified thermal blankets and reinforced hinges, and entered service without incident. In 1994 both satellites suffered attitude control failures as a result of solar activity, and although Anik E1 recovered Anik E2 did not.

FLTSATCOM 5's rough ride
On 6 August 1981 an Atlas–Centaur deployed FLTSATCOM 5, a communications satellite for the US Navy.[29] The satellite separated in a 'blind spot' over the eastern Atlantic, and was acquired when it came into view of a ground station. As was normal, the telemetry was at first garbled, since the signal was initially weak and the

decoding software had difficulty synchronising with the subcarrier's data. When it *remained* garbled, contingency procedures, including resetting some of the systems, made it readable. The bus voltage and current, and the battery charging current, were fluctuating wildly. The spacecraft was spinning and its thruster temperatures were elevated because they had been firing. However, the attitude dynamics were anomalous, implying that some mechanical damage had changed the moments of inertia. Bypassing the battery-charge regulator and placing the batteries directly onto the bus stabilised the voltages but not the currents. The power engineers concluded that the solar arrays had been damaged, and pinpointed the damage by comparing the telemetry with their wiring diagrams. The satellite was out of immediate danger, but time was of the essence. The low perigee of its transfer orbit caused rapid orbit decay, and each orbit took it twice through the van Allen belts. Before the Star 37F motor could be fired, the satellite had to be spun up from its present rate of 30 revolutions per minute to 60. As this was being done, a nutation suddenly developed at 45 revolutions per minute, and the operation had to be suspended while this unwanted motion was damped out. Once the satellite was spinning at the required rate, the motor had to be aligned for the burn to raise the perigee and reduce the inclination to 2 degrees, but the possibility of mechanical damage meant that the thrust vector might not be aligned with the satellite's spin axis, in which case the offset would sap the motor's performance by a cosine factor. A quick dynamic analysis of the spin telemetry indicated that the motor could be as much as 15 degrees off the spin axis. Hedging on the motor's performance, the controllers aimed for a 7-degree inclination, which the liquid propulsion system would subsequently be able to reduce to 2 degrees. If the misalignment was less than estimated, the burn would produce an orbit with a period in excess of 24 hours, but this too would be correctable. As events transpired, the result was an acceptable 'drift' orbit at 5.4 degrees. At this point, the satellite unfolded its solar arrays without incident, but the two UHF antennas failed to deploy and they resisted efforts to shake them loose. Measurements of the gain pattern of the omni antenna suggested that the UHF-transmit antenna was bent, but in fact the inner lining of the payload shroud had delaminated and struck the satellite. The UHF-receive antenna had

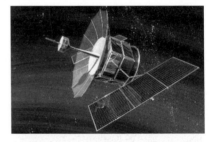

An artist's impression of a FLTSATCOM satellite.

broken at its base but remained attached during the initial operations. Only when the spacecraft was spun up to 45 revolutions per minute did it fly off, causing the nutation event. Fortunately, a signals-monitoring aircraft had been able to record telemetry during the 'blind spot', and when this was made available it enabled a variety of failure modes to be simulated. However, despite having been nursed to near-geosynchronous orbit, the satellite had to be written off.[30]

UoSAT 1's gravity-gradient boom

The UoSAT series of microsatellites generally achieved Earth-pointing by the use of gravity-gradient stabilisation – a technique that exploits the fact that in a spherically symmetrical gravitational field the centre of gravity of an asymmetrical object does not coincide with its centre of mass. An elongated object therefore has a pendulum stability, and will tend to orient itself along the local vertical. To give a satellite adequate stability, its moment of inertia in one or two axes requires to be made much greater than in the other(s). Some long cylindrical spacecraft, such as space stations, have this geometry, but for small satellites it is common practice to deploy a semi-rigid boom with a mass on its tip. However, pendulum stability has two weaknesses. First, being bi-stable, the inverted attitude is just as stable as the desired one. Second, it is undamped, and if the pendulum is established in an offset attitude it will swing back and forth without converging on the vertical; hence artificial damping must be introduced. On the UoSATs (and many others) this damping was to be done by magnetorquers, which were also used to set up the attitude and undertake other manoeuvres. The attitude of the satellite was sensed primarily by a 3-axis magnetometer that was mounted on the boom's tip-mass to minimise its sensitivity to the magnetic fields generated by the main body. On being released, UoSAT 1 adopted an appropriate attitude for gravity-gradient stabilisation, and the command given to deploy the bistem boom – a prestressed beryllium–copper tape that formed a cylindrical rod of 1 centimetre diameter when unreeled by an electric motor. The magnetometer telemetry was monitored as the boom came out. Suddenly a change in the signal indicated that the cable to the magnetometer, which had also to be drawn out, had snagged, causing the still-unreeling boom to buckle and bend. The only option was to retract the boom and operate the satellite in a much less convenient spin-stabilised mode.

Cabling and antennas

Many spacecraft use articulated antennas – dishes on gimbals to track the Earth from deep space, or to enable a satellite to track a ground station or a relay satellite. Accurate antenna pointing allows a narrower beamwidth, and therefore a higher data rate. In early operations of the Hubble Space Telescope, several disruptive safing events were traced to one of the two 1.3-metre-diameter high-gain antennas that were mounted on 5-metre booms for relaying via the geostationary satellites of NASA's Tracking and Data Relay System. A 2-centimetre-wide cable to the antenna had apparently become entangled. Restricting the range of travel of this antenna to 75 degrees (as opposed to the maximum of 90 degrees) eliminated the safing events at the expense of fewer downlink opportunities.[31]

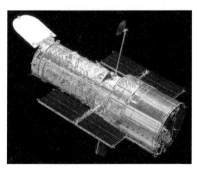

The Hubble Space Telescope, featuring one of the booms with a TDRS antenna.

A similar anomaly confronted the Mars

Global Surveyor spacecraft. The plan was to deploy the 1.5-metre-diameter high-gain antenna on its 2-metre boom immediately after achieving the mapping orbit, but difficulty encountered with a similar mechanism on Earth led to concern that the antenna on the spacecraft might become damaged during the deployment. In its stowed configuration the antenna was held tight against the side of the boxy spacecraft, and for it to be used the spacecraft had to be oriented to point it at the Earth, which required slewing the spacecraft back and forth to point its instruments at the planet and then to downlink the data. In its deployed configuration, the antenna was to rotate to maintain itself pointing at the Earth while the spacecraft maintained its instruments aimed at the planet. After three weeks of deliberation, on 28 March 1999 the antenna was commanded to deploy, which it did successfully. On

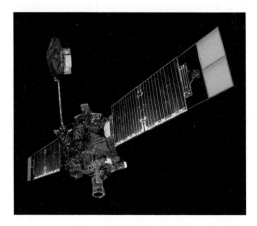

15 April, however, the azimuthal gimbal stalled, possibly as a result of a cable or thermal blanket snagging, or perhaps due to either a lubrication problem or a bent or broken gearwheel tooth.[32] After analysing the situation, the engineers concluded that the limited mobility of the antenna would facilitate unfettered mapping until February 2000. As the basic mapping mission was to last one year, this activity was able to be completed without significant loss.[33,34] After that time, the spacecraft had once again to divide its time between taking and transmitting data.

An artist's impression of Mars Global Surveyor with its high-gain antenna deployed.

'Remove before flight'

On being released by its Ariane 2 launch vehicle on 21 November 1987, Germany's TVSat 1 direct-broadcast satellite was to deploy one of the four segments on each of its two solar panels (which would ultimately span 20 metres and generate 3.5 kilowatts) for power during early operations. However, telemetry showed that only one array had deployed correctly. The other panel remained locked, and resisted attempts to free it by spinning and shaking the satellite. It was eventually realised that some of the hold-down clips that had been used to secure the panel during ground handling had not been removed, which left little scope for a recovery. There was some hope that the satellite might still be used with one panel deployed, on half power, transmitting a smaller number of channels than planned, but the uplink antenna by which the satellite was to receive the TV to be replayed was trapped behind the locked panel. Engineering tests were performed to verify the systems, but the communications mission was a total loss.[35,36]

Tape recorder failures

Often the rate at which a spacecraft can acquire data far exceeds the bandwidth with which it can transmit it to Earth, and so it requires some sort of onboard buffer. Prior to the advent of large solid-state magnetic bubble memory and the flash memory used in digital cameras – and even thereafter in some extreme cases – tape recorders were used. A tape recorder can be unreliable even in a terrestrial environment, and the interaction of a soft material (the tape) with moving, spinning parts under tension represents a serious challenge in space. The British high-altitude research satellite Ariel 3, launched in May 1967, had a tape recorder to store data for downlink; this began to behave erratically in October and failed completely four months later.[37] In 1971 the early Japanese satellite Shinsei lost its tape recorder after four months, but continued to send data in real-time.[38] Prospero, the first (and only) satellite to be launched by a British rocket (the three-stage Black Arrow) was put into orbit on 28 October 1971.[39] Its tape recorder failed in 1973, after 730 replays, but real-time telemetry continued to be received. Occasional direct telemetry was received up to 7 years later. The European satellite TD 1A, launched in 1972, lost its tape recorder.[40] The telemetry was compared with the performance of a recorder on the ground in an attempt to reproduce the symptoms and thereby diagnose the cause of the failure, which was interpreted to be a change in tape tension, possibly due to degradation of the tape lubricant. Tape recorder faults were not confined to the early days of space. Among other satellites to suffer was France's Earth-imaging satellite Spot 1, launched in February 1986.[41] One of its two Odetics tape recorders showed early signs of trouble and failed completely after about seven months. The Compton Gamma-Ray Observatory that was deployed by STS-37 on 7 April 1991 ran into trouble early the following year when first its primary and then its backup tape recorders failed. Its mode of operation had been to store data for 3 hours, then downlink it by S-Band through the geostationary satellites of the Tracking and Data Relay System. The only option was to downlink data in real-time. As the primary relay satellites provided coverage for only about 65 per cent of the satellite's orbit, the retired TDRS 1 was reawakened and relocated to plug this gap by relaying to a ground station which was set up at Tidbinbilla in Australia specifically for the purpose, and the satellite was operated in this way for the remainder of the decade.[42]

A spacecraft for which the tape recorder became invaluable was the Galileo mission to Jupiter. The 1991 failure to deploy the high-gain antenna severely compromised its data-return capacity. The only way to recover a significant fraction of the expected scientific results was to store data on the tape recorder and undertake onboard data compression to maximise the information content of the data that was sent back to the Earth at the very slow rate of the low-gain antenna. This required the spacecraft to be reprogrammed with special software, the writing of which took so long that it was not uploaded until shortly after the spacecraft had settled into orbit around its objective. By that time, the key to the plan, the tape recorder, had shown its own vulnerability. On 11 October 1995, as Galileo closed in on Jupiter, it was to start taking pictures to make a movie depicting the rotation of the planet. Having stored the first few images at 806 kilobits per second on the beginning of the tape (which required the tape to run through the system at 1.94 metres per second) the 900-megabit Odetics DDS-3100 recorder failed to stop and rewind.[43] Tests showed that the tape had slipped on its capstans (there were no pinch rollers) during the 15 hours between the rewind command being received and the motor being stopped by ground command. The spring to maintain tension on the tape was biased 'downhill' towards the middle of the tape, which meant that it was easier to move the tape in that direction. The rewind near the beginning of the tape was uphill, and thus less easy. The high acceleration required for the fast tape speed may have been a contributory factor to the problem. In view of concern that the tape may have become sufficiently weakened by such prolonged rubbing to break under dynamic operating loads, the tape was advanced to wrap the weak spot under 25 turns on the take-up reel. Careful tape management, combined with a restriction on the use of the high-speed mode, enabled the recorder to be successfully operated.[44] However, as the priority was to receive and store data from the atmospheric probe, it was decided not to risk running the recorder at high rate until this data had been downloaded to Earth, which meant that the plan to make observations of Io and Jupiter shortly prior to entering orbit had to be cancelled. Although the tape recorder did stick in January 1996 (believed to be due to a breakdown of the tape lubricant when the tape was moved quickly) it performed well throughout the rest of the mission.[45] An identical tape recorder on the ground was an important diagnostic tool. Intriguingly, this also failed on 11 October 1995, although in a different manner from the flight unit. A light-emitting diode and a photodiode were to sense the clear leader at the end of the opaque tape, but this failed and the recorder continued to run, with the result that the torque due to the tape-to-capstan friction broke the tape, which was exactly opposite to the tape-slippage problem that had occurred in space.

A tape-recorder fault on the Magellan spacecraft led to corruption of 50 gigabytes of its 700-gigabyte data return. The problem initially showed up on track number 2 of one of its two recorder, and within months had spread to the other three tracks. However, the corruption was a deterministic process, as the frame-synchronisation code on the tape was consistently changing a '1001' bit sequence to a '111' as one bit was flipped and another was lost totally. It was the loss of this bit that led to the synchronisation of the telemetry stream being lost.[46] Owing to the deterministic nature of the failure, an elaborate software search was able to identify and correct most of the bit-flips, with the result that data from 378 of the 415 corrupted orbits was able to be recovered.

Indian satellite mechanism failures

To test systems developed for its Insat series of satellites for communications, Earth-resources and weather forecasting, India built the 630-kilogram APPLE satellite and had it launched as a secondary payload on 19 June 1981 by the third Ariane test flight, and despite its solar panel jamming it was deemed successful.[47] For its operational satellites, India hired Ford Aerospace, which adopted an unconventional design for a geostationary satellite. Instead of symmetric solar panels, it used a single panel and, to counter the solar radiation pressure torque that this would impart, balanced it with a solar sail, which was a small structure on a mast opposite the solar panel. Unfortunately, when Insat 1A was launched by a Delta rocket in April 1982,

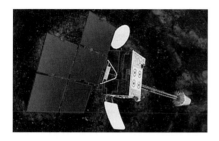

An artist's impression of Insat IA.

this mast failed to deploy.[48] (The design was modified, and the mast deployed correctly on Insat 1B.) Insat 1A also suffered anomalies in deploying its antenna. Initially the antenna failed to deploy completely, but after it had heated up in sunlight it was shaken loose by firing the thrusters. It soon became apparent that a faulty propellant-isolation-valve control circuit would cause the satellite to exhaust its attitude control gas long before its planned 7-year lifetime, but this became irrelevant in September 1982 when the satellite was lost. As it was coming out of eclipse, one of its Earth sensors was deliberately switched off because the Sun would be within its field of view. However, controllers failed to realise that as a result of a yaw error built up by the absence of the solar sail, the full Moon would intrude into the field of view of the *other* Earth sensor, which would protect itself by switching off. With both sensors off, the attitude drifted sufficiently to break the narrow-beam command link, and as the contingency logic did not include a switchover to an omni-directional antenna, the satellite was lost.

JERS 1 deployment problems

Foretelling mechanism and structure problems on later Japanese satellites, the radar-imaging antenna on the JERS 1 remote-sensing satellite that was launched by an H-1 on 11 February 1992 proved tricky to deploy. Evidently, one of six pins holding the 12 by 2.5-metre antenna had jammed. However, several weeks later, after the pin had been warmed by sunlight, it popped open. The mission was successful, although by 1998 two of its three gyros had failed and its batteries were degrading.[49]

The MARSIS antenna on Mars Express

The European Space Agency's Mars Express, which suffered a 30 per cent power loss *en route*, and delivered the ill-fated Beagle 2 probe, encountered another problem in orbit. The MARSIS subsurface sounding radar was to use much longer wavelengths than typical for an imaging radar, so as to enable its beam to probe the structure of the ice caps and penetrate the ground to seek subsurface water.[50] The 20-metre-long antenna booms were made of hollow cylindrical glassfibre segments that were folded up in the manner of an accordion, and once the box was commanded to open they were to deploy by using the elastic energy of their compressed segments. As this deployment mechanism was not easy to test on the ground, reliance was put on computer simulation of the deployment kinematics. The boom manufacturer, Astro Aerospace, employed an early version of the Automated Dynamic Analysis of Mechanical Systems (ADAMS) software. After the Mars Express system was delivered, work continued on a similar antenna for a related instrument for the 2005 Mars Reconnaissance Orbiter. During this work, refinements to the simulations indicated that the MARSIS antenna deployment might experience a previously unforeseen 'backlash' in which the deploying antenna might flip back and hit the spacecraft. The deployment had been scheduled for April 2004, several months after the spacecraft entered orbit, but it was postponed indefinitely. It was reasoned that as the power was constrained, it would be better to allow the other instruments – which were providing excellent results – a good long run before risking either damage or loss of control as a result of the deployment.[51],[52]

THERMAL FAILURES

The long-term temperature of a surface in space exposed to the Sun depends on what fraction of the incident sunlight it absorbs (i.e. its absorptivity, defined as unity minus its reflectivity) and how effectively it can radiate heat (i.e. its emissivity). The single most important parameter of a surface or coating is the ratio of absorptivity to emissivity. A large ratio means that the surface absorbs much sunlight, and needs to be hot to radiate that heat away. A low ratio is needed to keep cool in sunlight, with an efficient radiating surface that reflects most of the sunlight. Polished metal surfaces meet this requirement, as do specialised blankets such as plastic-coated foils in which the optically transparent plastic allows the foil beneath to reflect most of the light away, while the plastic gives a higher emissivity than the foil on its own. White paints have similar properties, but are less expensive. Another factor is that a surface should be electrically conductive in order to prevent electrostatic discharge. Usually a coating of indium oxide will be applied, but it must be borne in mind that doing so can alter the thermal properties of the material. Even when the intrinsic thermal properties of the surface are appropriate at launch, two factors are pertinent. First, charged particles and ultraviolet radiation degrade paints in space, particularly white paints,[53] and especially in the interplanetary (as opposed to low-orbit) environment, in which an increase in the ratio of absorptivity to emissivity of a factor of 2 due to ultraviolet exposure has been observed.[54] Second, if a surface should be covered with a thin layer of a contaminant (from outgassing, for example) this ratio will probably increase and warm the surface. In addition, a thermal coating can be damaged as a result of impacts. On encountering a micrometeoroid shower some years after its historic Mars flyby, Mariner 4 suddenly suffered 17 hits in the space of 15 minutes, its attitude was perturbed and its temperature fell by 1°C, implying that some of its thermal coatings had been degraded. The Giotto spacecraft suffered a veritable sandblasting during its flyby of Halley's comet and thereafter ran a little hot.

Various failures have been attributed to temperature – the accumulation of oxidiser in the cold propellant lines of Mars Observer being a good example. The life and capacity of a battery is highly temperature-dependent, and temperature control is an integral part of battery management. On the other hand, manipulating the temperature of systems via electrical heating and/or orienting a spacecraft to make use of either the heat of the Sun or the chill of deep space can coax a reluctant system into service – for example, controlling the temperature of a radio in order to manipulate its operating frequency, or controlling the viscosity of a damping fluid in antenna deployment. Temperature variations can cause parameter drift in electronic components (changing their capacitance, resistance, etc.) and consistently higher temperatures tend to accelerate ageing effects. High temperatures can induce outgassing from polymers and generally weaken materials, and batteries, being chemical systems, can degrade. Low temperatures in general are more benign than high ones, although freezing of battery electrolyte or propellant can be an issue, as can electronic parameter drift and material parameters such as viscosity or stiffness. And if the temperature becomes extreme in either direction, differential thermal expansion can be a problem for all systems, causing mechanical interference or optical misalignment. In general, temperature cycling is a bad thing, causing fatigue and failure, often leading to open circuits as a result of wire breaks or bond failures. Incorrect thermal design – or deficiencies in the implementation of a valid design – are sometimes the root cause of a problem. Japan's experimental Yuri 1 (BSE) satellite was placed into geostationary orbit in 1978 and some of its components overheated because the solar absorbance of some surfaces exceeded the design value by a factor of 3. This particularly threatened the battery, which required careful monitoring and frequent conditioning.[55]

Lunar Orbiters

Five spacecraft launched in 1966 and 1967 by Atlas–Agenas were inserted into lunar orbit to investigate possible landing sites for the Apollo crews, and many of the in-flight difficulties were related to thermal control. Lunar Orbiter 1 overheated *en route* and was tilted 36 degrees away from its nominal Sun-pointing orientation to improve its thermal control.[56] When its star tracker failed to lock on to Canopus, the Moon was used as an attitude reference. Once in lunar orbit, some pictures were degraded by imperfections in its photographic system (which was derived from the Corona reconnaissance system that was similarly affected) and the highest-resolution pictures were blurred by the failure of the image-motion compensation system.[57,58] A layer of S-13G white paint on its successor partially alleviated the overheating problem, and

the thermal control manoeuvres were better integrated into the flight plan. Although the spacecraft fell silent due to the failure of its travelling-wave-tube amplifier shortly before the film was fully replayed, the mission achieved its objectives.[59] When the film-winding mechanism misbehaved on the third mission it was decided to advance the playback schedule, but this was foiled by an electrical transient that burned out the motor with 72 of the 211 exposed frames remaining. The star

A model of a Lunar Orbiter spacecraft.

tracker on Lunar Orbiter 4 was spoofed, probably by glints from Sun and Earthshine. Whereas its predecessors had used low equatorial orbits, this one entered a polar orbit for general mapping. To counter the additional thermal stress, in excess of 500 quartz mirrors were installed on its underside. Barely had the photography begun than telemetry indicated that the thermal door of the camera section – which was opened to take pictures and then immediately closed – had only partially closed. If it remained open the camera might chill down and cause any gas inside the camera compartment to condense and fog the lens window, but if the door shut and failed to open again that would end the mission. The attitude of the spacecraft, and control of the door, had to be carefully judged to prevent fogging while minimising light leakage that could ruin the film. Of the photographs returned, 64 per cent were judged to be of excellent quality or had only modest fogging. Controllers also had to contend with some erratic signals from the readout encoder, which showed a tendency to halt the mechanism prematurely. The only problem suffered by the final spacecraft was an early difficulty in locking on to Canopus.

Mariner 10

The first (and so far only) spacecraft to visit the planet Mercury almost did not make it. Ironically, since many of the later problems were associated with overheating, one of the first problems was the failure of the heaters for its primary instrument, the television cameras. These heaters were to keep the optics (which obviously could not be pointed at the Sun) at a constant benign temperature. If the optics were to become too cold, thermal contractions might cause their components to become misaligned, unfocused, or even to crack. Despite the fact that the lifetime of the vidicon tubes of the cameras in space was unknown, and had been nominally qualified for only a total of a few weeks of use, it was decided to keep the cameras switched on so as to maintain their operating temperature. In December 1973, a drop by a factor of 4 in the signal output, believed to be due to the thermally induced failure of a joint in the feed for the high-gain antenna, threatened the television plans. Intriguingly, this went through several cycles of failing and repairing itself during the mission. The fact that Mariner 10 had more anomalies than its predecessors may well have been due to the higher thermal loads. On 8 January 1974 the power mysteriously and irreversibly switched to the back-up system. On 28 January, after a gyro

An artist's impression of Mariner 10.

malfunction, the attitude control system wasted 16 per cent of its supply of nitrogen propellant. The loss of so much gas so early was of great concern, as it was hoped that Mariner 10 would be able to perform three flybys of Mercury. The Venus flyby in February put the spacecraft on course for Mercury, but the gyro problem ruled out mid-course corrections. Nevertheless, the flyby on 29 March was a success. However, two days later an electrical short drained 90 watts from the power system and made the power electronics bay run very hot, which obliged the engineers to switch off the cameras to relieve the load. Shortly thereafter the tape recorder failed, which meant that future flyby data would have to be downloaded in real-time, presuming that the high-gain antenna held out. To make matters worse, part of the telemetry system was also lost. At one point, the scan platform on which the cameras were mounted threatened to fail. Despite these issues, the second flyby of Mercury was accomplished on 21 September. The gyro problems were compounded on 6 October by a bright particle nearby that caused the star tracker to lose its lock on the Canopus reference star. Because the gas supply was by this point perilously low, the engineers decided to tilt the solar panels in order to use solar radiation pressure (i.e. 'solar sailing') to help to control the attitude of the spacecraft. The failure of the cooler of the radio receiver at the Deep Space Communications Complex at Canberra in Australia, the only site with a clear line of sight to the spacecraft for the final flyby on 16 March 1975, resulted in the loss of much of the data. (This cooler kept the receiver noise down, and the unbuffered telemetry from the television cameras at the maximum data rate became unacceptably noisy.) A partial recovery was effected by sending only the central part of each image at one-fifth speed to yield a tolerably low error rate.[60],[61],[62] The attitude control system ran dry one week later. Overall, the mission was a great success, squeezing a vast scientific return from a modest spacecraft, in part by pushing the bandwidth to an unprecedented 117,600 bits per second (a decade earlier, Mariner 4 had transmitted its pictures of Mars at a mere 8 bits per second) and in part by exploiting a resonant orbit to set up three flybys of Mercury. The stress on the operations teams in nursing the ailing spacecraft along to yield a total of 2,400 useful pictures of Mercury and 3,400 of Venus was nevertheless severe.

Pioneer Venus
On 9 December 1978 one large and three small Pioneer Venus probes penetrated the hot, dense, corrosive atmosphere of Venus. Only the large probe had a parachute to slow its descent. None of the small probes was expected to survive reaching the surface, but one did, and transmitted for over an hour. Its silence was prompted not by the depletion of its battery, but a component failure as its internal temperature soared. While this is hardly a failure to meet a specification, it is a good case of 'heat death'. All of the small probes suffered sensor anomalies due to thermal decomposition of shrink tubing on external wires, releasing hydrogen fluoride gas that corroded the wires' insulation.[63]

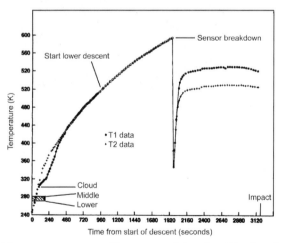

The temperature profile measured by the 'Northern' Pioneer Venus probe during its descent through the Venusian atmosphere.

Galileo's probe

As it approached Jupiter in 1995, the Galileo spacecraft released a probe to report on conditions in the planet's atmosphere. Unlike the probes of the Pioneer Venus mission, which were hermetically sealed, only the individual electronics boxes within the Galileo probe were sealed; the shell was open. Its thermal design assumed that temperatures would rise at a certain rate, but actually the temperatures inside the probe rose rather more rapidly and this led to significant calibration problems with the instruments – as much a result of the large rate of temperature change and the resultant gradients of temperature within the equipment as to the high temperatures themselves. The probe transmitted for almost an hour, during which it descended by parachute some 165 kilometres, extending from about 25 kilometres above to 140 kilometres below the 1-bar reference level. It fell silent when its communications system succumbed to the heat and pressure.[64] The pressure at failure was 22 bars, which was well beyond the original specification of 10 bars.

Rescuing GOES 7

After a trouble-free launch by Delta on 26 February 1987, the GOES 7 satellite was to use a solid apogee motor to circularise into geostationary orbit. The safe–arm mechanism and the explosive transfer assembly (the part that apparently failed on Hipparcos) were qualified to operated in the range 15°C to 25°C, but telemetry showed that the temperature of the safe–arm mechanism had risen through 35°C early on. Thermostat-controlled heaters were on each of the safe–arm and transfer components. The flight controllers turned off the primary heater, but when the temperature continued to rise they also turned off the backup on the safe–arm mechanism, at which point the mechanism began to cool. The heater was reactivated for a short period while the satellite was out of communication, and on reacquiring the signal it was found that although the safe–arm mechanism was satisfactory, the explosive transfer assembly had warmed to 35°C, which prompted the theory that the heaters on the two thermostats had been cross-wired. It was decided to make the circularisation burn on the satellite's second apogee rather than on the fourth, and this was accomplished successfully.[65,66]

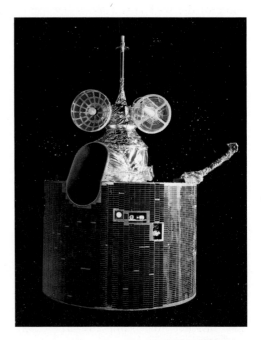

An artist's impression of GOES 7.

Alexis

After its remarkable recovery, the Alexis satellite had some later problems. As the Earth moved around the Sun, the satellite's orbit fell into the dawn–dusk plane so that it was fully illuminated, with no eclipses. It seems that some component in its radio transmitter began to overheat, and contact was lost for a time in August 1993. At first, a problem with the solar array voltage regulator was suspected, but the solution to that – ironically, switching on the heaters in order consume as much power as possible – did not help. As the radio system lacked temperature sensors, the problem could not be fully diagnosed. Nearby equipment reached 37°C, so the sensitive component was probably much higher. However, in subsequent dawn–dusk periods, shutting down the instrument payload kept the radio system cool enough to remain functional. Interestingly, the sensitivity of this component was not caught during system-level thermal vacuum testing because the transmitter had failed early in the test sequence, and the remainder of the testing was done using a cable. After the transmitter itself was repaired, it was thermal vacuum tested but a second full system test was not done.[67]

Magellan's thermal problems

Some of the problems suffered by the Magellan mission to Venus may have been exacerbated by poor thermal control. Much of the spacecraft was protected by blankets of Astroquartz,[68] but large elements such as the propulsion module and the high-gain antenna had been painted by a special inorganic water-based white paint developed by the Goddard Space Flight Center to reflect the intense heat and also resist discoloration. Even so, action was required to prevent excess heating of spacecraft components.[69] The 37-minute mapping swath at each periapsis was shortened by 10 minutes to enable the spacecraft to readopt the playback attitude, resulting in a slight loss in mapping data. In addition, the solar panels were rotated 90 degrees for about 5 minutes at the start and end of each mapping pass in order to reduce the reflection of sunlight off the mirrored solar cells onto the body of the spacecraft. It would appear then – at least on the basis of these few cases – that spacecraft using white paint have suffered more thermal difficulties than those using other coatings.

Genesis miswired

The Genesis spacecraft was launched on 8 August 2001 to spend several years in the company of the SOHO spacecraft in a halo orbit centred on the Lagrange point some 1.5 million kilometres up-Sun of the Earth, where it was to collect particles of the solar wind in ultra-pure collectors that were to be returned to Earth in September 2004 inside a small sample-return capsule. When the backshell of the capsule was opened nine days after launch to initiate sample collection, its temperature rose much more than expected, and threatened to cook the battery that was to run the sequencer and fire the mortars to deploy the parachute when it entered the Earth's atmosphere. To counter the large ratio of absorptivity to emissivity of the collector surfaces, and to cool the capsule, most of the non-collecting surfaces had highly reflective thermal paint. Thermal-vacuum testing had suggested that after a day or so

An artist's impression of the Genesis spacecraft with its sample collectors exposed. On returning to Earth, the capsule failed to deploy its parachute and embedded itself in the ground (bottom).

the temperatures would equilibrate, but it seemed that the thermo-optical properties of the capsule had significantly degraded. The capsule's cover was partially closed to shade the interior while the situation was assessed. Tests on an identical battery on the ground suggested that, despite exposure to elevated temperatures, it should indeed have the requisite capacity, and even a little margin.[70] The re-entry trajectory on 8 September 2004 was perfect for a descent to the Utah Test and Training Range, which was selected for being a large, flat and controlled area. To preclude the delicate contents being damaged on landing, the capsule was to be caught in mid-air by a helicopter flown by a Hollywood stunt-pilot, but the parachute was not deployed, and the capsule smashed into the desert at 300 kilometres per hour, broke open and compromised the integrity of the solar wind samples inside.[71] "For the velocity of the impact, I thought there was surprisingly little damage," said Roy Haggard of Vertigo Incorporated of Lake Elsinore in California, who was one of the first to inspect the capsule *in situ*. "I observed the capsule penetrated the soil about 50 percent of its diameter. The shell had been breached about three inches and I could see the science canister inside and that also appeared to have a small breach." It is hoped that even though the samples of solar wind (on ultra-pure wafers of silicon, gold and germanium) were contaminated by exposure to air and dirt, the precise composition may nonetheless be partially recovered because the solar wind particles should have been implanted some distance into the wafers, while terrestrial contamination should be confined to the outer few atomic layers of the wafers.

At the time of writing, the failure review board is still deliberating. However, it is known that the deployment pyrotechnics did not fire – indeed the presence of unfired ordnance on the capsule made its recovery hazardous – which was consistent with the battery failing due to its overheating, although other possibilities (such as a software error) were conceivable. If it had been believed that the battery would not survive the planned mission duration, then the spacecraft could have been returned early. In fact, the failure board identified a separate problem that would have prevented the parachutes deploying. A pair of *g*-switches were to have monitored the deceleration during entry, and, once the capsule had been slowed to a safe speed, triggered the pyrotechnics to deploy the parachutes. For redundancy, there were two

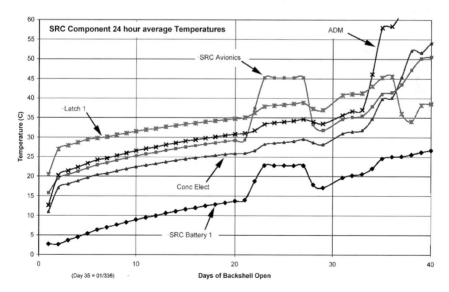

A plot of the temperature variations of the Sample Return Capsule (SRC) of the Genesis spacecraft showing the temporary anomalous heating of its battery and avionics.

pairs of switches. However, when the remains of the capsule were X-rayed, all four switches were found to be mounted the wrong way around – and it emerged that the engineering drawings specified their installation that way![72] The switches had been tested with a bidirectional shock, but this was insufficient to show their (incorrect) polarity.

After this incident the design of the Stardust spacecraft was re-examined, because its capsule employed the same entry system as Genesis, and the evidence confirmed that the g-switches had been correctly installed, so hopefully on its return in early 2005 it will be successfully recovered.

NOTES

1. 'The Alexis mission recovery', J. Bloch *et al.*, AAS-94-062.
2. 'Alexis, the little satellite that could: four years later', D. Roussel-Dupre, SSC97-IV-3, 11th AIAA/USU Conference on Small Satellites, Utah State University.
3. http://www.muses-c.isas.ac.jp/kawalab/astro/abst/1991_4_abst.txt
4. 'Aerobraking Magellan: Plan versus reality', D.T. Lyons, AAS/AIAA Spaceflight Mechanics Meeting, Cocoa Beach Florida, 14-16 February 1994. Paper AAS 94-118.
5. 'Magellan aerobraking at Venus', H. Curtis, *Aerospace America*, January 1994, p. 32.
6. *Aviation Week & Space Technology*, 15 September 1997, p. 26.
7. *Aviation Week & Space Technology*, 20 October 1997, p. 25.
8. *Aviation Week & Space Technology*, 3 November 1997, p. 24.
9. For a list of papers on MGS aerobraking, see: acquisition.jpl.nasa.gov/rfp/mro05/ AerobrakingReferences.pdf
10. http://www.jpl.nasa.gov/releases/99/mgsmag.html
11. *The Hubble Wars*, E. Chaisson, Harpercollins, 1994, p. 49.
12. 'In-orbit performance of ASCA satellite attitude control system', K. Ninomiya *et al.*, AAS 94-065 *Advances in the Astronautical Sciences*, p. 555.
13. 'Upper Atmosphere Research Satellite attitude disturbances during shadow entry and exit', M. Lambertson *et al.*, AAS 93-319, *Advances in Astronautical Sciences*, vol. 84, 1993, p. 1003.
14. *Aviation Week & Space Technology*, 7 July 1997, p. 31.
15. http://www.spaceflightnow.com/news/n0310/31adeos2/
16. *Investigation of ADEOS Midori's Functional Failure*, NASDA Report no. 64, October 1997.
17. *Orbital anomalies in Goddard spacecraft 1984*, E.F. Shockley, GSFC September 1985.
18. http://utenti.lycos.it/paoloulivi/mariner4.html
19. *Exploring space: Voyages in the Solar System and Beyond*, W.E. Burrows, Random House, 1990, p. 190.
20. http://llis.nasa.gov/llis/plls/ (0356)
21. 'The Ulysses mission', K.P. Wenzel, R.G. Marsden, D.E. Page and E.J. Smith, *Astronomy and Astrophysics*, Supplement Series 92, vol. 207, 1992, p. 207.
22. http://www.igpp.ucla.edu/pds3/ULY_5001/DOCUMENT/MISSION/MISSION.HTM
23. http://ulysses-ops.jpl.esa.int/ulsfct/nutation/presentation1/sld010.htm
24. *Solar System Log*, A. Wilson, Jane's, 1987.
25. *Uranus*, E.D. Miner, Springer–Praxis (2nd ed), 1997.

26. http://ringmaster.arc.nasa.gov/voyager/datasets/pps/insthost.html
27. 'Engineering the Voyager Uranus Mission', R.P. Laeser, *Acta Astronautica*, vol. 16, 1987, p. 75.
28. 'Anik E2 recovery mission', D. Wang and N. Martens, *Acta Astronautica*, vol. 29, 1993, p. 811.
29. 'Spacecraft Anomalies', E. Reeves, in *Cost-Effective Space Mission Operations*, D.G. Boden and W.J. Larson (eds), McGraw-Hill, 2000, p. 499.
30. http://www.astronautix.com/craft/fltatcom.htm
31. *The Hubble Wars*, E. Chaisson, Harpercollins, 1994.
32. *Aviation Week & Space Technology*, 26 April 1999, p. 85.
33. http://www.aqua.co.za/assa_jhb/new/Canopus/Can1999/c996jpl.htm
34. http://mars.caltech.edu/probe/mgs.html
35. *Aviation Week & Space Technology*, 30 November 1987, p. 24.
36. *Aviation Week & Space Technology*, 14 December 1987, p. 27.
37. http://www.skyrocket.de/space/doc_sdat/ariel-3.htm
38. http://centaur.sstl.co.uk/SSHP/micro/micro70s.html
39. http://www.geocities.com/CapeCanaveral/Launchpad/6133/blackarrow.html
40. *TD 1A investigation into the cause of the failure of the number 1 tape recorder*, G. Bodereau, ESRO TR-13, ESTEC, January 1973.
41. *Aviation Week & Space Technology*, 14 July 1986, p. 138.
42. http://project-tools.com/pages/grts1.htm
43. *Aviation Week & Space Technology*, 23 October 1995.
44. *Aviation Week & Space Technology*, 30 October 1995, p. 28.
45. *Aviation Week & Space Technology*, 29 January 1996, p. 33.
46. 'Magellan recorder data recovery algorithms', C. Scott, H. Nussbaum and S. Shaffer, Proceedings of the International Telemetry Conference, 1993.
47. *The Japanese and Indian Space Programmes: Two Roads into Space*, B. Harvey, Springer–Praxis, 2000, p. 147.
48. *Aviation Week & Space Technology*, 22 November 1982, p. 80.
49. *The Japanese and Indian Space Programmes: Two Roads into Space*, B. Harvey, Springer–Praxis, 2000, p. 69.
50. 'The lightweight deployable antenna for the MARSIS experiment on the Mars Express spacecraft', G.W. Marks, M.T. Reilly and R.L. Huff, Proceedings fo the 36th Aerospace Mechanisms Symposium, Glenn Research Center, 14–17 May 2002.
51. ESA Science News Release SNR 7-2004, 29 April 2004.
52. 'Mars Express update: MARSIS radar boom deployment good to go', E. Lakdawalla, (www.planetary.org) 17 May 2004.
53. 'Evaluation of thermal control coatings degradation in a simulated geo-space environment', J. Marco and S. Remaury, High Performance Polymers, vol. 16, 2004, p. 177.
54. *Spacecraft Systems Engineering*, P. Fortescue and J. Stark (eds), Wiley, 2003.
55. 'Satellite broadcasting experiments and in-orbit performance of BSE', S. Shimoscko, M. Yamamoto, M. Kajikawa and K. Arai, *Acta Astronautica*, vol. 9, 1982, p. 499.
56. http://www.brainyencyclopedia.com/encyclopedia/l/lu/lunar_orbiter_1.html
57. *To a Rocky Moon: A Geologist's History of Lunar Exploration*, D.E. Wilhelms, University of Arizona, 1993, p. 155.
58. *Destination Moon: A History of the Lunar Orbiter Program*, B.K. Byers, NASA, TM X-2487.
59. *Solar System Log*, A. Wilson, Jane's, 1987, p. 41.

60. *Solar System Log*, A. Wilson, Jane's, 1987, p. 80.
61. *Journey into Space*, B. Murray, W.W. Norton, 1989.
62. *Exploring Mercury: The Iron Planet*, R.G. Strom and A.L. Sprague, Springer–Praxis, 2003.
63. *Pioneer Venus 12.5 km Anomaly Workshop Report*, A. Seiff *et al*, NASA Conference Publication No 3303, 1995.
64. *Jupiter Odyssey: The Story of NASA's Galileo Mission*, D.M. Harland, Springer–Praxis, 2000.
65. *Aviation Week & Space Technology*, 9 March 1987, p. 266.
66. http://www.boeing.com/defense-space/space/bss/factsheets/376/goes/goes.html
67. 'Alexis, the little satellite that could: four years later', D. Roussel-Dupre, SSC97-IV-3, 11th AIAA/USU Conference on Small Satellites, Utah State University.
68. *The Magellan Venus Explorer's Guide*, C. Young (ed), JPL Publication 90-24, 1 August 1990.
69. NASA Release, 28 February 1991.
70. 'Genesis: The middle years', N.G. Smith, K. Williams, R. Wiens and C. Rasbach, IEEE Aerospace Conference, Big Sky, MT, 2003.
71. Press Release: 2004-219, 8 September 2004.
72. 'Flawed drawings caused spacecraft crash', *Nature* online, 18 October 2004.

15

Failures on the ground

CONSTRUCTION FAILURES

While many examples are covered elsewhere (e.g. the Genesis failure, magnetorquer polarity errors, etc.), it is interesting to assemble a collection of failures caused by mistakes or accidents prior to launch. Spacecraft are delicate systems, and are generally safer in orbit than they are on the ground. Some of these examples illustrate the importance of assuring the reliability not only of the high-tech spacecraft themselves, but also of low-tech infrastructure on the ground: water pipes, cables, and leaky roofs can all threaten a spacecraft. Some of the examples are cases of things simply being built incorrectly. Still other failures can be traced to failures in testing. Even the silliest of mistakes have been committed during construction – including at least one case on a V-2 research flight in the early days when an experimenter neglected to remove the lens cap from the camera that was to photograph cosmic-ray tracks in what was otherwise a functional cloud chamber.[1]

Even more frustratingly, the Venera 11 and Venera 12 spacecraft launched by the Soviet Union successfully touched down on Venus in December 1978, but neither could eject the lens cap of its camera. By the sheerest bad luck, in March 1982, when Venera 14's lens cap was ejected with a more powerful spring, it came to rest at the very point at which the surface-sampling arm swung down, with the result that the sensors determined the physical characteristics of this litter, rather than the surface.[2]

When Venera 13's camera ejected its lens cap this fell nearby, but in the case of Venera 14 (bottom) the cap came to rest at the spot where the surface sampling arm deployed.

The Galileo probe's g-switch

Like many entry probes, the probe released by the Galileo spacecraft to penetrate the atmosphere of Jupiter in 1995 used the deceleration profile to time the deployment of its parachute. To assure aerodynamic stability, and proper inflation of the parachute without either the dynamic loads or heating damaging it, deployment must occur within a certain range of Mach number and dynamic pressure. Even if the atmosphere were to be slightly thicker or thinner than expected, or if the probe's entry angle was slightly off-nominal, simulations had indicated that the parachute would deploy properly at a pressure level of 0.1 bar, as determined by the interval between the g-switches. Specifically, G1 was to go 'high' at 6 g and 'low' at 4.5 g, while G2 would trigger at 25 g and reset at 20 g. The expected sequence would therefore be G1 switching on first, then G2, then G2 switching off and then G1.[3] However, the telemetry showed G2 triggered first because the switches had been cross-wired. Despite this construction error, the logic triggered the deployment in an environment that allowed the parachute to inflate safely.[4] However, the 53-second delay resulted in the instruments sampling at a greater depth than intended, beginning at a pressure of 0.35 bar rather than 0.1 bar. In some respects, this meant the loss of about a fifth of the probe's science, missing, for example, the chance of detecting an expected ammonia cloud deck that forms the visible surface of the planet, because by the time the instruments were activated the probe had already plunged into the clouds.[5]

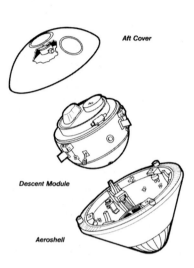

A schematic of the major components of the Galileo spacecraft's atmospheric probe.

NOAA 13 screwed

Two weeks after its launch by an Atlas-E on 9 August 1994, the NOAA 13 weather satellite failed, unable to charge its battery. A clue as to the cause was telemetry showing a temperature rise in the electronics box shortly prior to the failure. An inspection of an identical unit on the ground strongly suggested that a 1.25-inch screw that projected too far had disabled the charger by penetrating the insulation and causing an electrical short onto a metal radiator plate.[6] The unit, which had been designed 22 years previously, and flown on 16 satellites without incident, had therefore been treated as a standard unit, but it was an unforgiving design and after an assembly technician at Martin Marietta Astro Space had pushed the screw too far the quality checks failed to recognise the problem.

FUSE misaligned

While checking out the Far Ultraviolet Spectroscopic Explorer on top of its Delta II at Canaveral in 1999, engineers were concerned by a 5.5-degree attitude error in a ring-laser gyro. The obvious conclusion was that the gyro was faulty, but an inspection of the cables running between the satellite and the launch vehicle prompted the realisation that the satellite was incorrectly mounted! It had an unusual design that precluded using the alignment pins of the standard adapter. Instead, it was decided to align the satellite with reference to marks on the mating rings, but when the view of these marks was masked by ground support equipment a certain amount of guesswork had been used, introducing the gyro offset.[7]

The FUSE satellite being mated with its launch vehicle.

Fit for function

Incredibly, the first flight of the Delta IV was delayed for several weeks when it was discovered that the two strap-on solid rockets would not fit onto their attach points, and technicians were called out to the pad with tools to "grind and slightly reshape" the core vehicle to accommodate the boosters.[8]

THE HAZARDOUS GROUND ENVIRONMENT

Despite the harsh temperatures, the radiation hazard, and all the other hazards of the space environment, this is the environment in which a spacecraft is designed to function, and invariably everyone involved breathes a collective sigh of relief when the product of their labours survives its ride into space. But first it must survive routine movements, connections, and testing on the ground. Ground-handling failures make shocking reading – sometimes verging on the comical.

Magellan

The Magellan spacecraft suffered a number of traumas. First, during processing at the Kennedy Space Center a technician tripped on his lab coat and damaged the nozzle on its IUS stage, which had to be replaced. Next, a technician miswired the solid motor that was to insert the spacecraft into orbit around Venus, but fortunately this error was detected.[9] Another incident occurred during a battery test. As the connectors were behind a thermal blanket, out of sight, the technician had to undertake a 'blind mating', and in doing so the harness connector shorted out the battery. As this discharged through the cables, the harness and thermal blanket caught fire, in part because the harness used kapton-insulated wires.[10,11] Although the connectors were keyed or 'polarised' such that the connection could not be made unless they were co-aligned, it transpired that they could be jammed together sufficiently to permit a short. The harness and connector had both to be replaced. One complication was that no procedures had been established for the removal of fire-extinguishing foam, requiring the development of an *ad hoc* procedure. The incident cost a week in a tight schedule.

Lightning strikes

Several weeks prior to its scheduled launch on an Atlas IIAS in August 1998, both of the telemetry encoders on JCSAT 6 – which was one of the unlucky HS-601 series built for Japan Satellite Systems – began to behave erratically during tests at an Astrotech satellite processing facility adjacent to the Kennedy Space Center. A check of Air Force records of electrical activity in the area indicated that lightning had struck within 2 kilometres of the facility. Even though the test equipment was heavily shielded against lightning, the satellite was affected.[12,13] After being repaired, the satellite was launched on 16 February 1999. (It is not just satellites that are at risk: at a Huygens meeting in this area a year or so earlier, one of us, RL, observed a laptop fail during a thunderstorm; it later worked under mains power, but only after the damaged battery had been removed.)

Insat 1D, unlucky twice

India's multipurpose geostationary satellite, Insat 1D, was to have been deployed by a Shuttle, but after the loss of Challenger it was reassigned on the final vehicle in the early Delta series. When it was on the pad on 19 June 1989, a crane hook tangled with cables and smashed into the satellite, destroying its C-Band antenna. It was initially feared that the satellite would have to be written off, but Ford Aerospace said that it was repairable at the cost of $10 million. While it was undergoing repair at Palo Alto in California the facility was shaken by an earthquake, which added another $150,000 to the repair bill! After finally being launched on 12 June 1990, Insat 1D operated for over 5 years.[14]

Galileo's ribs

One celebrated – if that is the appropriate word – example of ground-handling failure is the high-gain antenna of the Galileo spacecraft: when commanded to deploy on 11 April 1991 it jammed. An Anomaly Team was established by JPL, drawing upon mechanical, electrical, thermal, and materials specialists, as well as staff from the Harris Corporation, which had built the 4.8-metre-diameter umbrella structure using the same design as the antennas of the Tracking and Data Relay

An artist's impression of an IUS with the Galileo spacecraft, its high-gain antenna in its stowed configuration.

Satellites. The deployment sequence called for an electric motor to release the pins that held the 18 graphite–epoxy ribs of the structure alongside the central support tower. The ribs were to draw a thin molybdenum mesh into a parabolic dish. The pin release motor was immediately beneath a small circular shield that had been installed on the tip of the tower to protect the antenna's structure from the Sun while the spacecraft was inside the Earth's orbit. It was clear that this command had been received and that the mechanism had been activated, because the motor had drawn an electrical current. However, telemetry showed that the motor had operated at a higher power than expected, indicating that it had encountered an unexpected frictional force. Another anomaly was the microswitch that was to have deactivated the motor after a few seconds, by which time the pins ought to have released, did not trigger,

indicating that at least some of the pins had failed to release. Another clue was that Galileo had assumed a slight wobble, which implied that the antenna was partially deployed in an asymmetrical state. Finally, in certain orientations, a Sun sensor's field of view was masked by one of the ribs of the antenna. In all likelihood, several ribs had jammed as a result of etching of their surfaces once excessive friction between the stand-off pins and sockets had eroded the molybdenum disulphide dry lubricant. The cause was vibration experienced on repeated cross-country trips: after man-

An artist's impression of the Galileo spacecraft in the Jovian system with its partially deployed high-gain antenna.

ufacture in Florida the antenna had been driven over to JPL by truck in a special container; in 1985 it was taken to the Kennedy Space Center for integration and launch; on the loss of Challenger and the cancellation of the Centaur deployment stage in 1986 the spacecraft was driven to JPL for modification and returned to Florida in 1989. Extensive analysis determined that the problem existed at launch and went undetected. It was not related to the revision of the flight plan to include flybys of Venus, which required that the tip shield be added, nor to the consequent delay in the deployment of the antenna until the spacecraft was beyond the Earth's orbit and in the environment for which it had been designed.[15,16]

This was a potential 'show-stopper' for the mission, since the high-gain antenna was to have been able to sustain X-Band transmissions at a rate of 134 kilobits per second. If it could not be coaxed out, when Galileo reached the Jovian system it would be unable to report its findings! On 20 May, to determine whether the thermal expansion of the tower would release the pins, the spacecraft was turned 40 degrees from its usual Sun-pointing attitude to remove most of the antenna tower from the shadow of the tip shield. The pins did not release. On 10 July, with the spacecraft 1.84 AU from the Sun, it was turned almost all the way around to put the tower into shadow for 32 hours, to induce thermal contraction, and then the spacecraft was spun-up to apply centrifugal force to the ribs in the hope that this would release the pins. The pins did not release. On 13 August, 1.98 AU from the Sun, the antenna was returned to shadow, this time for 50 hours. Again, the pins did not release. On 29 October, the spacecraft flew by asteroid 951 Gaspra and stored its observations on tape. A further cold soak at the 2.3-AU aphelion in December was fruitless. While Galileo was near the Earth a year later, it transmitted the Gaspra data over its low-gain antenna. On 29 December 1992, exploiting the fact that the spacecraft would never again be so close to the Sun, it baked the high-gain antenna for 20 hours to induce thermal expansion and then the deployment motor was pulsed 2,000 times to 'hammer' the pins free. Over the following month, the motor was pulsed a total of 13,320 times, but to no avail. Whatever had fouled the antenna, it had established a very secure bond. On 10 March 1993, as part of a test of the procedure for the deployment of the atmospheric probe, Galileo was spun up to 10 revolutions per minute to measure the 'wobble' resulting from the asymmetric configuration, and to allow a plan to be devised to counter this when the probe was eventually released. While the spacecraft was spinning, the antenna motor was pulsed in the hope that the increased centrifugal force on the ribs would release the pins. Having tried every trick in the book, the managers finally accepted that the mission would not be able to proceed as planned. It had been intended to return 50,000 images during two years of exploration of the Jovian system, but this would be impracticable over the low-gain antenna. To put the task into context, consider that in 1965 Mariner 4 had taken 10 days to transmit two dozen images of Mars at 8 bits per second. Each of Galileo's images contained *16 times* as many pixels, so how could it possibly explore the Jovian system using an antenna that would be able to operate at a rate no higher than 10 bits per second? It would take two weeks to return a single image together with its associated engineering and navigational data. Unable to downlink its imagery in real-time, the spacecraft would have to save it on tape, but even if the

Deep Space Network could be devoted to the task 24 hours per day it would take *3 years* to replay a tape full of data!

Les Deutsch of the Telecommunications and Missions Operations Directorate at JPL was confident that something could be done by reprogramming the onboard computer to do image compression, and by enhancing the manner in which the downlink was encoded. He led a feasibility check with the Deep Space Network engineers. An engineering team then spent 3 months fleshing out the details of what could be done on the ground and on the spacecraft. Although Galileo would not

The 70-metre-diameter dish at Goldstone in California, which is part of the Deep Space Network.

approach Jupiter until late in 1995, it was evident that developing the new software would be a race against the clock.[17,18] The data processor, the Command and Data Subsystem (CDS), used six 8-bit microprocessors of 1970s vintage that had a total memory of 384 kilobytes. The Attitude and Articulation Control Subsystem had a faster 16-bit computer, but with only 64 kilobytes of memory, nearly all of which had to remain as it was. Most of the instruments were controlled by single 8-bit microprocessors with very limited memory. The constraints of memory, processing time, and data bus traffic were excruciating. The vast volume of data that the science instruments and engineering systems would create would somehow have to be fed through the narrow 'data pipe' provided by the low-gain antenna. One problem was locating programmers still fluent in the archaic assembly language of the CDS operating system. A 'science virtual machine' was created and integrated into the existing operating system to support editing, compression, and 'packetisation' of the science data. Some 80 kilobytes of memory in this virtual machine were assigned to a multipurpose buffer into which data for downlinking would be loaded and prepared for transmission. The science data would be collected and processed at high speeds, relative to the speed of the data pipe. It would not necessarily be transmitted in the order that it was received, so the new software had to prefix data packets with headers identifying the data type, its size, and its collection time. Before imaging data could be inserted into playback packets, it would have to be compressed by ratios of up to 15 to 1, compared to its format on the tape. The particles and fields data would have to be compressed by a factor of 100, since this type of data was to be put into real-time packets. To enable the Deep Space Network to detect and to rectify transmission errors, which would be critical when using data compression, the packets would have to be encoded. Every innovation was used to refine, compress, and package data so that only the most valuable information entered that slim pipeline to Earth – or to put it another way, to ensure that every bit transmitted carried the maximum amount of information. Combined with an upgrade to the Deep Space Network, these measures would boost the 'effective information rate' of the low-gain antenna by a factor of about 100, which would be sufficient to return 10 per cent of the planned 50,000 images – which was much better than initially feared.

The dropping of NOAA N-Prime

A classic example of ground-handling failure occurred on 6 September 2003, when the nearly completed NOAA N-Prime weather satellite was badly damaged at Lockheed Martin's plant at Sunnyvale in California. The 1.8-tonne, 6-metre-tall boxy structure was in being rotated from vertical to horizontal on its handling cart when it slipped off the mounting plate and fell to the floor. Normally, it would have been secured to the cart's turn-over fixture by an adaptor plate with 24 bolts, but two days previously the bolts had been removed by a technician of another crew to enable the fixture to be used by another project, and before he actually removed the adaptor plate the technician had been told to stop work as the fixture was no longer needed. Because the fixture was in storage, there was no requirement for him to replace the bolts or to document their removal. When the NOAA N-Prime crew retrieved the fixture in order to turn their satellite over, they failed to verify that the bolts were still in place – after all, the adaptor plate was on the fixture, exactly where they had left it. Interestingly, one report suggested that one of the engineers, who was not normally assigned to the project but was helping out, noticed that there were empty bolt holes in the adapter plate, and on pointing this out to the regular engineer was told that there should be some empty holes in the mounting plate as it was designed to be used on more than one fixture. Once the satellite was mounted on the plate, the engineer and a quality representative both signed the documentation stating that the fixtures were correctly configured, although that was not so. As the satellite was tilted on the fixture, it slid, and toppled onto the ground. Fortunately, no one was hurt, and a nearby satellite was not affected. The immediate priorities were to safe the batteries, to depressurise the propulsion system, to prevent further movement of the spacecraft, and to impound the records for the investigation.[19],[20] The damage to the $233-million satellite was put at $135 million, to be paid out of the profits that the company had earned, or was to earn, on the Polar-orbiting Operational Environmental Satellite programme.[21]

The NOAA N-Prime satellite, which was allowed to slip off its work platform.

Spainsat dropped

NOAA N-Prime was not the first satellite to be dropped. In 2001 Hisdesat of Madrid ordered a 1300 series X-Band communications satellite from Space Systems/Loral, but when it was time for thermal vacuum testing, the vacuum chamber at Loral's plant in California was occupied by the Japanese government's MTSAT 1R, so the new satellite was shipped to a Boeing facility at Kent in Washington. However, as it was being loaded into a container for return to Loral a cable snapped, causing it to fall about 1 metre to the ground and damaging it sufficiently to delay its launch to supplement the Hispasat 1A and 1B satellites.[22]

Genesis near-drop
The payload canister for NASA's Genesis spacecraft was also dropped while being transferred from a transportation dolly to a thermal vacuum chamber. In this instance, one of three threaded attach fittings had worked loose. It was found that these were only hand-tight and had no positive retention cable or other backup. In this case, however, the canister dropped only 5 centimetres and personnel were able to cushion the fall, with the result that no damage was sustained.[23]

TOPEX–Poseidon's spin
In 1993, as the TOPEX–Poseidon satellite was being lifted for placing into a thermal vacuum chamber, it began a slow end-over-end motion and rotating through 135 degrees before snagging on a suspension cable. The crew visually determined that a spreader bar that had been inherited from the Galileo project had been damaged and was at risk of failing, so they promptly lowered the spacecraft to the floor before it could move further and sustain damage.[24]

Galileo's drenching
When the Galileo spacecraft was driven cross-country to the Kennedy Space Center, the radioisotope heater units (encapsulated pellets of plutonium, each providing 1 watt of heat) that were to prevent vital subsystems from freezing in space were already installed. Since it would have overheated if a simple sealed container had been used, a custom container fitted with an air-conditioner to chill the internal environment to 10°C was used. On arrival, the container was placed in a controlled-environment facility and the pressure seal relieved. When the container was opened some weeks later, there were puddles of water on the floor, and signs of corrosion on the exterior surfaces of the spacecraft. Evidently, the moist warm ambient air (recall that Florida has a balmy climate) had mixed with the chilled air inside the container, leading to supersaturation and condensation. Fortunately, cleaning of the external surfaces was able to remove the damage.[25]

Huygens's insulation blown apart
A related problem at Kennedy Space Center delayed the launch of Cassini–Huygens by two weeks. The Huygens probe incorporated 34 radioisotope heater units to prevent its systems freezing after its release to coast to Titan in 2005, and therefore was provided with forced ventilation to keep it cool on the ground. On the pad, however, this airflow was set three and a half times too high, with the air streaming from the nozzle at 300 kilometres per hour, which shredded the internal insulation foam. An inspection using a fibre-optic borescope showed a tear in the kapton foil

that had been wrapped around at least one of the foam blocks. In weightlessness, the particles of foam that littered the interior of the probe might easily migrate to critical areas. The replacement of the blanket panels would be straightforward, as would the cleaning out of fragments from within the probe, but gaining access required that the whole Cassini spacecraft (with Huygens mounted on its side) be removed from the launch vehicle and transported to the Payload Hazardous Servicing Facility several kilometres from the pad. This clean-up process, and the subsequent verification, encroached on a launch window that lasted only a few weeks.[26,27,28] Luckily, other than losing two days to the weather, no other major problems occurred during the launch preparations and, in the end, the launch occurred in the middle of the window, which had the beneficial effect of increasing the spacecraft's propellant margin.

HST payload sandblasted

In 1993, as STS-61 was being prepared for the first servicing mission to the Hubble Space Telescope, technicians found the COSTAR and WFPC-2 canisters coated by fine particulate material from sandblasting to remove corrosion on the pad. Some of the grit had been blown in through the roof of the payload facility by high winds. Although the double-walled containers should have prevented direct contamination of the payloads, it was felt necessary to double check their integrity. This underscores a general issue with failures of the mishandling type: the schedule disruption in making sure that systems are all right can be significant even if the actual damage is minimal.[29]

Dirty Shuttle tanks

An even more impressive accident occurred in 1999 at Lockheed Martin's Michoud facility near New Orleans, where the external tanks for the Shuttle are fabricated. A 5-centimetre-diameter compressed-air pipe buried about 40 centimetres beneath the floor ruptured. Because the factory, built during World War II, was on a swamp, the floor was supported by river sand, which the airflow sprayed throughout the factory. One estimate was that everything within 60 metres of the rupture received a coating of sand several millimetres thick. Lockheed drafted hundreds of workers to clean by hand the tooling and flight hardware for a dozen external tanks in various stages of production.[30]

Mars Odyssey's water leak

The Mars Odyssey spacecraft arrived at Kennedy Space Center in January 2001 and a webcam was installed in the clean room to enable the public to monitor the preparation of the spacecraft for launch in April. While checking the camera on 19 January, JPL public affairs staffer Ron Baalke noticed "a brown liquid spread out across the floor, including underneath the spacecraft and the gamma-ray spectrometer instrument sitting nearby". As the technical crew had retired for the evening, the spacecraft was unattended. Baalke immediately called the project manager, and within 10 minutes technicians had stemmed the leak and begun to mop up. The liquid had leaked from a rust inhibitor used in the spectrometer's cooling unit, and spilled 500 gallons of water onto the floor of what was meant to be a 'clean room'. Because the spacecraft was on a dolly, several centimetres off the floor, it was relatively safe, and the other ground support equipment was undamaged. While humidity monitors would eventually have detected the leak, Baalke's fortuitous observation made the clean-up easier and reduced the delay to the test schedule to just a few days.[31]

GPS satellite soaked

On 8 May 1999 a GPS Block-IIR in the clean room atop Pad 17A at Canaveral was damaged during a thunderstorm.[32,33] The investigation found that rain water had entered the clean room through an open fastener hole caused by a sheared screw on the roof. The fact that the room was not watertight had been known for several years, and satellites awaiting the mounting of the Delta II's shroud were routinely protected. It was standard procedure to withdraw staff from the pad upon the approach of a storm due to the risk of lightning strikes, but in this case the technicians working on the satellite inadequately assembled the protective plastic sheeting. The rain ingress was particularly severe, and water pooled in a hollow in the material, overwhelmed the tape that fastened it, opening a gap, and dripped onto the satellite causing $2.1 million worth of damage.[34] Colonel Edwin E. Noble, who led the investigation, was critical of the fact that "the risk of this condition has been accepted by all parties".[35]

Eutelsat W1 soaked

On 18 May 1998, the Spacebus 3000B2 satellite that was being built by Aerospatiale at its factory at Cannes in France was undergoing transmitter tests within an indoor antenna range – a large room with foam-lined walls to absorb the radio waves and prevent any reflections that might corrupt the measurements. On a test at the full power of several tens of watts, the microwaves heated the foam sufficiently for it to ignite. The sprinkler system drenched the satellite, Eutelsat W1, which, ironically, was to have been shipped the following day. Aerospatiale placed a $50 million insurance claim. The satellite, which was then dubbed the 'Wet One', was refurbished and launched as Eutelsat W5 by a Delta IV on 20 November 2002.[36,37,38] Aerospatiale was to have made all four satellites for the W-series, but after the fire Eutelsat elected to order the direct replacement from Matra Marconi Space, and this was launched as Eutelsat W1 by an Ariane 4 on 6 September 2000.[39]

Building collapse

On 12 May 2002 several workers were killed when the roof of Assembly Building 112 at the Baikonur cosmodrome collapsed. This massive structure, some 250 metres in length and 60 metres in height, was built for the N-1 moonrocket and later used to lift the Buran shuttle onto its Energiya launch vehicle. The investigation blamed poor upkeep. A 50-centimetre-thick blanket of kerazmit on the roof had absorbed three days' worth of heavy rain immediately prior to the collapse. This gravel-like material was supposed to repel water, but had lost this ability over the years. Ironically, the collapse was triggered by construction workers placing a load of 18 tonnes of 'rubberoid' on one section of the roof in preparation for repairs. The mishap severely damaged an Energiya launch vehicle and Buran shuttle (apparently the one that actually flew in orbit) both of which had to be scrapped.[40] No flight hardware was affected, but the mishap threatened the schedule for commercial Soyuz launches, including the launch of the European Space Agency's Mars Express spacecraft.

VAB damaged
Over the space of a few weeks in the summer of 2004 hurricanes Charley, Frances and Ivan hit Florida, and the winds damaged the roof of the Vehicle Assembly Building at the Kennedy Space Center and tore 850 aluminium panels from its walls, exposing about 20 percent of the interior to the environment.[41,42] While none of the flight hardware inside was damaged, the need to repair the building would probably delay the return to flight of the Shuttle following the loss of Columbia.

Woodpeckers
Even wildlife can cause problems for space missions. The insulation on the external tank of the Shuttle, being foam, is soft. After STS-70 had been rolled out to the pad in the summer of 1995, several woodpeckers of the yellow-shafter flicker species tried to nest on it, and in doing so pecked 75 holes up to 10 centimetres across. The Shuttle had to be returned to the Vehicle Assembly Building for repairs.[43]

Mars Polar Lander
One sequence in a system-level functional test of the Mars Polar Lander spacecraft simulated the mission profile for immediately following landing. The software programmed into the vehicle commanded the medium-gain antenna to exercise its full gimbal range. On realising that a collision was imminent because a previous test sequence had left the solar panels in their stowed configuration, the test conductor attempted to cancel the test, but the uplink path was being reconfigured from cable to radio frequency. Furthermore, the emergency power-off switch only affected the ground support equipment, since the spacecraft was running off its battery. When the antenna drove into the undeployed solar panel, this cracked the composite reflector dish and scratched the substrate of the panel. In essence, this was a failure of configuration control, in that, prior to ordering the antenna to move, there was no check that the entire spacecraft was in its landed configuration, with the panel deployed.[44] Another valuable lesson was: Always have a means of disabling an autonomous system!

Hubble smash
In 1989, about a year before STS-31 was to launch the Hubble Space Telescope, it was realised that no one had verified that the secondary mirror was securely attached to its mounting structure. This was supposed to have been done nearly a decade earlier, but the documentation for the test was spectacular by its absence and none of the technicians could recall doing it. An elaborate clean-room operation was ordered in which the huge satellite was tilted horizontally and a structure similar to a diving

board inserted through the aperture of the telescope to enable a (small) technician to crawl along the board to inspect the secondary, which was indeed correctly affixed. While there was considerable scope for mishap, no damage was done. However, when the clean-room crew swung the satellite back to its vertical storage orientation they inadvertently slammed its rear into the wall, destroying one of its low-gain antennas.[45]

Murphy's law

More so than many other chapters, these tales seem to support the notion that if anything can go wrong, it will.

TESTING FAILURES

There are two types of testing failure. First is the failure of the test objective, in that the test is a measurement that proves to be inaccurate – an improperly conducted test may fail to detect a manufacturing or design error, or introduce a manufacturing error as a result of suggesting that (inappropriate) modifications are required. The second kind of testing failure is one in which the test causes damage. In conventional engineering practice, several spacecraft models may be produced in advance of assembling the unit(s) that are to fly. The first is usually a thermal and structural model whose rôle is essentially to serve as a pathfinder, to get the metalwork right. Since this lacks the electrical and computational sophistication of the flight unit, it is relatively cheap to construct. This is followed by the engineering model, which is a functionally representative near-replica of the flight unit, but may well trade-off structural accuracy for convenience of access to its systems. After the flight unit is constructed, the engineering model will serve as a diagnostic and corrective testbed. In addition to the flight article, there may be a flight-ready spare, or at least all the components needed to assemble a complete spacecraft at short notice in the event that something catastrophic befalls the original.

The purpose of testing is to verify the functionality of a system in a given set of circumstances. Because a test that 'pushes' a component to a point close to failure is (by definition) one that stresses the component, the subsequent reliability of that component may be compromised, for instance by fatigue in the case of stress testing, or growth of intermetallics in high-temperature tests of electronics. Parts or systems that are intended to be flown should therefore *not* be stressed to the extreme conditions in which they will be required to function. Two different test levels are usually applied, 'qualification' for a test or engineering model, and a significantly less stressful 'acceptance' level for a flight unit. The qualification level is the design performance level, and usually represents the expected design flight levels, with some margin – i.e. the qualification levels are often more arduous than those expected in flight, although in some cases actual levels can be more severe due to environment modelling errors or launch problems. Thus a system that meets qualification levels can reasonably be expected to function properly in flight. The acceptance tests are based on the assumption that the components in the flight system are sufficiently similar (e.g. from the same production batch) that passing an acceptance test that should detect some gross deficiency implies that the performance should be the same as those that were subjected to the qualification test. That, at least, is the theory.

A bad case of the shakes

A good example of *over-test* failure is the High Energy Solar Spectroscopic Imager (HESSI) developed for NASA by the University of California at Berkeley. In a vibration test at the Jet Propulsion Laboratory on 21 March 2000 its solar panels sustained over $1 million of damage. The satellite was mounted on a vibration table to be subjected to short bursts of periodic oscillations – a so-called sine-burst vibration test. The plan was to start with six bursts at an amplitude of 1.88 g, and then increase the amplitude by steps to a maximum of 7.5 g. However, even the first vibration level exceeded this maximum, and before the operator could intervene to abort the test the satellite had been subjected to 21 g. The investigation discovered that the magnesium table to which the satellite had been affixed, which was to have slipped frictionlessly over the granite block built into its mounting, had actually been sticking to it. The slip-table, block, and the armature of the shaker (a device similar to a large loudspeaker) were misaligned. Earlier in the day, the operators had noticed a "different sound", but had not realised the cause. With the table sticking, the accelerations experienced by the satellite were not those programmed. The *average* motions may have matched the desired sinusoidal motion, but the stick-and-slip motion introduced peak accelerations that were much greater.[46,47] One review argued that staff turnover, and the consequent loss of experience, may have contributed to the failure.

Hubble's abberation

The optical flaw in the Hubble Space Telescope must surely rank as one of the most profound (and illuminating) cock ups in the history of spacecraft engineering. Following its release by STS-31 on 25 April 1990, the telescope's systems were methodically checked. The first image returned on 20 May was of an open star cluster, and the astronomers were delighted to note that it resolved many previously unsuspected double stars. It was slightly fuzzy, but that was because the optics had yet to be focused. Unfortunately, the telescope was unable to focus. Only 15 per cent of the light from a star formed a point-like image, the remainder was smeared into a surrounding smudge. Computer modelling proved that the 2.4-metre-diameter primary mirror had spherical abberation!

As the mirror was ground extremely precisely, the failure cannot be described quite as a manufacturing error. Its shape was verified during the grinding process by recording the light pattern generated by the interference of a laser beam which bounced between the mirror and a test fixture. NASA impounded the manufacture and test facility, and all the documentation, such as it was. Because the test fixture had been mothballed once the contractor shipped the mirror, it was available for inspection. It was found that when the fixture, a 'reflective null', was built in 1981, it had been assembled incorrectly, with the result that the interference pattern sought in the grinding process did not correspond to the intended mirror shape. In fact, the mirror was 2 microns flatter than it ought to have been. In effect, some 3 cubic centimetres of glass had been erroneously ground away! Shockingly, two simple and independent tests (an inverse null and a reflective null) had indicated that the shape of the mirror was *incorrect*, but these tests, one of which was repeated dozens of times, were deemed to be in error due to confidence in the test fixture. Ironically, the test fixture had been machined correctly. The flaw was that several cheap washers had been inserted, displacing a lens by 1.3 millimetres, to make the reflective null match up with the laser reflection from the end of a precision-length metal bar that was traceable to the National Bureau of Standards. The bar had a small black-painted cap with a hole through which the laser was to shine in order to reflect off the rod. However, in a test for a mirror for a reconnaissance satellite, a small fleck of black paint had been knocked off the surface of the cap, and this exposed a shiny surface 1.3 millimetres from the end of the rod. As a result, the laser beam bounced off the cap, instead of the rod itself, and the test fixture was jury-rigged to match, with the later result that the mirror was ground to the wrong shape. This accurate failure diagnosis facilitated the design of a 'corrective optics' package to restore the optical properties of the flawed mirror, and this was installed in December 1993 by STS-61.[48,49]

In the years that the telescope was used in its flawed state, the fact that the images were blurred in a deterministic way enabled post-processing to recover some of the lost resolution. The raw image was convolved with a point-spread-function to yield a blurred image. The intensity value of a given pixel in the blurred image represents a weighted sum of contributions from the original pixels adjacent to the original. The blurred image therefore provided a set of simultaneous equations. While these equations could not be *solved* in an absolute sense because information had been lost, it was possible to make 'maximum likelihood' estimates of the original image. A variety of algorithms (such as Fourier methods, Maximum Entropy and the Lucy–Richardson algorithm) were employed to perform this deconvolution. Although deconvolution techniques were used by radio astronomers, they had largely been ignored by their optical counterparts. Their 'forced' introduction in response to the abberation on the Hubble Space Telescope led to deconvolution becoming a familiar tool to astronomers.

In fact, all of this need never have happened. The Hubble Space Telescope was to have been launched in the summer of 1986, but the loss of Challenger rendered this plan obsolete. While the satellite awaited launch, its data system was subjected to a rigorous end-to-end test, but the optics was not. Perkin–Elmer and Eastman Kodak had bid for the contract to supply the main mirror. Kodak's $100-million bid had included provision for an end-to-end optical test. Perkin–Elmer had won the contract with a bid of $70 million (although the credibility of this bid is strained by the fact that the eventual cost was $450 million) that did not include an end-to-end optical test. Nevertheless, the *ad hoc* use of washers to shim out the reflective null, and the subsequent tests that revealed a problem with the mirror seem like a set of warning signs that really ought not to have been ignored.

The shocking truth about Mars Polar Lander

Arriving at Mars several months after the appalling loss of Mars Climate Orbiter, the Mars Polar Lander was subjected to an intensive review. This mission was acknowledged to be very demanding – it was to deliver a payload comparable to Mars Pathfinder but using a smaller launch vehicle, on an extremely tight schedule in the cost-constrained 'faster-better-cheaper' regime. Unfortunately nothing was heard from the lander after the cruise stage turned to release it, nor indeed from the two Deep Space 2 penetrators that were to be released a few seconds after the lander. The investigation was hampered by the fact that neither the lander nor the penetrators were equipped to transmit telemetry during entry, descent and landing. A remote possibility was that the mechanical properties of the strange 'layered terrain' in the south polar region, where all three were directed, had enabled it to swallow them whole, before they could start to transmit. Initial suspicion fell on the separation event, because a failure in this critical sequence could have affected all three vehicles. If the deployment had occurred as planned, then the vehicles must have suffered independent failures. As regards the lander, one 'killer' failure mode was determined in the following months. A design error that could have been (but was not) caught in testing would have caused the spacecraft to erroneously sense touchdown when its three landing legs were deployed at an altitude of 40 metres, which in turn would have prematurely switched off the engines, leaving the vehicle to crash. Hall-effect (magnetic) sensors were meant to detect the motion of the legs in response to contacting the surface. The computer was to read these sensors 100 times per second and interpret two consecutive positive signals from any leg as indicating touchdown. Unfortunately, the act of deploying the legs could cause the sensors to generate transient signals. Later tests showed that there was a 47 per cent to 93 per cent chance of producing a transient that lasted sufficiently long to result in a misleading positive signal, and this was for each leg – it translated into *at best* a 1 in 8 chance that the lander would *not* erroneously sense touchdown as the legs deployed. At that altitude, the lander would have been descending at 13 metres per second. In free-fall, it would hit the ground at 22 metres per second with an impact energy some 40 times greater than that for which it was designed. Since this drop would have lasted only a few seconds, the lander would probably have come to rest in an upright position. Many components may have survived the impact, but a sidewall to which the delicate waveguides of the communications system were attached would almost certainly have buckled, rendering it silent.[50,51]

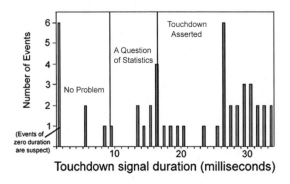

The durations of transient signals from the touchdown sensors on the legs of the Mars Polar Lander, from 47 deployment tests.

This failure mode derives from several separate aspects of the development. It had been recognised that Hall-effect sensors can generate transients (indeed, one had occurred during a leg-deployment test in 1997) and therefore a requirement was introduced to ignore the signals until the lander was near the ground, but instead of expressing this in a way that could be tracked and verified as a mission critical item, such as "the system shall reject landing sensors until close to the ground", the wording was expressed as a case of "shall not", and the intent of the requirement was not captured when the requirements flowed down to the flight software specification. The problem could have been caught during an entry, descent and landing test on the actual spacecraft, but a leg was found to have been miswired during the test. Either as a direct result of this miswiring, or because the test was concluded before reaching that part of the sequence (the specific objectives of the test had been met, after all) the problem was not caught, and the tight schedule had not permitted the test to be re-run. Another aspect was that the leg-test code might logically only be run during the final part of the descent, when it was needed, but a requirement was introduced that all flight software objects (i.e. processes) start up prior to entry, in order that a phased startup did not produce spikes in CPU utilisation that might reset the flight processor during entry, descent and landing, as that would have been catastrophic.

A cartoonist's suggestion for what befell the Mars Polar Lander.

Such a small process as the leg-check code was unlikely to cause this problem, but it was easier to track and implement a 'blanket' requirement than to evaluate the implications of some exceptions. This is therefore an instructive example of how one safety improvement in a complex system can introduce new risks. Finally, while the original version of the code was insensitive to such transients, it was revised in 1998 to accommodate a requirement to include the 'touchdown flag' in channellised telemetry – ironically, a feature that would have made the problem easier to diagnose if the spacecraft had survived the crash! The new software did not clear this flag, as there was no software requirement to do so.

The lesson learned from the ambiguity over the fate of Mars Polar Lander, was that future landers must provide at least basic telemetry during entry, descent and landing.

NOTES

1. *Beyond the Atmosphere: Early Years of Space Science*, H.E. Newell, SP-4211, NASA, 1980, p. 43.
2. http://www.vectorsite.net/taxpl_4.html#m6
3. 'Galileo atmospheric entry probe mission description', N.S. Vojvodich, R.J. Drean, R.W. Schaupp and D.L. Farless, AIAA-83-0100, AIAA 21st Aerospace Sciences Meeting, 10–13 January 1983, Reno, Nevada.
4. 'The Galileo spacecraft architecture', W.J O'Neil, in *The Three Galileos: The Man, the Spacecraft, the Telescope*, C. Barbieri *et al.* (eds), Kluwer, Amsterdam, 1997, p. 75.
5. 'Galileo probe: in-situ observations of Jupiter's atmosphere', R.E. Young, M.A. Smith and C.K. Sobeck, *Science*, vol. 272, 1996, p. 837.
6. http://www.met.fsu.edu/explores/Guide/Noaa_Html/noaa13fail.html
7. *Aviation Week & Space Technology*, 5 July 1999.
8. *Aviation Week & Space Technology*, 22 July 2002, p. 51.
9. *Aviation Week & Space Technology*, 24 April 1989, p. 42.
10. http://llis.nasa.gov/llis/plls/ (entry 0386)
11. http://llis.nasa.gov/llis/plls/ (entry 0053)
12. *Aviation Week & Space Technology*, 17 August 1998, p. 31.
13. *Aviation Week & Space Technology*, 17 August 1998, p. 48.
14. *Two Roads into Space: the Japanese and Indian Space Programmes*, B. Harvey, Springer–Praxis, 2000.
15. http://www2.jpl.nasa.gov/galileo/hga_fact.html
16. http://www2.jpl.nasa.gov/galileo/hgafaq.html
17. 'Galileo's telecommunications using the low-gain spacecraft antenna', J. Statman and L. Deutsch, in *The Three Galileos: The Man, the Spacecraft, the Telescope*, C. Barbieri *et al.* (eds), Kluwer, Amsterdam, 1997.
18. http://www.lpi.usra.edu/publications/newsletters/lpib/lpib76/gal76.html
19. http://www.spaceref.com/news/viewsr.html?pid = 10299
20. http://www.rocketmanblog.com/2003/11/noaan_prime_acc.html
21. http://www.space.com/spacenews/businessmonday_041011.html
22. *Space News*, 15 December 2003.
23. http://llis.nasa.gov/llis/plls/ (entry 0914)
24. http://llis.nasa.gov/llis/plls/ (entry 0267)
25. http://llis.nasa.gov/llis/plls/ (entry 0376)
26. *Aviation Week & Space Technology*, 8 September 1997, p. 30.
27. *Solar System News*, no. 20, September 1997.
28. http://helio.estec.esa.nl/ssd/public/ssn/20/main.html
29. *Aviation Week & Space Technology*, 8 November 1993, p. 27.
30. *Aviation Week & Space Technology*, 4 January 1999, p. 31.
31. JPL Press Release, 2 February 2001.
32. *Aviation Week & Space Technology*, 10 May 1999, p. 29.
33. *Aviation Week & Space Technology*, 17 May 1999, p. 27.
34. *Aviation Week & Space Technology*, 26 July 1999, p. 27.
35. *Aviation Week & Space Technology*, 23 August 1999, p. 34.
36. http://www.selkirkshite.demon.co.uk/analoguesat/w1info.html
37. http://www.frontierstatus.com/fs0101.shtml
38. www.spaceandtech.com/spacedata/logs/2002/2002-051a_eutelsat-w5_sum.shtml

39. www.spaceandtech.com/spacedata/logs/2000/2000-052a_eutelsat-w1_sumpub.shtml

40. http://www.spacetoday.net/Summary/915

41. http://www.nasa.gov/home/hqnews/2004/sep/HQ_04301_FrancesRecovery.html

42. http://www.msnbc.msn.com/id/5926976/

43. *Aviation Week & Space Technology*, 5 June 1995, p. 66.

44. http://llis.nasa.gov/llis/plls/ (entry 0885)

45. *The Hubble Wars*, E. Chaisson, Harpercollins, 1994, p. 155.

46. http://hesperia.gsfc.nasa.gov/hessi/news/latest.html

47. http://www.gsfc.nasa.gov/news-release/releases/2000/h00-80.htm

48. 'Diagnosing the optical state of the Hubble Space Telescope', A. Vaughan, *Journal of the British Interplanetary Society*, 1992, p. 487.

49. *The Hubble Wars*, E. Chaisson, Harpercollins, 1994.

50. 'Low-cost, lightweight Mars landing system, R.W. Warwick, IEEEAC Paper 1473, IEEE Aerospace Conference, Big Sky, Montana, March 2003.

51. 'Report on the loss of the Mars Polar Lander and Deep Space 2 missions', J. Casani *et al.*, JPL Special Review Board, JPL D-18709, 22 March 2000.

16

Operator and software errors

OPERATOR ERRORS

As operator errors are among the most avoidable errors, and can involve culpability, they are often the most vaguely reported of incidents, whether for professional courtesy or out of legal considerations. Hence, specific details are rather difficult to find. On the other hand, it should be recognised that operators, who often work long stressful hours, are the heroes who have nursed spacecraft back from a variety of system failures. As an example of a simple mistake, on 5 December 1984 controllers mistakenly uploaded the NOAA 7 satellite with the ephemeris for NOAA 6; the satellite lost attitude control, but was recovered a day later.[1]

The loss and recovery of Olympus

When the Olympus satellite developed by British Aerospace for the European Space Agency was launched by an Ariane 3 on 12 July 1989, it became the most sophisticated communications satellite in geostationary orbit with four payloads in different frequency bands for several rôles, including serving as a relay for other satellites.[2] It nominally used one of two redundant infrared Earth sensors to control its roll and pitch axes, and one of four redundant gyros to control yaw. It also carried a pair of radio-frequency sensors that were nominally to adjust the pointing of the communications antenna reflectors but were also capable of Earth sensing. It had an additional set of gyros and Sun sensors for use in the transfer orbit, and these offered a further degree of backup. No sooner had Olympus achieved its operating station than it was discovered that the primary Earth sensor lost its function for several hours after local midnight. At that time, the Sun was at its closest to the Earth, as seen from the satellite, the Earth-facing sensor was heated, and differential thermal expansion eliminated the clearance between its oscillating and fixed parts, preventing it from functioning until it cooled down. The redundant sensor began to suffer the same fault after 21 months in orbit. To sense the Earth, the operators based in Fucino in Italy began to use the radio-frequency sensors to monitor a

Preparing the Olympus satellite.

beacon transmission from the ground as a reference, but the system was sensitive to radio noise from the ground. When the radio-frequency sensors lost lock on 29 May 1991, the satellite adopted the Emergency Sun Acquisition mode. Unfortunately, "certain procedural and operational errors" while recovering from this Sun-pointing mode were made by the controllers, leaving the satellite tumbling. With the solar arrays no longer facing the Sun, the batteries soon discharged, the satellite cooled down, and the propellant froze. Worse, autonomous thruster firings during the tumble had introduced a delta-V of 15 metres per second, which had nudged the satellite off its orbital slot.[3]

The only transmission from the satellite was an unmodulated carrier. Fluctuations in the signal strength indicated that Olympus was spinning with a period of 89 seconds. As the spin axis would remain fixed with respect to the stars, the hope for recovery was that the illumination would become favourable in the coming weeks or months, as the Earth travelled around the Sun. Determining the spin axis was thus crucial. Monitoring efforts required enlisting other assets. A radar at Millstone Hill in Massachusetts was used to determine the satellite's orbit: its longitude was drifting at five degrees per day. Since the satellite would soon be out of range of the Fucino station, the assistance of ground stations in Perth in Australia and Goldstone in California was obtained. It had to be assumed that the delta-V had been imparted along the Z-axis at the time of the anomaly. At the times in which the solar arrays were able to power the transmitter, the variation in the output as the satellite rotated enabled its attitude to be estimated. The interference and reflection effects that caused the antenna radiation pattern of the telemetry transmitter to be non-uniform provided an independent estimate. Although both methods agreed to within a few degrees, the actual direction of the X-axis was ambiguous. It was evident that the power situation would improve substantially by mid-September, but eclipses later in the month would introduce complications.[4] By 19 June, the bus voltage was sufficient to allow the satellite to receive ground commands. By the end of the month, heaters had been activated to warm the nickel–hydrogen battery, to enable it to be recharged, but there was not yet sufficient power to run the heaters and the charger simultaneously. The temperature of the thruster nozzles – when heated by the Sun – resolved the ambiguity of the X-axis. This indicated that the north solar array should be driven round to improve the power situation further (the rotator for the opposite array had failed in January 1991). With near-continuous power now available, it was feasible to accelerate the recharge and to re-establish full telemetry coverage. However, the power situation was still precarious, and precise attitude determination was crucial. This was complicated by the change in the mass properties of the spacecraft, since the arrays were in new positions and the propellant used during the anomaly had produced a slight misalignment between the reference axes (defined, for example, by the Sun sensors) and the spin axis, such that special filtering methods had to be used to process the sensor data. Also, because the arrays were canted asymmetrically, solar radiation pressure exerted a torque that was both increasing the period of the spin by 0.4 second per day and precessing the axis sunward at about 0.5 degree per day. The next milestone was early August, by which time the satellite would have completed a circle of the Earth and, all being well, it would be able to resume its assigned station. In July, the propellant was thawed. The final steps began on 26 July, with the thawing of the thrusters. Once the spin rate was reduced to 2 degrees per second, the satellite was able to go autonomously into Sun-acquisition, and thence gingerly reacquire Earth-pointing orientation. A burn on 13 August halted the longitude drift in order to resume the original geostationary point. The recovery took thousands of telecommands, as well as extensive rehearsal using a satellite simulator.[5] By 19 August, the recommissioning of the payload began. Operations then resumed, but were disrupted by the Perseid meteor shower in 1993.[6]

Rescuing SOHO

The Solar and Heliospheric Observatory was a joint project, with the European Space Agency providing the spacecraft for an international suite of instruments and NASA the Atlas IIAS launch vehicle. After launch on 2 December 1995, it was manoeuvred into a 'halo orbit' centred on the Lagrange point 1.5 million kilometres up-Sun of the Earth, where on 14 February 1996 it oriented itself to enable its instruments to monitor the Sun at a variety of wavelengths for a nominal 3-year mission. On 25 June 1998, two months into its *extended* mission, the satellite was lost as a result of maintenance activity. One of its three gyros, Gyro B, which was intended to measure the roll rate, had been left in a high-gain setting, causing it to give a reading 20 times greater than normal, and this prompted the spacecraft to adopt its Emergency Sun Reacquisition mode at 19:16 local time at Goddard Space Flight Center, the control centre in Maryland. In this mode, the flight computer responded by switching over to Gyro A. When the software was written it had been

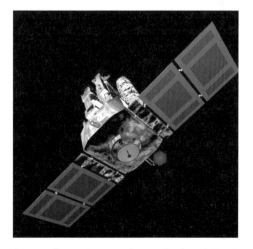

An artist's impression of the SOHO spacecraft.

presumed that both gyros would be operating. However, as similar gyros on the BeppoSAX satellite had suffered failures, it had been decided to switch off a gyro when it was not in use in order to preserve its life. Unfortunately, the software did not recognise that Gyro A was actually off, and proceeded to calculate a roll rate by subtracting 'bias' values from the sensor's reading. Since the sensor had no output, this bias subtraction led the software to perceive that the spacecraft was in a roll, which it tried in vain to correct by firing the thrusters. At 20:35, by which time some of the controllers had gone home, the spacecraft went into a deeper safe mode. On the basis of the discrepancy between gyros A and B, the skeleton crew of operators made a diagnosis (which, despite their having operated the spacecraft for several years, would ultimately be shown to be incorrect) that Gyro B was faulty, and at 23:45 they switched it off. However, as SOHO tried to recover using the inactive Gyro A, it began to wobble. At 00:38, now spinning at 7 revolutions per minute and coning about its X-axis at 60 degrees, it adopted a third safe mode. Contact was lost within 5 minutes, which in retrospect was fortunate since the ground intervention was only making matters worse.[7]

On 23 July the 300-metre-diameter Arecibo radio telescope in Puerto Rico was used to locate SOHO. Acting as radar, the telescope emitted microwave energy with a power of 560 kilowatts, but by the time this blast reached the spacecraft the power density had fallen to less than 1 *micro*watt per square metre, and a tiny fraction of this was reflected back. After the 70-metre dish at Goldstone had integrated this echo for 60 seconds the signal-to-noise ratio was about 10, indicating that the radar cross-section of the satellite was about 15 square metres. As the satellite was 6 metres wide this meant that it was being viewed obliquely (whereas in normal service it would have had its main axis on the line of sight). The Doppler broadening of the echo indicated that it was spinning at 1 revolution per minute. The satellite's exact orientation was not known but its angular momentum vector should still be pointing more or less at the Sun, and in the absence of active control it would be rotating around its axis of maximum moment of inertia, which would set its solar arrays edge-on to the Sun, so it was in serious trouble. A prolonged period spent in this orientation would not only freeze the hydrazine propellant but would also possibly damage the instruments, or even the core systems. On the other hand, the European Space Agency had gained experience in reactivating Olympus from a similar spinning coma.[8] On 3 August SOHO was coaxed into activating its transmitter.[9] Only the carrier signal was received, but this was a start. Over the next few days the telemetry was activated. The spectrum of the transmission showed subcarrier bands, indicating that the signal was being modulated with data, but the transmission was too brief for the Deep Space Communications Complex in Australia to lock up to recover it. Meanwhile, because SOHO had become a major asset in the prediction of space weather and the anticipation of hazards to satellites and astronauts from solar activity, NASA began to consider installing some of its instruments onto Triana, another satellite that was intended for this up-Sun point (although political considerations ultimately prevented its launch). After commanding a different charging strategy for the batteries, the power situation began to improve.[10] By 11 August, the Deep Space Network was able to lock onto the telemetry signal to extract the temperatures, voltages and Sun-sensor output. Although the tankage was only slightly below the 2°C freezing point of hydrazine, some of the pipework was much colder. Regaining control would require the propulsion system to be thawed, because the spin rate was too rapid for the momentum wheels to counter. One communications session that lasted some 100 minutes established that the batteries were charging. As the power was restored, the spacecraft was able to be thawed out.[11] The response of the satellite to commands indicated that its command and data-handling system still worked, but a backup receiver had failed. When Gyro A was commanded to spin up it did not do so. It was determined that the deep cold had caused a wire inside it to break. Since Gyro C was suffering intermittent faults, the Emergency Sun Reacquisition was performed using Gyro B.[12] Clearly, with no redundancy for a type of unit that was prone to failure, the spacecraft's future was open to question, but when Gyro B failed on 12 December 1998, 'gyroless' control modes were already under development. Despite some of its instruments having been chilled down to –122°C, the spacecraft was not only able to continue, but at the time of writing was still providing excellent results.

This recovery prompted a major review, which, in addition to determining the direct failures, indicated several deeper contributing factors. Among these were understaffing of operations, pressure to maintain the flow of scientific data at the expense of diluting engineering support, and poor data displays (matters were exacerbated by the transfer of operations to a new multi-mission control centre at Goddard).[13,14]

Mars Climate Orbiter

In what is sure to be a prime example for science teachers worldwide in stressing the need to state the units in which numbers are quoted, NASA suffered an appalling embarrassment in 1999. Mars Climate Orbiter was to join Mars Global Surveyor to recover more of the science from the failed Mars Observer. However, instead of entering orbit around Mars, a simple unit conversion error caused the spacecraft to penetrate the atmosphere and burn up. A data file that was routinely sent from the contractor, Lockeed Martin, to the engineers at JPL who were to navigate the spacecraft's trajectory, was supposed to be expressed in metric units, per the software interface specification, but had actually been written in Imperial (so-called 'English') units, the continued use of which by the aerospace industry of the USA is a perplexing anomaly. The impulses for propulsive manoeuvres had been expressed as pound-seconds instead of newton-seconds – units which differ by a factor of 4.45. As a result, trajectory errors had built up which resulted in the spacecraft flying too close to the planet for its orbit insertion burn. The situation was complicated somewhat by the design of the

An artist's impression of the Mars Climate Orbiter spacecraft.

spacecraft. Mars Global Surveyor (with which the operations teams had gained much useful experience in the preceding years) had two solar panels in a symmetric configuration, but Mars Climate Orbiter had only one, and this caused the pressure of solar radiation to induce a significant torque during the interplanetary cruise. As this torque built up, the spacecraft's reaction wheels spun faster to absorb it and maintain the appropriate pointing, until they became saturated, at which time the thrusters performed an angular momentum dump. In addition to providing torque, the thrusters caused slight delta-Vs that, for several reasons, went largely undetected. First, software interfaces had errors in the first four months of the cruise. Second, the only way to measure the delta-Vs was by Doppler tracking, and the Sun–Earth–spacecraft alignment was such that most of the delta-V was in the undetectable direction orthogonal to the Earth–Spacecraft line. Finally, with one spacecraft operating in orbit and Mars Climate Orbiter and Mars Polar Lander *en route*, the operations teams were understaffed due to the budgetary constraints imposed by the 'faster-better-cheaper' strategy. Nevertheless, when all is said and done, there is little excuse for a contractor failing to meet a simple units specification.

The warning signs began to appear in the week prior to Mars arrival, after the fourth trajectory correction. The plan called for a closest point of approach of 226 kilometres at Mars Orbit Insertion, to enter a highly elliptical capture orbit. At the apoapsis, a burn would drop the periapsis to 210 kilometres for the aerobraking that would gradually lower the apoapsis. However, while initial trajectory determinations indicated an altitude of 150–170 kilometres, this figure migrated down to 110 kilometres as the planet's gravity drew in the spacecraft. If this discrepancy had been detected earlier in the cruise then an economical burn could have been made, but this became ever more expensive as the planet loomed. One mystery was that Doppler-only trajectory determinations tended to predict lower altitudes than those that used Doppler and range information. However, it was believed that an altitude as low as 80 kilometres would be survivable, and the first solid indication that something was wrong was when the spacecraft slipped behind the planet's trailing limb 49 seconds before the predicted moment. At that time it was already firing its engine to enter orbit. It should have emerged from occultation 21 minutes later, safely in orbit, but nothing was heard from it. The investigators soon identified the units error, and when the mission was recalculated it was found that the trajectory would have taken the spacecraft to within 57 kilometres of the planet's surface, which was too deep into the atmosphere for survival.[15]

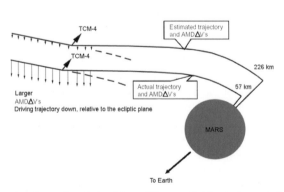

A schematic of the Mars Climate Orbiter's intended and actual approaches to Mars (not to scale).

The irony is that unit foul ups had happened previously.

In 1985, in a high-precision tracking experiment for the Strategic Defense Initiative (also known as 'Star Wars'), an Air Force laser on a Hawaiian mountain was aimed at Shuttle mission STS-51G to be reflected by a 20-centimetre-diameter mirror mounted in the window of the side-hatch. However, on the first attempt the mirror was pointed *away* from the Earth because the altitude of the laser site had been entered into Discovery's computer as a number in feet instead of in nautical miles, as the software expected![16]

Soviet Mars missions

On 10 May 1971 the Block-D upper stage of a Proton launch vehicle was placed into low orbit. It was to re-ignite about an hour later to send its payload to Mars but because an operator had entered the eight-digit firing command in reverse order the manoeuvre did not occur.[17,18] The orbiter/lander payload was disguised as Kosmos 419. The Soviets followed up later in the month with Mars 2 and Mars 3, both of which dropped probes into the Martian atmosphere, but to no effect.[19]

Two spacecraft that were to orbit Mars and investigate Phobos, the larger of its two moons, were successfully dispatched in July 1988. Unfortunately, the first of these craft was lost in late August to a command error. A sequence, some 20 or 30 pages long, had been uplinked to Phobos 1 to be executed while it was out of communication with the Earth. Unfortunately, the last digit was omitted and the spacecraft's computer interpreted this final command as an instruction to deactivate the attitude control thrusters, which it duly did. In the absence of attitude control, it started to rotate, lost solar power, and was never heard of again.[20,21] Normally an upload would be verified by a ground computer (or a spacecraft simulator, such as the engineering models used for this purpose in the West), but this was out of commission and rather than wait for software verification the technician overrode this stage and sent the sequence to the spacecraft; he may well have been working to a tight schedule. Roald Sagdeev, then director of the Space Research Institute, IKI, has speculated that the difficulties may also have been associated with the transfer of control of the spacecraft from a command centre in the Crimea to one just outside Moscow.[22] The second spacecraft entered Mars orbit on 30 January 1989, but was lost while manoeuvering close to its namesake moon on 27 March.[23]

The mystery of Beagle 2

Beagle 2 was a very compact Mars lander developed on a tight schedule and on a shoestring budget by a consortium led by Britain's Open University with Astrium Ltd as the prime contractor. Nevertheless, 400 man-years of effort were invested in this project.[24] The 93-centimetre-diameter, 69-kilogram probe was carried on the side of the European Space Agency's Mars Express spacecraft, which was launched by a Soyuz–Fregat on 2 June 2004. The probe was released by a spring-loaded ejector that spun it up for stability on 19 December, for Mars arrival on Christmas day. It had a tiled entry shield similar to that of the Huygens probe, but with tiles made of a cork composite.[25] An accelerometer was to trigger the deployment of initially a drogue and then the primary parachute. A radar altimeter was to inflate a trio of air bags immediately prior to impact. On the ground, Beagle 2 was to analyse the atmosphere for traces of organic materials, and to deploy an innovative 'mole' device to burrow into the ground to retrieve soil from a depth of about 1 metre – something that had never been done before. Initial communications were to be made via NASA's Mars Odyssey orbiter, and Mars Express was to serve as the primary relay once it had settled into its operating orbit. When Mars Odyssey received no signal, the 76-metre-diameter Jodrell Bank radio-telescope in England attempted in vain to detect the transmission directly.

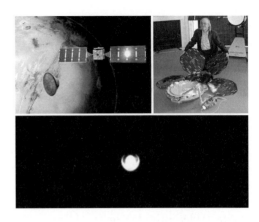

An artist's impression of Mars Express deploying the Beagle 2 lander. Colin Pillinger with a model of Beagle 2 in its surface-deployed configuration. A picture taken by the spacecraft of the departing lander (bottom).

As in the cases of Mars Polar Lander and the Deep Space 2 probes, therefore, there was essentially no information from which to deduce the failure mode. It is clear that the project's schedule was extremely tight, and uncertainty about funding at the beginning did not help. Mass and energy margins were slim, and there was no redundancy in the systems. While it was ambitious and remarkable, and employed many innovative design features, Beagle 2 had no resilience. It could therefore have been disabled by any of numerous single-point failures. One possible environmental cause, however, would have been the delayed deployment of its parachutes as a consequence of dust having warmed the Martian atmosphere more than the values to which Beagle was designed, resulting it hitting the ground at too great a speed. Even if the parachutes worked, it was belatedly noticed that in the images taken by Mars Express just after the probe's separation there appeared to be an icy deposit on the rear of the capsule, suggesting that the ammonia gas that was to have inflated the air bags might had leaked away in space.[26] Unfortunately, the truth will never be known.

SOFTWARE FAILURES

As spacecraft become more complex, the behaviour of software becomes ever less deterministic – that is, the number of potential states available, the number of combinations of processes, modes and stimuli, become so large as to be untestable within the time and budgetary constraints of a project. The result is unexpected behaviour. For example, a task to run the meteorology instrument on the Mars Pathfinder lander clashed with a bus data distribution task, causing the computer to reset. It was noticed that this failure (for which a patch would be straightforward) only occurred when the system was being used heavily, and the 'window' for this fault was a mere 15 computer instructions in length.[27] The Magellan spacecraft suffered a similar asynchronous sensitivity in which a sequence of instructions that was intended to be uninterruptible was sometimes interrupted.[28] On the Galileo spacecraft, a data compression program on its CDS A processor took too long to execute, forcing the processor to shut down and put the spacecraft into a safe mode – during which it was nonetheless able to perform a critical orbit manoeuvre.[29]

A case of overspin

An intriguing software specification failure came to light with the group of Microsat satellites that were placed into low orbit for the Department of Defense by a Pegasus on 17 July 1991. The seven 25-kilogram cylindrical spinning satellites were to use a small nitrogen thruster system to adjust their orbits so as to maintain their relative positions. Each satellite had a magnetometer and an Earth-horizon sensor to determine its attitude, and was to spin at the nominal rate of 3 revolutions per minute with its spin axis aligned so that it would roll along its orbits in the manner of a wheel. Unfortunately, thrust misalignments of the nitrogen jets led to excessive spin-up of some of the satellites, which were designed with an arbitrary limit of attitude control up to 7 revolutions per minute. However, since the software sampled and controlled the attitude at a rate of 0.66 hertz, aliasing could occur for high spin rates and, indeed, the direction of rotation could be misread. Specifically, when the nitrogen thrusts caused the spin to jump to above 7.5 revolutions per minute, the software control law acting to achieve the desired difference between successive magnetometer readings in fact caused the spacecraft to spin up instead of to slow down. Indeed it spun up two of the satellites to the next speed at which the inferred spin-direction sign changed, namely 22.5 revolutions per second.[30] Although the control law was adequate for conditions within the bounds of the specification, the specification failed to recognise possible exceptions.

The loss of Clementine

In 1990 the Ballistic Missile Defense Organisation set out to test a suite of miniaturised sensors intended for detecting and tracking objects in space. The initial

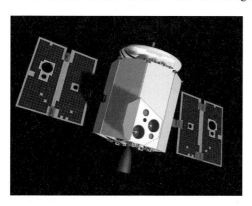

An artist's impression of the Clementine spacecraft.

plan to have one satellite track another, but when the target was eliminated by budget cuts it was decided to rendezvous with asteroid 1620 Geographos. As a preliminary, the spacecraft, named Clementine, was to spend several months in lunar polar orbit and use its multispectral sensors to chart the composition of the surface. It was launched from Vandenberg on 25 January 1994 by a refurbished Titan II, entered lunar orbit on 19 February, functioned flawlessly, and departed on 3 May 1994 to rendezvous with its primary target in August. Unfortunately, four days later, during a 20-minute telemetry hiatus while preparing for a manoeuvre, a flaw in the software of the autonomous computer fired the attitude control thrusters for 11 minutes, exhausting the supply of propellant and leaving the spacecraft spinning at 80 revolutions per minute, which forced the abandonment of the mission.[31],[32]

Spirit's flash memory

After spending 18 'sols' on the surface of Mars, the Spirit Mars Exploration Rover fell silent for several hours on 21 January 2004. When contact was re-established the vehicle was found to be in a fault mode, and it continued to suffer anomalies in a pattern that baffled its terrestrial controllers. The problem was traced to the 256-megabyte flash memory on which science data was stored – a much more reliable and lighter option than a tape recorder. A 128-megabyte random access memory was used for working storage, and held the software that managed the file system for the flash memory. Unfortunately, the file manager was consuming too much of this memory and causing the computer to reset repeatedly. Lax file management had contributed to the problem. Some 1,000 files were left on the file system from the interplanetary cruise, and when these were deleted on 30 January the problem was greatly relieved, but not totally eliminated. Ultimately, completely reformatting the 224-megabyte partition restored

An artist's impression of a Mars Exploration Rover about to roll off its landing platform.

nominal operation without side effects, although engineers had to be careful that the reformatting did not corrupt the other 32-megabyte partition that contained the flight software![33] Subsequent operations managed the files much more carefully. Although this glitch took 14 days to diagnose and fix, thereby wasting one-sixth of the nominal mission duration, Spirit then performed for a considerably longer time than expected and, indeed, was still active at the time of writing.

NOTES

1. *Orbital anomalies in Goddard spacecraft 1984*, E.F. Shockley, Goddard Space Flight Center, September 1985.
2. http://www.wtec.org/loyola/satcom/c5_s2.htm
3. *Spaceflight*, November 1991, p. 384.
4. 'Olympus recovery: attitude aspects', P. de Broek, Proceedings of the ESA Symposium on Spacecraft Flight Dynamics, Darmstadt Germany, September–October 1991, ESA SP-326 (December 1991), p. 221.

5. 'Theoretical description and operational experience of a new gyro-based Earth-pointing control mode for the Olympus satellite after spacecraft recovery', N. Wiengarn, M. Pecchioli and N. Mardle, AAS 93-280, in Spaceflight Dynamics 1993, *Advances in the Astronautical Sciences*, vol. 84, p. 461.

6. http://www.selkirkshire.demon.co.uk/analoguesat/olympus.html

7. *Aviation Week & Space Technology*, 6 July 1988, p. 32.

8. *Aviation Week & Space Technology*, 27 July 1998, p. 28.

9. *Aviation Week & Space Technology*, 3 August 1998, p. 30.

10. *Aviation Week & Space Technology*, 10 August 1998, p. 29.

11. *Aviation Week & Space Technology*, 17 August 1998, p. 32.

12. *Aviation Week & Space Technology*, 28 September 1998, p. 31.

13. 'An analysis of causation in aerospace accidents', K.A. Weiss, N. Leveson, K. Lundqvist, N. Farid and M. Stringfellow, Space 2001, Albuquerque, New Mexico, August 2001.

14. http://sunnyday.mit.edu/accidents/

15. *Mars Climate Orbiter Mishap Investigation Board Phase-I Report*, 10 November 1999.

16. *Space Shuttle Log*, T. Furniss, Jane's, 1986.

17. *Solar System Log*, A. Wilson, Jane's, 1987.

18. *The Difficult Road to Mars*, V.G. Perminov, NASA Monographs in Aerospace History, no. 15, 1999.

19. http://www.skyrocket.de/space/doc_sdat/mars-71.htm

20. *Aviation Week & Space Technology*, 7 November 1988, p. 27.

21. UPI Report, Houston, 10 September 1988.

22. *The Making of a Soviet Scientist*, R. Sagdeev, Wiley, 1994.

23. http://heasarc.gsfc.nasa.gov/docs/heasarc/missions/phobos2.html

24. Managing Mars, B. Kirk, *Aerospace*, September 2003, p26.

25. *The Guide to Beagle 2*, C.T. Pillinger, M.R. Sims and J. Clemmet, December 2003.

26. An official Commission of Inquiry was set up by the British National Space Centre and the European Space Agency. However, its report in May 2004 was not released to the public – a policy that met with widespread criticism as implying a 'cover-up' or whitewash. Separately, the Beagle 2 team later (August 2004) released a 276-page report of their own *Beagle 2 Mars Mission Report*, M.R. Sims (ed), University of Leicester 2004.

27. *Aviation Week & Space Technology*, 21 July 1997, p. 24.

28. *The Evening Star: Venus Observed*, H.S.F. Cooper, Johns Hopkins Press, 1994, p. 46.

29. *Aviation Week & Space Technology*, 3 August 1998, p. 31.

30. 'Microsat attitude control system on-orbit maintenance and performance', P. Carruthers, C. Mills and K. Reiss, AAS 92-071, Guidance and Control conference 1992, *Advances in the Astronautical Sciences*, vol. 78, p. 491.

31. http://www.lpi.usra.edu/expmoon/clementine/clementine.html

32. http://www-phys.llnl.gov/clementine/status/status.html

33. *Aviation Week & Space Technology*, 9 February 2004, p. 33.

17

Conclusions

FALSE ECONOMIES

Half a century into the Space Age, launch vehicles and spacecraft are still prone to failures for all manner of reasons. In marked contrast to other fields of high technology, the combination of high cost and huge risk seems to be intrinsic to the endeavour.[1] In the decade leading up to 1995, advances in technologies and data compression improved the economic performance of a commercial communications satellite (as measured by its lifetime, power, bandwidth and throughput) by a factor of 25, yet during this time launch vehicle capability remained static. The result was that launches and insurance often cost more than the payload. A rocket is essentially a controlled explosion. In the drive for the greatest performance, rockets are always on the 'edge' of technology. Yet despite their apparent robustness, they are delicate vehicles that are easily torn apart by aerodynamic stress. Many failure modes can be overcome by the provision of backup systems but this increases the mass, complexity and cost. It had been intended to phase out 'expendable' rockets upon the introduction of the Shuttle, but following the loss of Challenger in 1986 their production was resumed and their designs modified to drive down costs. In early 1999 a review by the Aerospace Corporation of 60 significant launch failures since 1990 by all vendors concluded that often underlying the specific cause was a case of "forgetting a lesson learned somewhere along the way".[2] The report opined that in attempting to introduce new launchers while maintaining their existing ones, companies were spreading their experienced personnel over too many projects. In fact, the industry appeared to be rife with endemic poor quality control in manufacturing – as recently evidenced by the failure of the Centaur of a Titan IVA to deploy a Milstar, the failure of an IUS of a Titan IVB to deploy a DSP satellite, the stranding of Orion 3 by a Delta III, and the catastrophic loss of several Protons. But it was not just launch vehicles, poor coordination between teams had doomed the Mars Climate Orbiter, and inadequate testing had resulted in the loss of Mars Polar Lander. The underlying commonality was the relentless pressure to drive down costs.[3]

PATTERNS OF FAILURE

One conclusion that can be drawn is that many mistakes were committed repeatedly – if lessons were being learned, then either they were not being learned well or they were being forgotten. We hope that this book goes some way to help to remedy this. The fact that a generation of spacecraft engineers has now passed since the start of the Space Age might be a contributing factor to the loss of collective experience. On the other hand, the rapid pace at which (generally smaller) planetary missions were pursued in the late 1990s has created a new generation of veterans. Some of the failures were well-known aspects of spacecraft engineering: the poor reliability of gyros, articulating parts, and so on. Others were subtle, due to an unexpected environmental effect or the use of a new component. One might wonder why components that show good reliability do not become universal, but of course components may be discontinued for commercial or other reasons. One recent concern is that while environmental legislation is discouraging the use of lead in wire coatings, in space the lack of lead allows tin-based solders to grow tin 'whiskers' – narrow crystals that can cause short circuits.[4] Also, as technology improves, substitution of more modern parts will probably improve performance, albeit at the risk of introducing new issues. While ground testing can expose a spacecraft to hazards, it seems more likely that tests will uncover potential problems. Unfortunately, the mantra of "test what you fly, and fly what you test" is rarely achieved. When a schedule slips, it is often the testing campaign that is trimmed back in order to meet the deadline.

Failure is not an option: problem-averse psychology

In view of the loss of Mars Climate Orbiter, the Mars Polar Lander mission – which was in flight – was intensely scrutinised, but this risk review failed to detect the flaws in the spacecraft. There may be a psychological factor at work: reviewing a project is a thankless task. If (as might be hoped) the project has been well managed, then the risk review is unnecessary. A review may be perceived as disruptive to ongoing work, and the review board treated with some hostility. By undertaking a review, an engineer is gaining 'an opportunity to fail' – if the *risk review board* fails to anticipate a failure, it itself becomes culpable. On the other hand, a *failure review board* is more attractive – there is a clearly defined requirement, and it is almost guaranteed success because it will either identify the cause of the failure or, if it does not do so, it can conclude that there was insufficient data to narrow down the cause, or imply that perhaps the contractor was insufficiently cooperative. Of course, no project sets out to fail, and each no doubt believes that it is pursuing a 'right first time' strategy. Nevertheless, failures occur for the most preventable of reasons, even when it is glaringly obvious (in retrospect) that there was a problem: consider, for example, the flaw in the mirror of the Hubble Space Telescope and the unceremonious toppling of the NOAA N-Prime satellite. A manager requires a certain strength of will to carry a project through. This is particularly the case when challenges need to be overcome. There may therefore be a reluctance to accept indications of a problem, since dealing with it may introduce delays or increase costs. However, it should be the duty of engineers and scientists to make cautious nuisances of themselves and call attention to issues of concern. Indeed, there have been cases – on the COBE mission, for example – when the manager offered a reward to anyone who could identify a serious flaw. It is better to raise a potential problem and have it promptly dismissed, and risk being ridiculed by your colleagues, than to remain silent and then have to live with knowing that you could have saved a mission. Sometimes institutional issues, such as the diffusion of responsibility, or poor information flow, can prevent problems from being seen, or acted upon.[5,6] Perhaps, then, a less formal but earlier exposure of a project to outside scrutiny may help.

Risk: you get what you pay for
Engineering is the art of doing with one dollar that which any damn fool can do with two. A spacecraft can be made arbitrarily (although not completely) robust, but at a cost. Additional fuel, ample solar array capacity, spare gyros or other systems would all make a spacecraft more resilient, but would add to its cost and mass. In a sense, therefore, risk is just a different manifestation of cost, and the degree to which risk is perceived will determine how much resilience the customer will pay for.

There is a cycle in engineering, as the perception of risk ebbs and flows. There is a continual pressure to improve, to pare margins and hence to perform better or cheaper. This pressure is difficult to resist because it is imposed by customers who give their business to the most competitive offer. On the other hand, the perception of risk is dramatically increased by an instance of failure, and to some extent this pressure restores the margins. A classic example is the Forth Rail Bridge, which spans the 2.5-kilometre-wide Firth of Forth north of Edinburgh in Scotland. Opened in 1890 after seven years of construc-tion, it was the largest span bridge in the world, the first to be constructed purely of steel, and the largest civil engineering project of the nineteenth century.[7] Its formidable structure is the exemplar of Victorian over-engi-neering. Competitive pressure is not evident, because it was designed soon after the disaster in which, on the stormy night of 28 December 1879, the iron bridge over the Firth of Tay near Dundee collapsed as a train was

The Forth Rail Bridge.

crossing it, plunging 75 people to their deaths. That bridge had been built on a shoestring budget. It was the worst British structural engineering failure. Although the contract for the Forth Rail Bridge had been given to the same designer, the increased perception of risk prompted a redesign in which the margins were greatly increased, and the budget increased accordingly.[8]

Up through the Viking missions in 1975, NASA planetary missions were launched in pairs. This, in part, reflected the uncertainty of a single probe succeeding. But because the recurring cost of a spacecraft is considerably lower than the total cost, making several reduces the unit cost. The Soviets also initially had a strong serial production philosophy, reused elements for various types of spacecraft, and made multiple launches per opportunity. NASA's successful Mariner series was similarly cost-efficient, but an attempt to repeat the idea with Mariner Mark II failed – many of the systems for the Cassini spacecraft were to have been reused on a spacecraft for a Comet Rendezvous and Asteroid Flyby, but this was cancelled. However efficient they might be, combined projects like these, and series of spacecraft, entail expensive total project costs. In the United States, at least, the hurdle is usually the budget for the coming fiscal year, and the political temptation to 'descope' multiple missions is often irresistable.

Large projects render themselves vulnerable to risks – a launch failure or some other problem can cripple a large spacecraft or a small one, but the damage to the programme is greater in the first case. Spreading payloads between many smaller satellites is, in this sense, more resilient, but in another sense is less efficient. The originally stated intent of the Discovery programme was to accept a higher level of risk. However, when failures and overruns occurred in the Discovery and Mars programmes, the perceived level of risk increased, and conservatism progressively increased, culminating in the dispatching of two vehicles for the Mars Exploration Rover (MER) mission – and because both were successful, two sites on Mars were able to be investigated for a small additional increase in the overall cost.

PREDICTING FAILURE

While some degree of random failure is inevitable, there is at least empirical support for the notion that 'stretching the envelope' leads to failure. The Aerospace Corporation, which has conducted lengthy studies into spacecraft costs and failures, has developed a 'complexity index' in an effort to gauge the technical difficulty of a mission. This index incorporates such factors as the spacecraft power, the pointing accuracy, and the type of propulsion. Of course, a single metric such as this is unlikely to adequately represent all of the factors that can contribute to failure, but it is at least a start. When the complexity indices of various missions were plotted against the development time and cost, successes occupied the upper left of the graph (indicating that long development and/or comfortable budgets for a given level of complexity yielded success) and failures tended to fall in the lower right (too fast or too cheap).[9]

One conclusion from this study is that the schedule may be a stronger factor than the cost. Cheap or expensive missions can both succeed, but rushed missions almost always fail. Historically, planetary missions tend to take 27 per cent longer to prepare than planned, and because they usually have inflexible launch dates they become schedule-constrained.[10] They also have a higher failure rate, although at least part of this must be due to less well-characterised environments. Interestingly, the predictive power of these relationships is not infallible. The recent MER mission, for example, lies well in the 'failure region'.[11] Perhaps psychology is again at work: after the losses of Mars Climate Orbiter and Mars Polar Lander, NASA simply *had*

FACTORS CONTRIBUTING TO COMPLEXITY INDEX CALCULATION

FACTOR	UNIT	MINIMUM	AVERAGE	MAXIMUM	EXAMPLE	
Bus Cost	(FY97$M)	1	25	135		
Development Time	(MOS)	11	35	84		
					value	sub-index
Satellite Launch Mass	(kg)	25.9	252.4	890.0	494.0	69%
Design Life	(mos)	0.3	23.2	96.0	36.0	59%
Max Distance from Earth Orbit	(au)	0	0.13	2.7	0.0	0%
Beginning-of-Life Power	(W)	12	294	2600	400	52%
End-of-Life Power	(W)	3	252	2302	350	52%
Solar Array Area	(m2)	0.2	3.3	12.2	5.0	60%
Solar Cell Type		Si	—	GaAs	GaAs	100%
Array/Antenna Configuration		body-fixed (B)	deployed (D)	articulated (A)	D	50%
Battery Type		lead-acid	NiCd, SNiCd	NiH2	SNiCd	66%
Battery Capacity	(A-hr)	1.0	15.4	119.6	35.0	59%
Structures Material		Aluminum	—	Composite	Al	0%
Attitude Control Type		None	Grav-Grad, Spin	3-axis	Spin	66%
Number of P/L Instruments		1	3	10	2	23%
Pointing Accuracy	(deg)	0.005	3.9	35.0	1.0	87%
Pointing Knowledge	(deg)	0.002	2.1	20.0	0.7	83%
Number of Thrusters	(#)	0	3	20	6	59%
Propulsion Type		None, Cold Gas	Monoprop	Biprop, Ion	Biprop	75%
Downlink Communications Band		UHF/VHF	S-band	L/X-band	S-band	50%
Max Downlink Data Rate	(Kbps)	0	552	2250	128	12%
Solid State Recorder Memory	(Mbytes)	0	321	4000	2000	73%
Thermal Control Type		passive	semi-active	active	active	100%
Mean Complexity Index		**2%**	**41%**	**80%**		**59%**
Normalized Complexity Index		**0%**	**50%**	**100%**		**79%**

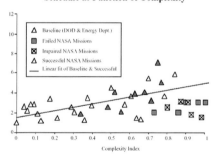

Schedule as Function of Complexity

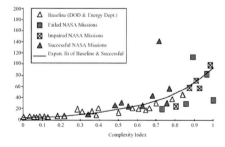

Spacecraft Cost as Function of Complexity

Complexity Index. (The Aerospace Corporation, reported in *Aviation Week & Space Technology*)

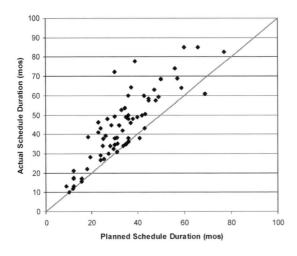

Schedule overruns. ('The effect of schedule constraints on the success of planetary missions', R.E. Bitten *et al.*, IAA Conference on Low-Cost Planetary Exploration in 2003.)

to succeed, with the result that vast manpower resources were assigned and, despite the constraints, the collective will assured success. Or perhaps the complexity index metric is faulty and overestimates the challenges. Or perhaps MER just got lucky!

Our prognosis, then, is not particularly rosy. In as much as failures are the means by which the level of effort or margin on a project are determined (the 'error signal' in control theory), a non-zero failure rate is inevitable. Otherwise how do we know we are not overpaying? Unlike consumer electronics items manufactured by the million in the same plant, spacecraft are often 'one-off', and so whether an individual mission fails or not gives no statistically significant information.

FINAL THOUGHT

Our impression on writing this book is of growing surprise that anything works at all! Failure is the norm: something goes wrong on almost every project. The measure of a spacecraft is perhaps not whether it fails, but rather how well it tolerates the inevitable failures. When flight controllers for the Apollo programme were invited to participate in purchasing decisions for spacecraft hardware, they did not ask the potential vendor how a product worked, they were more interested in how it failed! How robust a spacecraft is depends not only on engineers employing technical skill, ingenuity and care, but also on their being sufficiently canny to manage to build systems which are, one way or another, more robust than the customer was willing to pay for.

NOTES

1. *Aviation Week & Space Technology*, 13 December 1999, p. 108.
2. *Aviation Week & Space Technology*, 3 May 1999, p. 31.
3. *Aviation Week & Space Technology*, 24 May 1999, p. 27.

4. Eurocomp no. 7, Autumn 2004, p. 3.

5. 'An Analysis of causation in aerospace accidents', K. Weiss *et al.*, Space 2001.

6. http://sunnyday.mit.edu/accidents

7. http://www.chrishobbs.com/johnfowlerforthrail.htm

8. http://www.undiscoveredscotland.co.uk/queensferry/forthrailbridge/

9. *Aviation Week & Space Technology*, 12 June 2000, p. 47.

10. 'The effect of schedule constraints on the success of planetary missions', R.E. Bitten, D.A. Bearden, N. Lao and T.H. Park, IAA Conference on Low-Cost Planetary Exploration, Noordwijk, September 2003.

11. 'Perspectives on NASA's "Faster Better Cheaper" experiment: when is a spacecraft development too fast and too cheap?', D.A. Bearden, Presentation to NNSA Future Technology Conference, 17–19 May 2004.

Index